AF538320

Macromodeling with SPICE

J. Alvin Connelly
Georgia Institute of Technology, Atlanta, Georgia

Pyung Choi
Kyungpook National University, Korea

PRENTICE HALL, Englewood Cliffs, New Jersey 07632

Library of Congress Cataloging-in-Publication Data

Connelly, J. Alvin.
Macromodeling with SPICE / J. Alvin Connelly, Pyung Choi.

p. cm.
Includes bibliographical references and index.
ISBN 0-13-544941-3
1. Electric circuits–Computer simulation. 2. Electronis circuit analysis–Data processing. 3. SPICE (Computer program) I. Choi, Pyung. II. Title.
TK454.C68 1992
621.319'2'0013–dc20 91-26959
CIP

Acquisitions editor: Elizabeth Kaster
Production editor: Irwin Zucker
Copy editor: Judy Ashkenaz
Prepress buyer: Linda Behrens
Manufacturing buyer: David Dickey
Editorial assistant: Jaime Zampino
Supplements editor: Alice Dworkin
Cover design: Wanda Lubelska Design

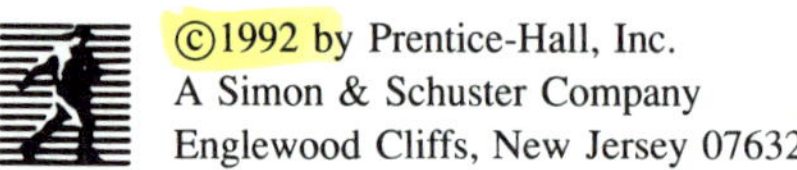

A Simon & Schuster Company
Englewood Cliffs, New Jersey 07632

Printed in the United States of America

10 9 8 7 6 5 4 3 2 1

ISBN 0-13-544941-3

Prentice-Hall International (UK) Limited, *London*
Prentice-Hall of Australia Pty. Limited, *Sydney*
Prentice-Hall Canada Inc., *Toronto*
Prentice-Hall Hispanoamericana, S. A., *Mexico*
Prentice-Hall of India Private Limited, *New Delhi*
Prentice-Hall of Japan, Inc., *Tokyo*
Simon & Schuster Asia Pte. Ltd., *Singapore*
Editora Prentice-Hall do Brasil, Ltda., *Rio de Janeiro*

To Mary Nelle and Yeong Ae,
and to
Tim, Jeff, Stephen, Brian, Leota, Sharon,
Yun Jung (Christine), and Jun il (David)

Contents

FOREWORD ***ix***

PREFACE ***xi***

1 INTRODUCTION ***1***

1.1 History of SPICE 1

1.2 SPICE Capabilities 2

1.3 List of Symbols and Notation 3

1.4 Subcircuits 6

1.5 Extending Simulation Capabilities 7

1.6 Summary 14

References 14

2 MACROMODELING METHODOLOGY ***15***

2.1 Introduction 15

2.2 A General Procedure for Macromodel Development 16

2.3 Arithmetic Functions 17

2.4 Piecewise Linear Approximations 23

2.5 Behavioral Level Filter Macromodels 30

2.6 Summary 40

References 40

3 DROP-IN MACROMODELS 42

3.1 Introduction 42

3.2 Behavioral Macromodels 43

3.2.1 Voltage Summer, 43
3.2.2 Voltage Multiplier, 45
3.2.3 Voltage Divider, 47
3.2.4 Squarer, 49
3.2.5 Ideal Transformer, 51
3.2.6 Square Root Extractor, 52
3.2.7 Triangle-to-Sine Wave Converter, 54
3.2.8 Phase Shifter, 57
3.2.9 Integrator, 59
3.2.10 Differentiator, 62
3.2.11 Absolute Value, 64
3.2.12 Peak Detector, 66
3.2.13 Frequency Multiplier, 68
3.2.14 Frequency Divider (by 2), 71
3.2.15 Frequency Adder/Subtractor, 73
3.2.16 Phase Detector, 76
3.2.17 Delay Line, 79
3.2.18 Schmitt Trigger Simulator, 80
3.2.19 Sample-Hold, 83
3.2.20 Pulse Width Modulator, 86
3.2.21 Amplitude Modulator, 88
3.2.22 Log Amplifier, 91
3.3.23 The Nth Root Extractor, 94
3.2.24 Noise Sources: Voltage and Current, 97
3.2.25 Random Noise Generator, 103
3.2.26 Voltage Controlled Oscillator (Sine, Square, and Triangle Waveforms), 107
3.2.27 Frequency Counter: Frequency-to-Voltage Converter, 111

3.3 Functional Macromodels 117

3.3.1 Voltage Comparators, 117
3.3.2 Operational Amplifiers, 120

3.4 Summary 132

References 132

4 THE PHASE LOCKED LOOP MACROMODEL **134**

4.1 Introduction 134

4.2 Electrical and Performance Parameters to Be Modeled 135

4.3 Choosing the Drop-in Models 138

4.4 Summary 165

References 165

5 USING SPICE TO DESIGN, ANALYZE, AND TEST A PLL SYSTEM **167**

5.1 Introduction 167

5.2 Setting the Model Parameters 168

5.3 Testing the Analog PLL System Macromodel 170

5.4 Accuracy of the APLL Macromodel 195

5.5 Applications of the APLL Macromodel 209

5.6 Summary 213

References 213

Further Readings 213

6 AUTOMATIC GENERATION OF MACROMODELS **215**

6.1 Introduction 215

6,2 Data Required and Output File Produced 216

6.3 Computer Software 217

6.3.1 Batch File: MACRO.BAS, 217
6.3.2 Analog Phase Locked Loop Macromodel: APLL.BAS, 219
6.3.3 Macromodels for Continuous-Time Domain Transfer Function: CFFM.BAS, 223
6.3.4 Peak Detector Macromodel: PEAK.BAS, 225
6.3.5 Sample-and-Hold Circuit Macromodel: SAMPLE.BAS, 226
6.3.6 Bistable Network Macromodel: SCHMITT.BAS, 228
6.3.7 Noise Source Model: NOISE.BAS, 229
6.3.8 Macromodels for Piecewise Linear Circuits and Systems: PWLM.BAS, 231
6.3.9 Macromodels for Discrete-Time Domain Transfer Functions: ZFFM.BAS, 234
6.3.10 Macromodel for Operational Amplifier: AMP.BAS, 236

Appendices

A OVERCOMING CONVERGENCE PROBLEMS IN SPICE **241**

A.1 Bias Point Calculation 241
A.2 Transient Analysis 242
A.3 Example of an Orderly Procedure for Correcting Problems 242
A.4 Error Messages and What They Mean 246
Recommended References 247

B GETTING THE MOST OUT OF PROBE **248**

B.1 PSpice Probe 248
B.2 Versions Available 248
B.3. Arithmetic Functions 249
B.4 Control Keys 249
B.5 Special Tricks 250
B.6 Combining Probe Plots 253
Recommended References 254

C QUICK SPICE **256**

C.1 Elements 256
C.2 Semiconductor Devices 262
C.3 Subcircuits 267
C.4 Control Statements 267

INDEX **272**

Foreword

Simulation of electronic circuits has replaced breadboarding as a means of verifying and analyzing the performance of complex circuits or systems. Unfortunately, simulation is only as good as the models used to represent the components of the circuit or system which is being simulated. The well-known and extensively applied simulation program SPICE was based on efficient and accurate models for electronic devices, including the pn junction diode, the bipolar junction transistor, and the field-effect transistor. Although these device models have become very sophisticated in their capability and efficiency, the number of devices contained in today's electronic circuit or system has far outstripped the ability of SPICE-like simulators.

One solution to the limitations of SPICE-like simulators is to raise the level of the model from the device level to a higher level. Two categories of higher level models are available. One uses SPICE primitives, and the other uses SPICE primitives plus other simulator attributes not found in most SPICE-like simulators. The first category, known as macromodeling, is the subject of this book. The second category is simply called higher level modeling or behavioral modeling. The advantage of macromodeling is that it is compatible with SPICE-like simulators and does not require that the user learn a new simulator.

Macromodels are not new. They were first introduced in 1974 by Boyle, Cohn, Pederson, and Solomon. As computational power increased, the demand for macromodels did not grow. However, the size and complexity of electronic circuits has rapidly outstripped the increases in computational power, resulting in a crisis in simulation when using device-level models. It is not unusual to spend days on a dedicated high-speed computer in an attempt to simulate a complex circuit or system. If the simulation time becomes unreasonable, then the complex circuits or systems are partitioned and simulated individually. This method has the disadvantage of not being able to model high-level behavior, such as stability or timing.

The crisis in simulation has led to a renewed interest in macromodeling. This timely book offers one solution to the inability to simulate large circuits and systems adequately. It is the first book devoted totally to the subject of macromodels. A methodology for the development of macromodels is given, along with many examples. A complete library of drop-in macromodels along with programs for automatically generating macromodels, is included. The application of these macromodels to design, analyze, and test a phase locked loop illustrates how macromodels can be used to obtain a simulation of a complex electronic system. The phase locked loop is impractical to simulate at the device level.

The authors have extensive experience in the development of macromodels and their application to complex circuits and systems. The macromodels presented have been applied and debugged in the classroom. The authors also include many useful techniques for optimizing the use of SPICE. The subject material in this book will be useful to both practicing engineers and students using SPICE. This book will be useful to those who know nothing about macromodeling as well as those who are familiar with macromodeling. It will give the reader an increased capability of using SPICE based on knowing the proper level of modeling to be used in a given simulation. The authors are to be commended on their effort and for making an important contribution to the field of electronic circuit and system simulation.

Phillip E. Allen,
Schlumberger Professor of Electrical Engineering
Georgia Institute of Technology
Atlanta, Georgia

Preface

This book will be useful for individuals having an intermediate to advanced knowledge of SPICE and its various derivative circuit analysis programs. Electrical engineering students enrolled in upper-level courses as well as graduate students represent the primary academic market. Practicing professional engineers engaged in circuit and system design and analysis also will find this book very useful. In both cases it is assumed that the reader has some previous experience with the SPICE circuit analysis program. This familiarity can come from having used SPICE under the guidance of the built-in instruction manual included on the program available from Berkeley. Other good references for the beginning SPICE user are *SPICE A Guide to Circuit Simulation and Analysis Using PSpice* by Paul W. Tuinenga (Prentice Hall, 1988), *Computer-Aided Circuit Analysis Using SPICE* by Walter Banzhaf (Prentice Hall, 1989), and *SPICE for Circuits and Electronics Using PSpice* by Muhammad H. Rashid (Prentice Hall, 1990). Many new textbooks have included chapters and/or appendices covering introductory aspects of SPICE utilization.

This book has the following unique aspects, which intermediate and advanced SPICE users will find very helpful. The first chapter defines the symbols and notation used throughout the book. Then subcircuits are explained and used to develop an operational amplifier macromodel. An example is given of how to use subcircuit macromodels in an active filter. Finally, some special tricks are presented for extending the built-in capabilities of SPICE by sweeping temperature and component values as well as modeling zener diodes. Chapter 2, "Macromodeling Methodology," describes the techniques that were used in the development of the "tool kit" of macromodels for many circuit and system functions. These techniques can be extended by readers to customize their particular macromodels by either modifying one of the given models from the later chapters, or creating a new macromodel using the techniques presented. This chapter guides the reader through the creation and development of an initial macromodel that approximates either experimental results or specification sheet parameters. Next, a simple, easy-to-use macromodeling technique for piecewise linear circuits is explained and illustrated. Behavioral-level macromodels are discussed and examples are given for both continuous-time (s-domain) analog filters and discrete-time (z-domain) sample data filters.

Chapter 3, "Drop-in Macromodels," is a collection of over thirty behavioral, functional, and commercial macromodels that have been developed, simulated, tested, and verified. The macromodels are called *drop-in* because they can easily be used singly or in groups to perform virtually any analog signal-processing function. These macromodels can be used much as functional blocks or modules of laboratory equipment are used to simulate and evaluate the performance of circuits and systems. Using these drop-in macromodels with various test signals allows a circuit and/or system designer rapidly and easily to determine gains, impedance levels, noise, distortion, spectral content, and the like as functions of time and frequency without solving equations or performing tedious calculations. Furthermore, the designer can simulate a large system using the macromodels to identify which block or stage is the critical one for limiting the overall performance. Then a more complete, device-level model can be inserted for this critical block while continuing to use the other drop-in macromodel functions. The tool kit of available drop-in behavioral macromodels includes the following: summer, multiplier, divider, squarer, ideal transformer, square root extractor, triangle-to-sine converter, phase shifter, integrator, differentiator, absolute value, peak detector, frequency multiplier, frequency divider, frequency adder and subtractor, phase detector, delay line, hysteresis (Schmitt trigger) simulator, sample-hold, pulse-width modulator, amplitude modulator, log amplifier, Nth root extractor, voltage and current noise sources including 1/f frequency effects, random noise generator, voltage controlled oscillator (for sine, square, and triangle waveforms), and a frequency counter. Functional macromodels are given for voltage comparators and operational amplifiers. The description given for each drop-in macromodel includes the functional block diagram, input–output equations, design equations and methodology, a circuit diagram, the SPICE input control file, and at least one practical example of each macromodel with actual simulation results.

Chapter 4 explains how several of the behavioral drop-in macromodels are interconnected to form an APLL (Analog Phase Locked Loop). A phase detector, VCO (voltage controlled oscillator), and low-pass filter are used to produce the complete APLL macromodel. The functional block macromodels are verified with simulations for various test signals.

In Chapter 5, an APLL system is designed to model the NE565 integrated circuit phase locked loop. Simulation results are compared with experimental measurements to show that the PLL accurately and efficiently models system operation in all four regions of operation: unlocked state, locked mode, acquisition, and loss of lock regions. To verify its practicality, a FSK (frequency shift key) network is simulated.

Chapter 6 describes a computer program written in BASIC that permits the user to select from a menu a macromodel to be constructed . The menu contains nine macromodels: analog phase locked loop, continuous time domain transfer function, peak detector, sample-and-hold circuit, Schmitt trigger, noise source, piecewise linear circuit, discrete time domain transfer function, and an amplifier. Once the user selects the desired model, the program prompts for the specific performance information required. Once all questions have been answered, the program creates a SPICE subcircuit macromodel file having the required topology, elements, and component values. Diskettes are included so readers can rapidly, accurately, and automatically generate macromodels using an IBM or compatible personal computer. Chapter 6 provides a complete listing for all nine macromodels which have been programmed for automatic computer generation plus the menu program. The chapter contains a complete listing of the source programs for these nine models. Users having some familiarity with BASIC programming can easily adapt and modify these programs to include many more of the models covered in the book or can extend these computer programs to their own specialized macromodels.

Any serious SPICE user no doubt has experienced frustration with convergence problems in SPICE. Appendix A explains an orderly procedure to follow for correcting the bias point and transient analysis convergence problems often encountered. Also, a listing of error messages and their interpretation is given.

The PROBE feature available in PSpice versions from MicroSim Corporation is a very powerful tool that allows quick and accurate determination of many key circuit performance indexes. The regular mathematical functions in PROBE are explained in Appendix B along with some special graphical tricks for plotting key asymptotic lines and curves. Another useful feature of the PROBE command is the ability to merge two or more PROBE plots from two separate simulations into a single graph. In short, this appendix shows how to extend the capabilities of the postprocessor PROBE function.

The last section, Appendix C, is intended as a quick reference for the four parts of the SPICE program—that is, the elements, devices, subcircuits, and control statements. Figures and examples are provided to help the reader quickly locate and specify the correct format for each of these parts of SPICE.

Many of the macromodels presented in this book have been used in a variety of classes at Georgia Tech. The s- and z- domain macromodels have enabled students studying operational amplifiers and active and switched-capacitor filters to simulate easily complicated and involved behavioral transfer functions. Graduate students in a Frequency Synthesizers course have been able to simulate and better understand how a phase locked loop functions in all four regions of operation. Furthermore, they have been able easily to change the type and characteristics of the low-pass filter and to determine the important effects this has on the loop dynamics, the lock and tracking ranges, and the overall stability. Students in a Low Noise Electronic Design course have been able to model amplifier noise performance using the included macromodels. Also, they can quickly determine the noise bandwidth and the equivalent input and output noise levels in their circuits using the integration capabilities of PROBE with PSpice. The multipliers, mixers, phase shifters, and so on, have been used to demonstrate various modulation techniques in communication courses. It is hoped and expected that there will be virtually no end to the applications for these macromodeling tools in the hands of clever students, professors, and practicing engineers.

ACKNOWLEDGMENTS

It is not possible to enumerate all those whose assistance contributed to completing this book, but there are some who deserve special acknowledgment. We wish to express thanks to the many students who helped in the development of this book. Special thanks go to Ahmad Dowlatabadi, Fred Pruner, Katherine Taylor, Jim Caravella, Tim Holman, Mark Thrower, and Craig Hollabaugh. Appreciation is expressed to fellow faculty members at the Georgia Institute of Technology, notably Doctors P. E. Allen, M. Brooke, G. Casinovi, R. K. Feeney, D. R. Hertling, and W. E. Sayle. The encouragement and assistance of many people, especially Dr. Roger P. Webb, was essential to the success of the book. We also recognize the help and cooperation of Doctors E. J. Kennedy of the University of Tennessee and Eugene Chenette of the University of Florida.

The guidance and leadership of our editor, Elizabeth Kaster, is gratefully acknowledged, and the suggestions of the many reviewers are very much appreciated, especially Doctors Paul D. Stigall of the University of Missouri (Rolla) and James E. Morris of SUNY, Binghamtom. We also appreciate the thoroughness of our production editor, Irwin Zucker. The financial support of the Semiconductor Research Corporation is certainly recognized.

We gratefully acknowledge the support, patience, and understanding of our wives, Mary Nelle and Yeong Ae.

J. Alvin Connelly
Pyung Choi

1

Introduction

1.1 HISTORY OF SPICE

SPICE is a powerful, general-purpose circuit analysis program that simulates analog circuits. SPICE (an acronym for Simulation Program with Integrated Circuit Emphasis) is by far the most popular analog circuit simulation program being used today by both practicing engineers and students. SPICE was developed by the Integrated Circuit Group of Electronics Research Laboratory and the Department of Electrical Engineering and Computer Sciences at the University of California, Berkeley, California in the late 1960s and was released to the public in 1972.[1] The person credited with originally developing SPICE is Dr. Lawrence Nagel, whose Ph.D. thesis describes the algorithms and numerical methods used in SPICE.

Over the years, SPICE has gone through many upgrades. The most significant came with SPICE2, in which the kernel algorithms were upgraded to support advanced integrated system methods, many of which relate to IC performance. SPICE2 has virtually replaced SPICE1 as the SPICE choice. SPICE2 has been ported to numerous types of mainframe computers and personal computers, and to various operating systems. Because the development of SPICE2 was supported using public funds, this software is in the public domain, which means it may be used freely by any U.S. citizen. SPICE2 has become the industry standard and is simply referred to as SPICE. It is a large (over 17,000 lines of FORTRAN source code), powerful, and extremely versatile industry-standard program for circuit analysis and IC design.

Recently SPICE2 was upgraded to SPICE3. In the newer version, the program was converted from FORTRAN to C for easier portability. Also, several devices were added to the program library, such as a varactor, semiconductor resistor, and lossy RC transmission line models. However, the kernel algorithms were not changed, and the added components are not too significant because SPICE2 can simulate all the devices built into SPICE3 using external device modeling techniques.

Today there are more than 35 SPICE derivative programs, many known by associated acronyms such as HSPICE and RAD-SPICE (from Meta-Software), IG-SPICE (from A. B. Associates), I-SPICE from NCSS timesharing), PSpice (from MicroSim), IS_Spice (from Intusoft), SLICE (from Harris), ADVICE (from AT&T Bell Laboratories), Precise (from Electronic Engineering Software), and ASPEC (from Control Data Corporation).[2] In 1984 MicroSim introduced two versions of SPICE, called PSpice, that run on an IBM personal computer. The educational or student version of PSpice is distributed free by MicroSim. This has made SPICE available to vast numbers of students and has caused a rethinking of the way classes and laboratories are being taught. The student version of PSpice that is PC-based is limited to circuits with approximately 10 or fewer transistors. However, the professional (or production) version can simulate a circuit with up to 200 bipolar transistors or 150 MOSFETs.

SPICE and other simulation programs will be around and will be widely used for many years to come. Students will find SPICE an important tool for learning circuit analysis and design, and for testing electronic circuits in ways not easily done in most college laboratories.

1.2 SPICE CAPABILITIES

SPICE contains built-in models for passive elements such as resistors, capacitors, inductors, and transmission lines; semiconductor devices such as diode, bipolar transistor, junction field-effect transistor (JFET), and metal oxide semiconductor field-effect transistor (MOSFET); independent voltage and current sources, and dependent voltage and current sources including linear and nonlinear types (VCVS, VCCS, CCVS, and CCCS). By including control lines in an input file, SPICE can be made to perform many kinds of analyses of a circuit, as follows:

1. Nonlinear DC analysis, which determines the DC operating point of the circuit
2. Linear small signal analysis, which calculates the frequency response of the circuit
3. Transient analysis, which determines the response as a function of time over a specified time interval
4. Small signal DC transfer function analysis of a circuit from a specified input to a specified output (input resistance, output resistance, and transfer function)
5. DC small signal sensitivity analysis of one or more specified output variables with respect to every parameter in the circuit
6. Distortion analysis with AC analysis
7. Noise analysis with AC analysis, which determines the equivalent output and input noise at specified output and input nodes
8. Fourier analysis of an output variable when done with a transient analysis
9. Temperature analysis[3]

1.3 LIST OF SYMBOLS AND NOTATION

Independent sources. General form for voltage and current sources

```
Vxxxxxxx  N+  N-  <<DC>  DC/Tran Value>
                  <<AC> <ACMAG<AC Phase>>>

Iyyyyyyy  N+  N-  <<DC>  DC/Tran Value>
                  <<AC> <ACMAG<AC Phase>>>
```

Here are some examples.

```
VCC  10  0  DC  6
```

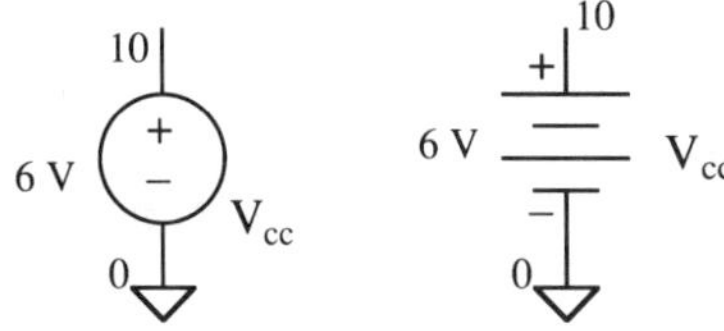


```
VIN  13  2  DC  0.001  AC  1  SIN(0 1 1MEG)
```

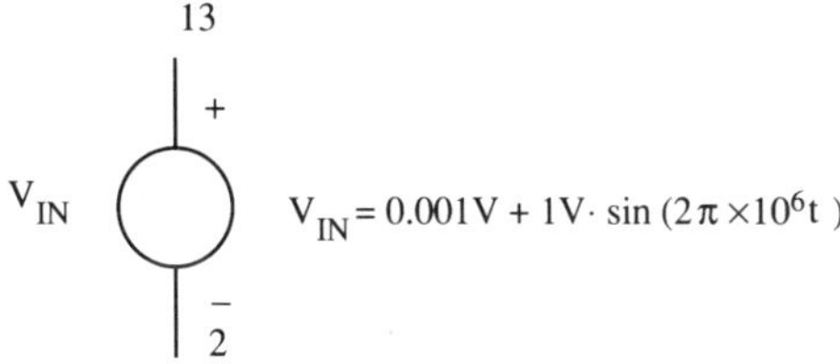


```
ISOURCE  40  30  DC  1A  AC  1MA
```

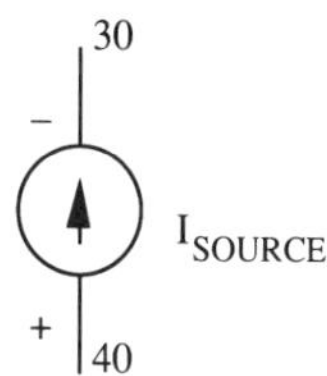

Note: Positive current is assumed to flow from the positive node, through the source, to the negative node. A current source of a positive value will force current to flow out of the N+ node, through the source, and into the N− node. Voltage sources, in addition to being used for circuit excitation, are the ammeters for SPICE. Zero-valued voltage sources may be inserted into the circuit for the purpose of measuring current.

They will have no effect on circuit operation because they represent short circuits. To find the AC current through a sample resistor use the following commands:

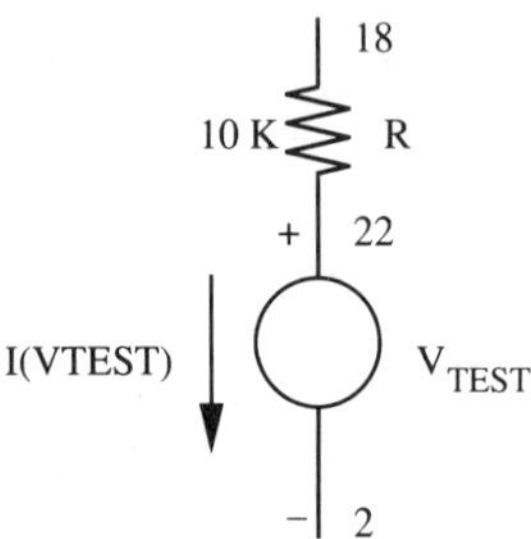


```
VTEST   22   2   DC   0V   AC 0V
.PRINT   AC   I(VTEST)
```

Dependent sources. Linear Voltage Controlled Voltage Source (VCVS)

```
Exxxxxxx  N+  N-  NC+  NC-  VALUE
E1  2  3  14  10  1.5
```

14, 10, 2, 3, E_1

$E_1 = 1.5[V(14) - V(10)] = 1.5[V(14,10)]$

Linear Voltage Controlled Current Source (VCCS)

```
Gxxxxxxx  N+  N-  NC+  NC-  VALUE
GTEST  4  12  100  16  8.4
```

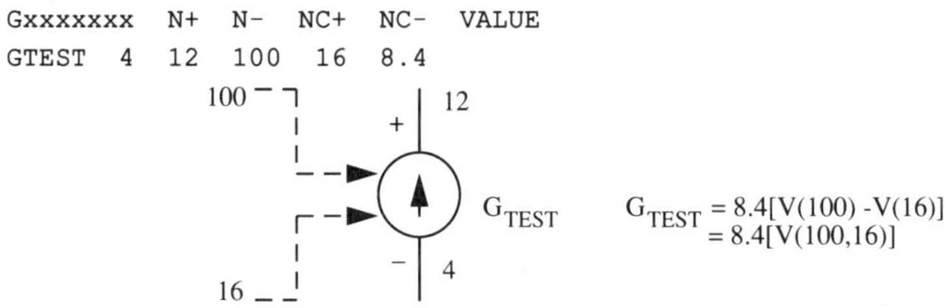

Linear Current Controlled Voltage Source (CCVS)

```
Hxxxxxxx  N+  N-  VNAME  VALUE
HSOURCE  9 16  VZ  0.5K
```

$I(V_Z)$, V_Z, 9, 16

$H_{SOURCE} = 0.5\ K\ I(V_Z)$

Linear Current Controlled Current Source (CCCS)

```
Fxxxxxxx  N+  N-  VNAME  VALUE
F1  13  7  VSENSE  5
```

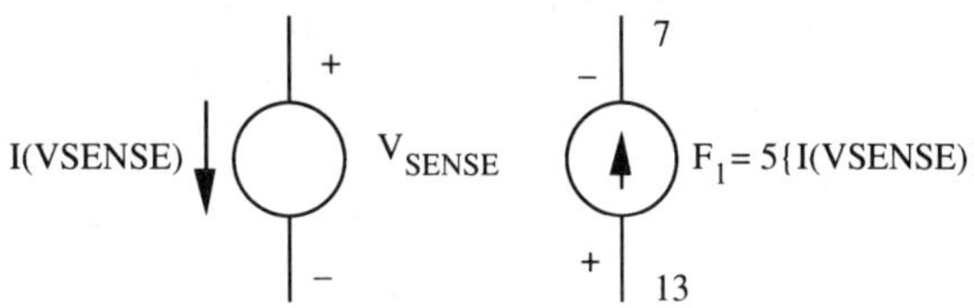

Nonlinear dependent sources (polynomial sources)

Symbols

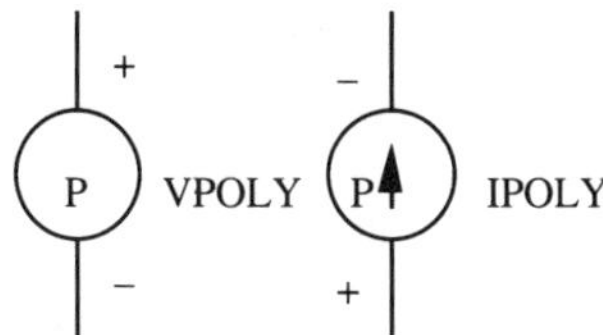

Dimensions

One : $fv = P_0 + P_1F_a + P_2F_a^2 + P_3F_a^3 + \cdots + P_nF_a^n$

Two : $$fv = P_0 + P_1F_a + P_2F_b + P_3F_a^2 + P_4F_aF_b + \\ + P_5F_b^2 + P_6F_a^3 + P_7F_a^2F_b + P_8F_aF_b^2 + P_9F_b^3 + \cdots$$

Examples

```
Gxxxxxxx  N+  N-  <POLY(ND)>  NC1+  NC1-  NC2+  NC2-  ···
       <P0  P1  P2  ···  >  <IC=   ···  >
GMLT  23  17  POLY(2)  3  5  1  2  0  1M  17M  3.5U
```

17

$GMLT = 10^{-3}[V(3,5)] + 17 \times 10^{-3}[V(1,2)] + 3.5 \times 10^{-6}[V(3,5)]^2$

+ 23

Summers

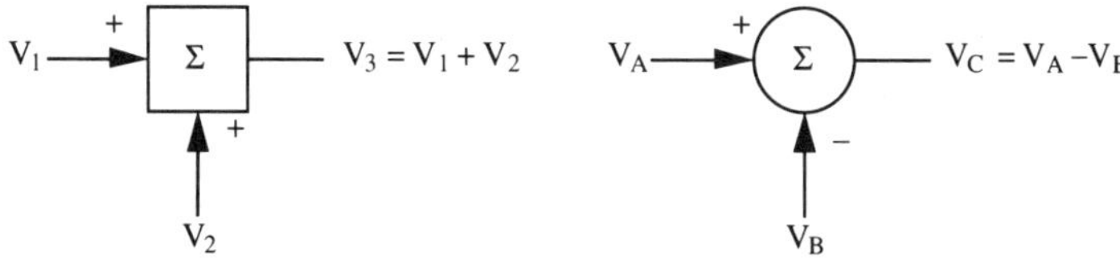

Multipliers

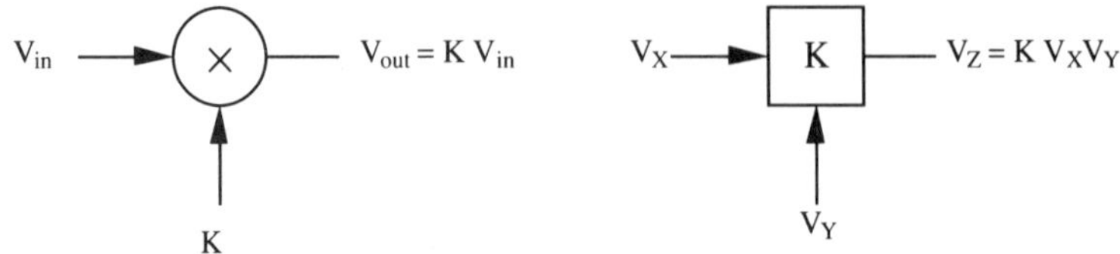

1.4 SUBCIRCUITS

A subcircuit statement or model looks like subroutine function in the FORTRAN program. This is a very useful and efficient way to duplicate a repetitive circuit in a large circuit or system. The definition of a subcircuit begins with the .SUBCKT statement and ends with a .ENDS statement. All the statements between a .SUBCKT and .ENDS are included in the definition. When the subcircuit is called by an X statement, all the statements in the definition replace the calling statement.

Subcircuit calls may be nested. That is, an X statement may appear between a .SUBCKT and a .ENDS. However, subcircuit definitions may not be nested. That is, a .SUBCKT statement may not appear in the statement between a .SUBCKT and a .ENDS.

Subcircuit definitions should contain only device statements (statements without a leading ".") and possibly .MODEL statements. Models defined within a subcircuit definition are available only within the subcircuit definition in which they appear. Also, if a .MODEL statement appears in the main circuit, that model is available in the main circuit and all subcircuits.

Node, device, and model names are local to the subcircuit in which they are defined. It is permissible to use a name in a subcircuit that has already been used in the main circuit. When the subcircuit is expanded, all its names are prefixed with the subcircuit identification name. For example, "Q13" becomes "X3.Q13" and node "5" becomes "X3.5" after expansion.

As an example of how to use the subcircuit feature and as an introduction to macromodeling, consider an operational amplifier (op amp) having the following characteristics:

Open loop voltage gain = OLG = 200K V/V
Differential input resistance = 2M Ω
Output resistance = 75 Ω
Gain-bandwidth product = GBW = 1 MHz

First find the location of the dominant pole assuming a one-pole frequency response.

$f_{dp} = \mathrm{GBW/OLG} = 10^6/200\mathrm{K} = 5$ Hertz $= 1/(2\pi^*\mathrm{CDP}^*\mathrm{RDP})$
Choose RDP $= 10^6\Omega$, which gives CDP $= 31.85$nF

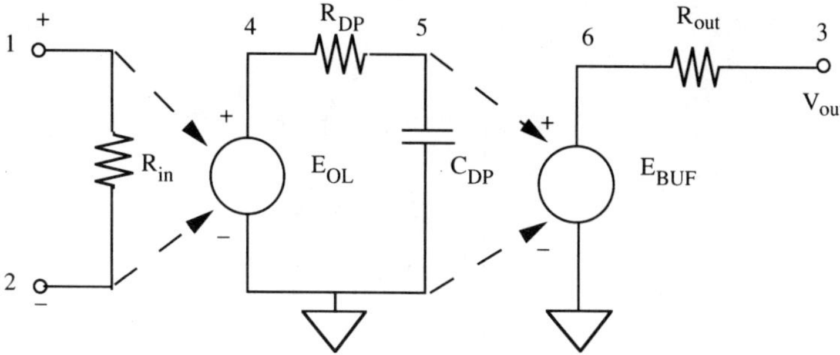

Figure 1.1 Op amp macromodel

The op amp model is constructed as shown in Figure 1.1, with its associated subcircuit model file.

SPICE code for Figure 1.1

```
*OP AMP SUBCIRCUIT
.SUBCKT OA741 1 2 3
* +IN  -IN    OUT
*   1       2     3
RIN 1 2 2MEG
ROUT 6 3 75
EOL 4 0 1 2 200K
RDP 4 5 1MEG
CDP 5 0 31.85NF
EBUF 6 0 5 0 1
.ENDS OA741
```

Next this op amp macromodel can be incorporated into a complicated active filter circuit to determine the frequency response. As an example consider the fifth-order low-pass filter shown in Figure 1.2. The subcircuit model is used once instead of using three different op amp circuits. The control file for Figure 1.2 is shown in Figure 1.3.

1.5 EXTENDING SIMULATION CAPABILITIES

Sweeping component values. In SPICE the .DC command can sweep model parameters for components and voltages or currents. This is accomplished by extending the definition of the source name to include a reference to the model parameter. The statement in an input PSpice file of

```
.DC  VIN 2 20 1
```

will sweep the DC voltage of the VIN source from 2 volts to 20 volts in 1 volt steps. As another example, the resistance multiplier in RMOD can be swept from 100 ohms to

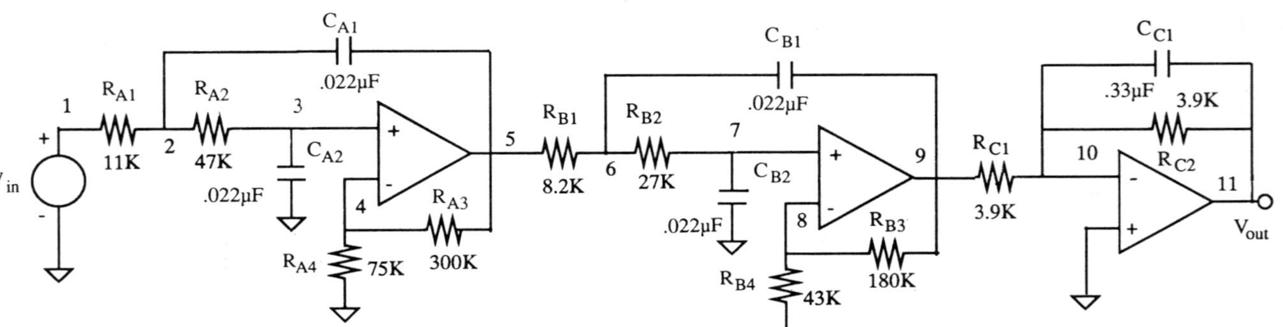

Figure 1.2 Fifth-Order active filter using subcircuits for the op amps.

```
MACROMODEL OP AMP ACTIVE FILTER EXAMPLE
* OP AMP SUBCIRCUIT
*
.SUBCKT OA741 1 2 3
* +IN  -IN    OUT
*   1       2      3
RIN 1 2 2MEG
ROUT 6 3 75
EOL 4 0 1 2 200K
RDP 4 5 1MEG
CDP 5 0 31.85NF
EBUF 6 0 5 0 1
.ENDS OA741
* FIRST STAGE
*
RA1 1 2 11K
RA2 2 3 47K
RA3 4 5 300K
RA4 4 0 75K
CA1 2 5 0.022UF
CA2 3 0 0.022UF
X1 3 4 5 OA741
* SECOND STAGE
*
RB1 5 6 8.2K
RB2 6 7 27K
RB3 8 9 180K
RB4 8 0 43K
CB1 6 9 0.022UF
CB2 7 0 0.022UF
X2 7 8 9 OA741
* THIRD STAGE
*
RC1 9 10 3.9K
RC2 10 11 3.9K
CC1 10 11 0.33UF
X3  0 10 11 OA741
* INPUT SIGNAL AND CONTROLS
*
VIN 1 0 DC 0 AC 1V
.TF V(11,0) VIN
.AC DEC 10 1 1MEGHZ
*USE PROBE COMMAND WITH MICROSIM PSPICE VERSION
*INCLUDE PRINT AND/OR PLOT STATEMENTS WITH
*REGULAR SPICE SIMULATIONS
.PROBE
.END
```

Figure 1.3 Input control file for active filter showing subcircuits

200 ohms in 5 ohm increments by using the following PSpice file:

```
RVARY 4 6 RMOD 1
.MODEL RMOD RES(R=100)
.DC RES RMOD(R) 100 200 5
```

Note that if other resistors in the circuit to be simulated are identified as RMOD, they will be swept as well. A technique for modeling time-dependent elements in SPICE is described in Connelly and Fisher.[4]

Sweeping temperature. The statement

```
.DC TEMP -55 125 5
```

will sweep the operating temperature from −55°C to +125°C in 5°C increments.

Modeling nonlinear temperature effects. Linear and quadratic functions of temperature are modeled in SPICE by using the command

```
R1 12 24 10K TC=0.001,0.015
```

The R1 resistor becomes a function of temperature according to

$$R1 = 10K[1 + 0.001(\text{TEMP} - \text{TNOM}) + 0.015(\text{TEMP} - \text{TNOM})^2]$$

Zener diode model. There is no special built-in model for a zener diode in SPICE. The regular diode model can be used for a zener by setting the reverse breakdown voltage, BV, and current, IBV, to match the desired zener characteristics. Note that the default value of IBV is 100 pA. The example in Figure 1.4 is for a 6.2 V zener with a reverse breakdown current of 1 μA.

SPICE code for Figure 1.4

```
DZENER 15 17 DZ
.MODEL DZ D(BV=6.2V IBV=1UA)
```

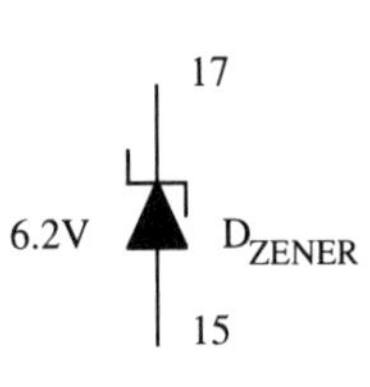

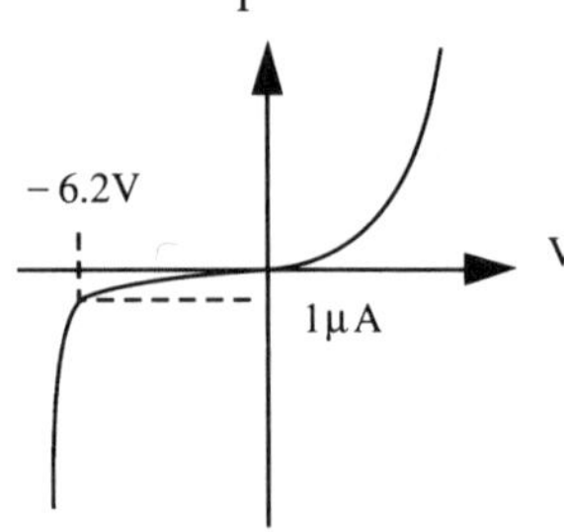

Figure 1.4 Regular SPICE zener diode model

The simulated V-I characteristics in Figure 1.5 show that the knee is not as sharp as is typical of most practical zener diodes. Expanded views around the forward and reverse voltage breakpoints indicate that the knees are almost identical.

An improved zener diode model is shown in Figure 1.6. The diode D2 is modeled as an "ideal" diode, and the VZENER supply sets the desired breakdown voltage. Resistor RF models the diode resistance in the zener region. The second diode D1 accounts for the regular diode characteristics in the forward direction. By setting N = 0.001 in the MODEL statement for DIDEAL, a sharp knee is produced as shown in Figure 1.7. If a model with more breakpoints is desired, see Section 2.4, "Piecewise Linear Approximations."

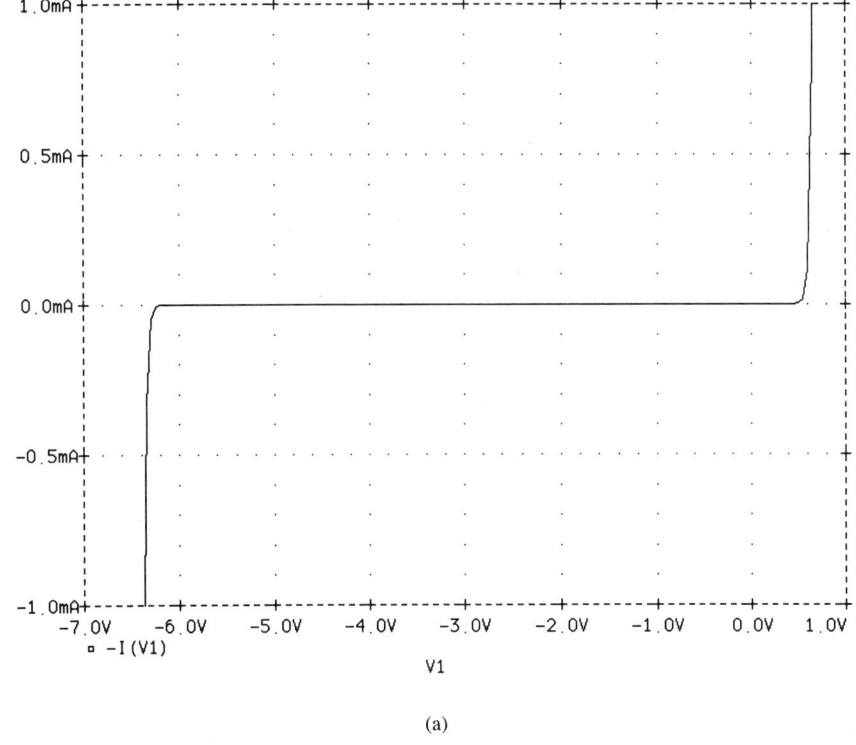

(a)

Figure 1.5 Simulation results for regular zener diode. (a) Forward and reverse characteristics. (b) Expanded reverse region. (c) Expanded forward region.

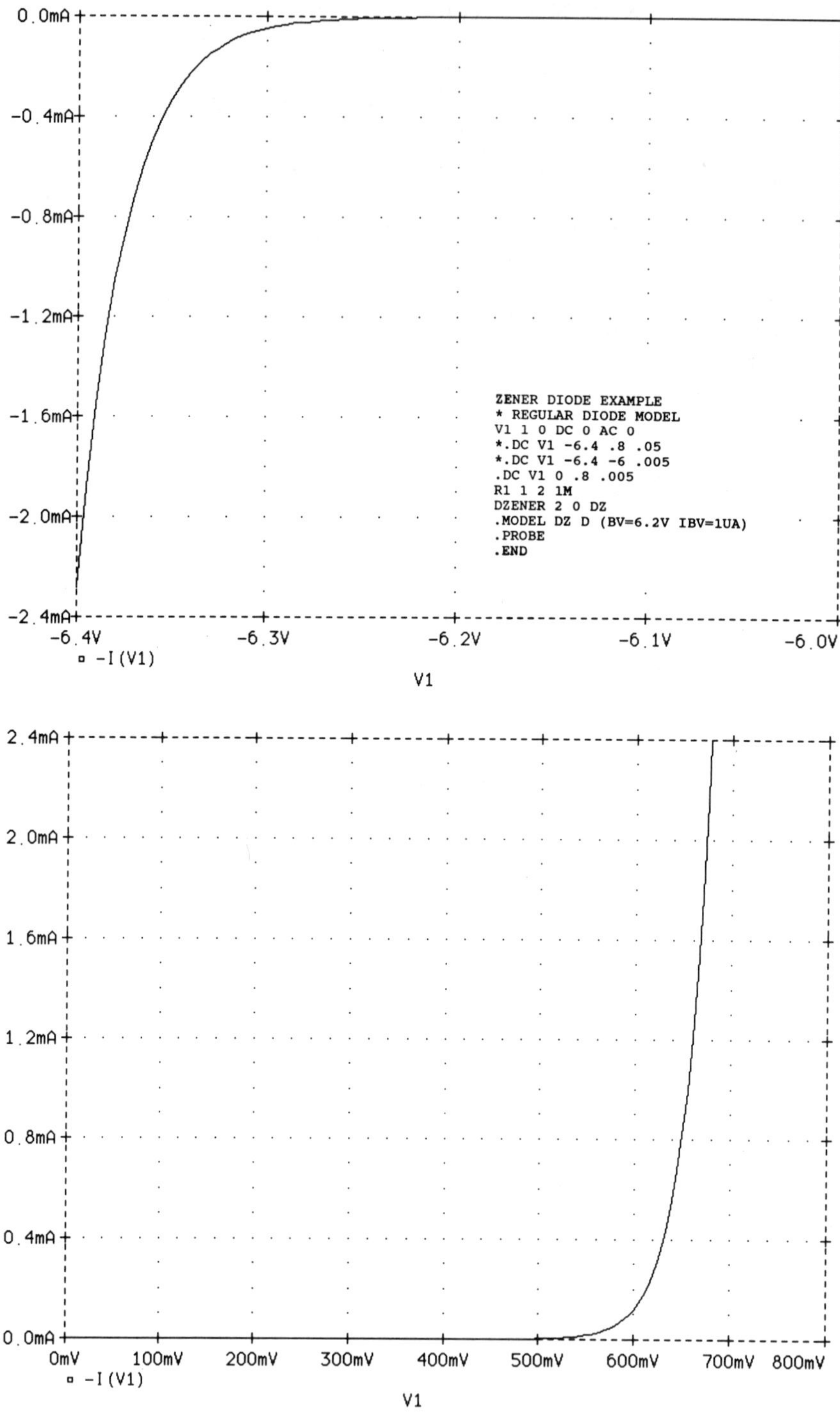

Figure 1.5(b) and (c)

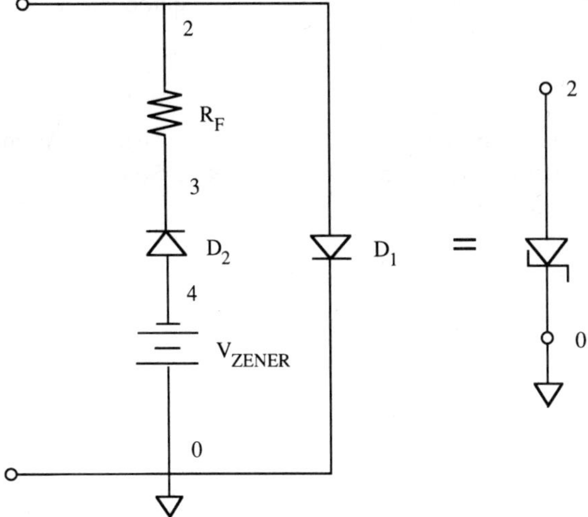

Figure 1.6 Improved zener diode model

SPICE code for Figure 1.6

```
D1 2 0 DREG
D2 4 3 DIDEAL
RF 2 3 10
VZENER 0 4 DC 6.2V
.MODEL DREG D
.MODEL DIDEAL D (N=0.001)
```

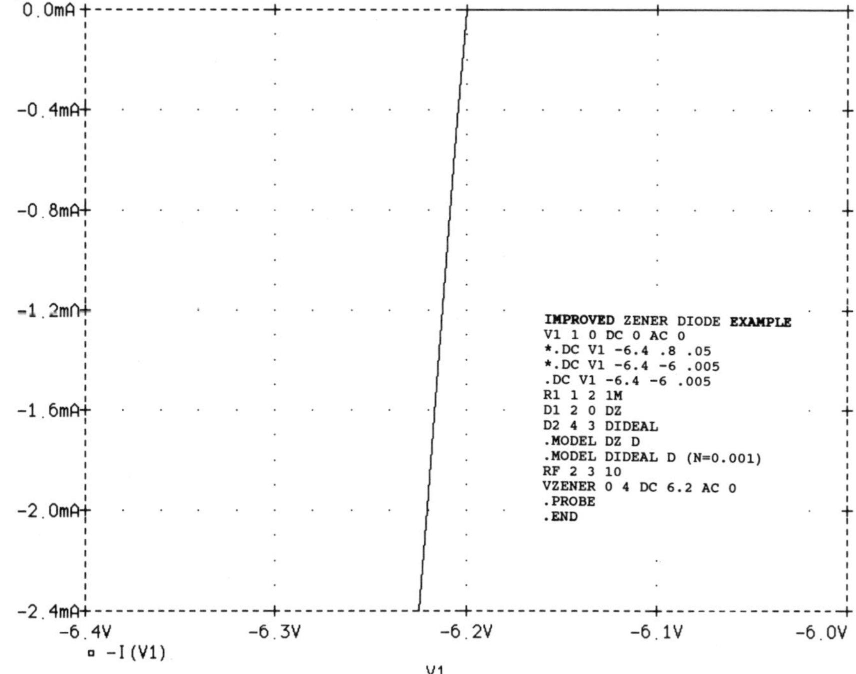

Figure 1.7 Simulation results for improved zener diode

1.6 SUMMARY

A short history of SPICE and its capabilities were reviewed in this chapter. An introduction was provided for using subcircuits, together with examples. Some ways to extend the simulation capabilities of SPICE were also presented.

REFERENCES

1. T. J. Byers, "SPICE—Breadboards Are Giving Way to PC-Based Circuit Simulation," *Radio Electronics*, November 1990, pp. 63–68.
2. Craig E. Smith and Ivin L. Holt, "Adding SPICE to the Electronics Industry," *Printed Circuit Design*, September 1990, pp. 39–47; Muhamid H. Rashid, *SPICE for Circuit and Electronics Using PSpice*, Englewood Cliffs, N.J.: Prentice Hall, 1990), pp. 1–3; Paul W. Tuinenga, *SPICE: A Guide to Circuit Simulation and Analysis Using PSpice*, Englewood Cliffs, N.J.: Prentice Hall, 1988), pp. xvi–xvii.
3. Paolo Antognetti and Giuseppe Massobrio, *Semiconductor Device Modeling with SPICE* (New York: McGraw-Hill, 1988).
4. J. A. Connelly and G. J. Fisher, "Modeling Time-Dependent Elements for SPICE Transient Analysis," *IEEE Transactions on Computer-Aided Design*, Vol. CAD-5, No. 3, July 1986, pp. 429–432.

2

Macromodeling Methodology

Macromodeling methodologies for algebraic functions, piecewise linear functions, series expansions, and s- and z-domain transfer functions are discussed. These methodologies not only are applicable to modeling complex analog circuits and systems, but also improve both simulation capability and flexibility of analog simulators without internal modifications. All equivalent circuit models developed by these methodologies are used directly for SPICE simulations, since the elements in the macromodels are standard SPICE primitives.

2.1 INTRODUCTION

Circuit simulation programs such as SPICE are useful tools for the circuit or system designer;[1] they allow the prediction of behavior performance of complicated circuits and systems without solving loop and node equations or building a breadboard. Computer simulation can save considerable time and money in the design cycle. However, the simulation of some complex analog circuits and systems frequently requires a circuit element or a functional block model not included in the simulator's standard category. Often these functional limitations prevent or retard the use of the simulator for the analysis of even a small system. Thus one of the important tasks in computer-aided analysis of electronic circuits and systems is the development of higher level models that are simple, computationally efficient, and sufficiently accurate.[2]

This chapter discusses generalized macromodeling methodologies for various analog functions. Included are algebraic functions and series expansions for analog signal processing, piecewise linear (PWL) approximations for nonlinear characteristics, and linear s- and z-domain transfer functions. These modeling techniques improve both the generality and capability of general-purpose circuit simulation programs and are applicable to large-scale integrated circuit analysis. The circuit buildup approach is employed to construct equivalent circuit models of the analog functions.

2.2 A GENERAL PROCEDURE FOR MACROMODEL DEVELOPMENT

A general procedure that is used to create an equivalent circuit structure macromodel is introduced. Figure 2.1(a) shows that the first step is to identify system performances and to select the electrical characteristics to be modeled. Selection of performance parameters and terminal behaviors is very important because the computational efficiency usually depends on the size of the macromodel. If a highly accurate model is required, it should include almost all characteristics. As a result, the size of the macromodel increases, which requires a long computational time.

For the functional description of the system, a careful partitioning is necessary since each functional block will become a submodel of the whole system macromodel. After partitioning based on performance functions, the system can be alternatively described by a functional block diagram, a pictorial representation of the functions performed by components and of the flow of signals. In this case each functional block includes its transfer characteristic and terminal behavior.

The second step in the general procedure is to understand the simulation program that is to be used as the modeling tool. Among the various kinds of analog circuit simulation programs, the model developer should choose a circuit simulator with good

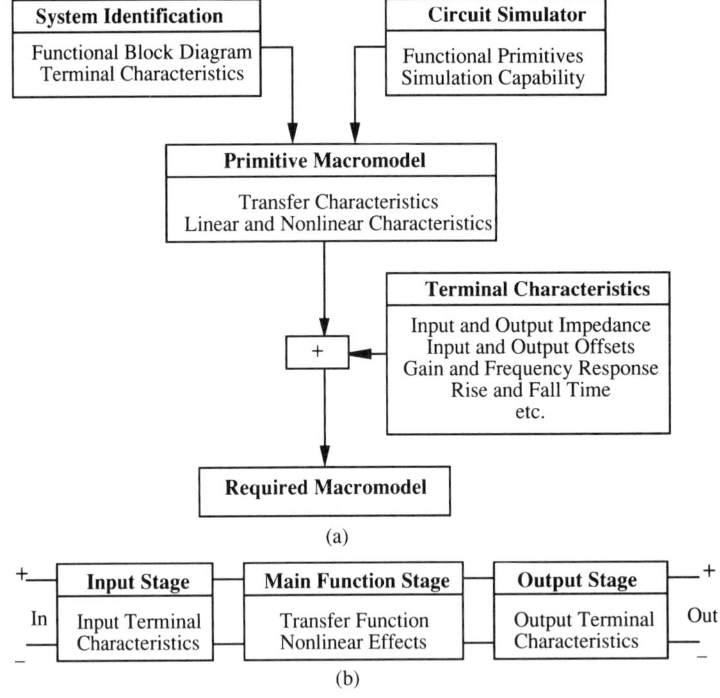

Figure 2.1 General procedure for macromodel development. (a) Developing steps. (b) Basic sections of macromodel.

flexibility, generality, and wide acceptance. The functional block diagram of the system gives basic information such as the required functional primitives, including passive elements, independent and dependent sources, and semiconductor device models. Furthermore, the circuit description language, analysis capabilities, CPU-run control file, and special facilities are also essential concepts and skills that must be mastered.

Combining the preceding two steps, a simple equivalent circuit (primitive structure) is created on the basis of the functional block diagram. The simulator-supplied primitives and the principles of circuits and system theory are normally used in this procedure. The primitive model includes the major transfer function, plus input and output impedances of the original system. Thus the basic structure of the primitive macromodel shown in Figure 2.1(b) is normally composed of three sections:

1. The input stage for implementing the input impedance
2. The function stage for the realization of the circuit's functional behavior or transfer characteristic
3. The output stage, which duplicates the output impedance

To improve the accuracy of the primitive macromodel, nonlinear and terminal characteristics of the original system are added. The PWL approximation and series expansion technique are ways to model the various nonlinearities. The terminal characteristics include input and output offset voltages and currents, rise and fall times, dynamic range, gain and frequency responses, signal delay time, and the like. Most of these parameters are modeled in the main function stage of the macromodel. But the model developer should note that the macromodel complexity is usually determined by the required accuracy and computational efficiency. If a highly accurate macromodel is required, the complexity and computational efforts increase. Hence, a compromise between accuracy and computational efficiency is necessary to obtain acceptable accuracy and reasonable simulation time.

2.3 ARITHMETIC FUNCTIONS

The development of macromodels from the functional block diagram frequently requires many basic mathematical operations such as addition, subtraction, multiplication, division, differentiation, integration, and the like. In the SPICE program, four different types of dependent sources, configured as linear and polynomials, can be utilized to model these basic mathematical operations.[3] All possible voltage and current combinations of the dependent sources are available and are characterized by following four equations:

- V = E(V); Voltage Controlled Voltage Sources (VCVS)
- I = G(V); Voltage Controlled Current Sources (VCCS)
- V = H(I); Current Controlled Voltage Sources (CCVS)
- I = F(I); Current Controlled Current Sources (CCCS)

Voltage addition/subtraction. Any number of input voltages can be added (or subtracted) through the linear dependent sources or the n-dimensional polynomial dependent sources by properly setting the dimension, controlling sources, and coefficients. For example, consider a circuit block whose function is the addition of n-input voltages as shown in Figure 2.2(a). The transfer function of the example system is given by

$$V_{out} = V_{in}(1) + V_{in}(2) + V_{in}(3) + \ldots + V_{in}(n) \tag{2.1}$$

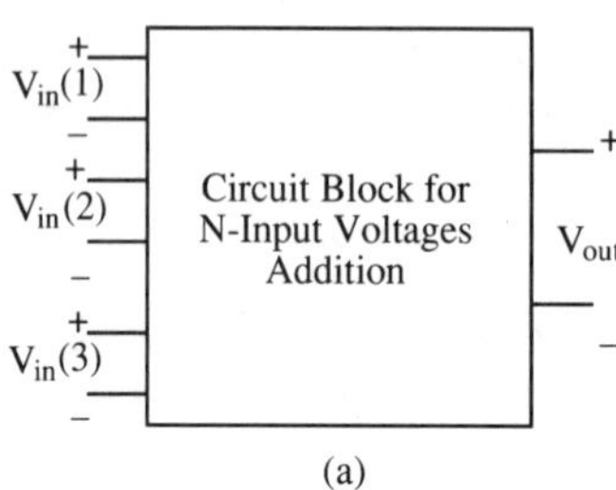

(a)

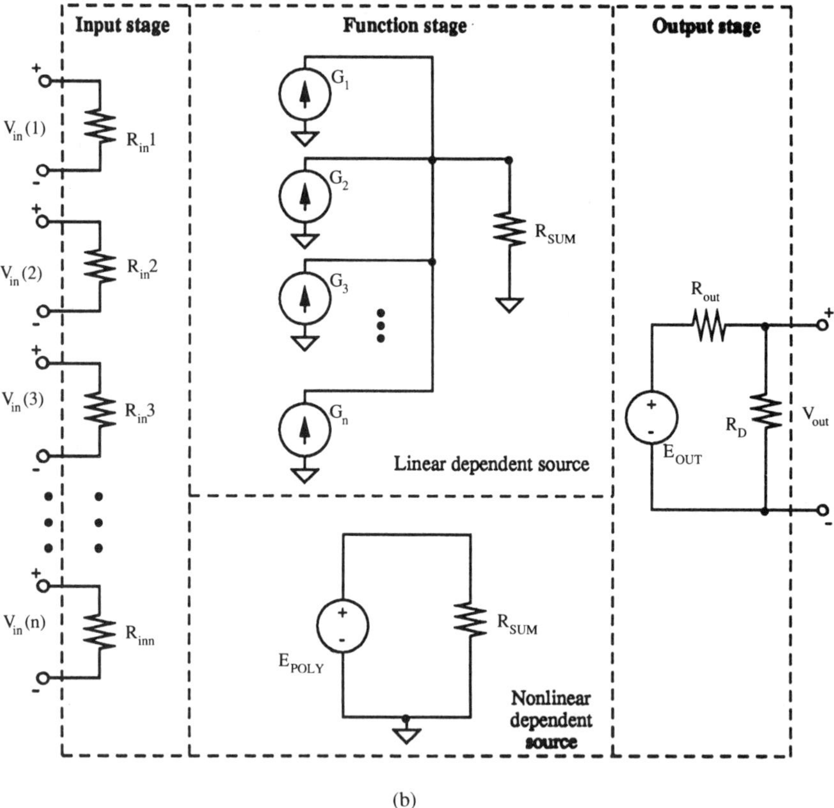

(b)

Figure 2.2 Addition/subtraction of n-input voltages. (a) Block diagram. (b) An equivalent circuit structure macromodel for (a).

Actually, there are many ways to model Equation (2.1) when linear dependent sources are used. Among many simple circuit configurations, the simplest equivalent circuit structure macromodel for Equation (2.1) is shown in Figure 2.2(b). It consists of three stages, input, function, and output just like those in the primitive macromodel. The resistor, R_{inn}, in the input stage is for modeling the input resistance of the n port in Figure 2.2(a). The voltage controlled current source, G_n, in the function stage is linear, and its controlling voltage is the voltage across at the input resistor, R_{inn}. The gain of each VCCS is unity. So the n-input voltage's addition function is modeled here by the proper combination of the linear dependent VCCSs in the function stage. The output resistance is modeled by the resistor, R_{out}, in the output stage, and resistor R_D is a dummy element to remove floating node.

The other type of equivalent circuit model uses the nonlinear (polynomial) dependent source such as VCVS or VCCS for the function stage instead of the linear dependent sources. The polynomial equation of the n-dimensional nonlinear dependent VCVS is given by

$$V_{out} = P_0 + P_1 V_{in}(1) + P_2 V_{in}(2) + \cdots + P_n V_{in}(n) + P_{n+1} V_{in}(1)^2 + \cdots \tag{2.2}$$

where P_n is the coefficient of the $V_{in}(n)$ input voltage. Thus, addition can be implemented by setting the P_n coefficients as follows;

$$P_0 = 0,\ P_1 = 1,\ P_2 = 1,\ P_3 = 1,\ \ldots,\ P_n = 1$$
$$\text{Rest of coefficients } = 0 \tag{2.3}$$

Voltage division. An equivalent circuit structure macromodel, using a linear dependent source in parallel with a polynomial dependent source, permits the division of two node voltages. Figure 2.3(a) shows a voltage division model that consists of the input, function, and output stages.

```
RIN1 1 3 [A]                 ; [A] = Input resistance I
RIN2 2 3 [B]                 ; [B] = Input resistance II
GDV1 0 4 1 3 [K1]            ; [K1] = Gain of GDV1
GDV2 4 0 POLY(2) 2 3 4 0
+ 0 0 0 0 [K2]               ; [K2] = Gain of GDV2
R1 4 0 1G
EOUT 5 7 4 0 1
R2 6 7 1G
ROUT 5 6 [C]                 ; [C] = Output resistance
```

In the function stage, the first VCCS, G_{DV1}, has a controlling voltage V_1 between nodes 1 and 3, and is linear. The second VCCS, G_{DV2}, is a two-dimensional polynomial with two controlling voltages. The first controlling voltage is the voltage between nodes 2 and 3, V_2, and the second is the voltage at node 4. Summing currents at node 4 gives

$$I(G_{DV1}) - I(G_{DV2}) - V(4,0)/R_1 = 0$$

$$K_1 \cdot V(1,3) - K_2 \cdot V(2,3) \cdot V(4,0) - V(4,0)/R_1 = 0$$

$$V(4,0) = \frac{R_1 \cdot K_1 \cdot V(1,3)}{1 + R_1 \cdot K_2 \cdot V(2,3)}$$

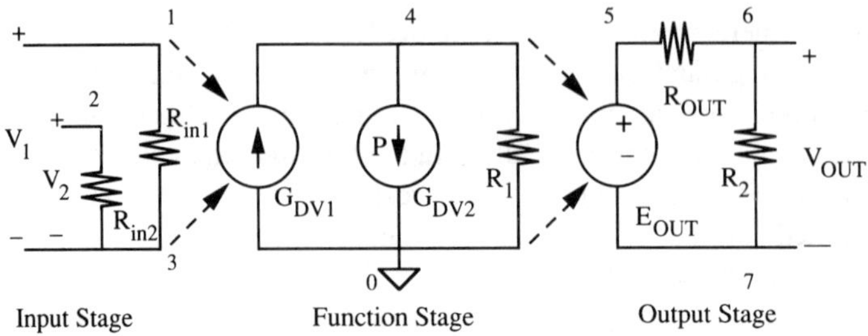

Figure 2.3 Equivalent circuit model for voltage division.

Making R_1 approach infinity forces the current $I(G_{DV1})$ to equal the current $I(G_{DV2})$ and makes the common voltage at node 4 approach the limit given by

$$K_1V(1,3) = K_2[V(2,3)V(4,0)]$$

$$V(4,0) = \frac{K_1 \cdot V(1,3)}{K_2 \cdot V(2,3)} \tag{2.4}$$

Voltage differentiation. Figure 2.4 shows an equivalent circuit model for performing differentiation. This model contains the same three stages: input, function, and output. The analog differentiator macromodel is formed by applying the input voltage across a capacitor. The current through this capacitor is converted to a voltage by forcing it through a resistance. Three VCVS elements are used to form the macromodel. The scaling factor K is the negative of the differentiator's time constant.

$$V_{out} = K \cdot \frac{d}{dt}(V_{in}) \tag{2.5}$$

SPICE code for Figure 2.4

```
RIN 100 101 [A]              ; [A] = Input resistance
E1 102 0 100 101 1
CTIME 102 103 [B]            ; [B] = See note
RTIME 103 104 [C]            ; [C] = See note
R2 103 0 1G
E2 104 0 103 0 -1G
EOUT 105 107 104 0 1
R3 106 107 1G
ROUT 105 106 [D]             ; [D] = Output resistance
```

Notes. To determine the values of [B] and [C], use the following equations:

$$[B] = [1E-3 * T]/[4 * \text{ Amplitude of the input signal}]$$

where T is the period of the input signal.

$$[C] = [1E3 * (\text{Desired amplitude of output})]$$

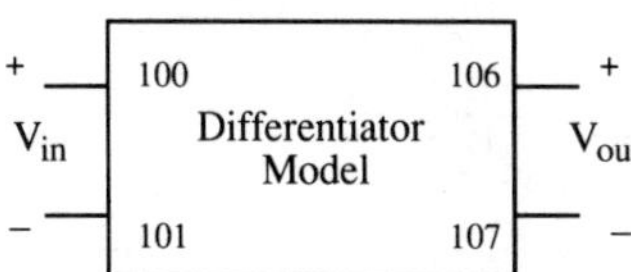

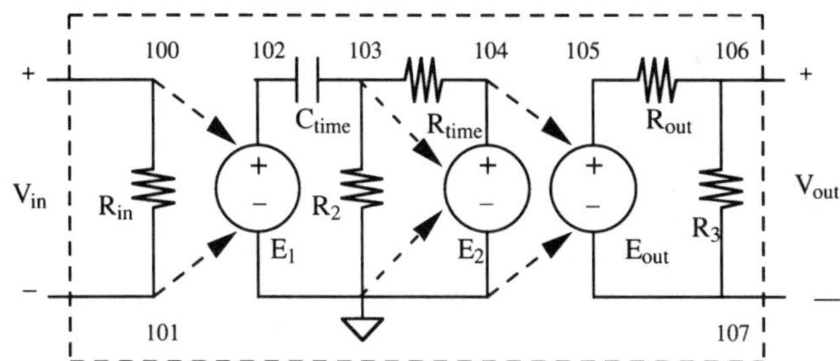

Figure 2.4 Differentiator macromodel: block diagram and equivalent circuit.

E_1 samples the input signal giving $E_1 = V_{in}$. E_2 and R_2 form an op amp model with the voltage gain of 1G and input resistance of $R_2 = 1G\Omega$. The $R_{time}C_{time}$ combination sets the time constant scaling factor. With R_2 and E_2 very large, the following equations describe the differentiator.

$$V(104,0) = (-sC_{time}R_{time})(E_1) \Rightarrow dV_{in}/dt$$

$$E_{out} = V(104,0) = -dV_{in}/dt. \quad (2.6)$$

For a further discussion and an example of this differentiator macromodel, see Section 3.2.10.

Voltage integration. Figure 2.5 shows a macromodel for an integrator. The analog integrator macromodel is formed by taking the voltage across a capacitor whose current is a linear function of the input voltage. Three VCVS elements are used to form the macromodel. The scaling factor K is the negative reciprocal of the integrator's time constant.

$$V_{out} = K \cdot \int V_{in} dt \quad (2.7)$$

SPICE code for Figure 2.5

```
RIN 100 101 [A]                ; [A] = Input resistance
E1 102 0 100 101 1
RTIME 102 103 [B]              ; [B] = See note
R2 103 0 1G
CTIME 103 104 [C]              ; [C] = See note
RF 103 104 [D]                 ; [D] = See note
E2 0 104 103 0 1G
EOUT 105 107 104 0 1
ROUT 105 106 [E]               ; [E] = Output resistance
R3 106 107 1G
```

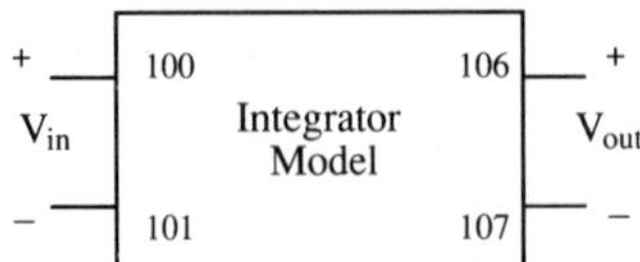

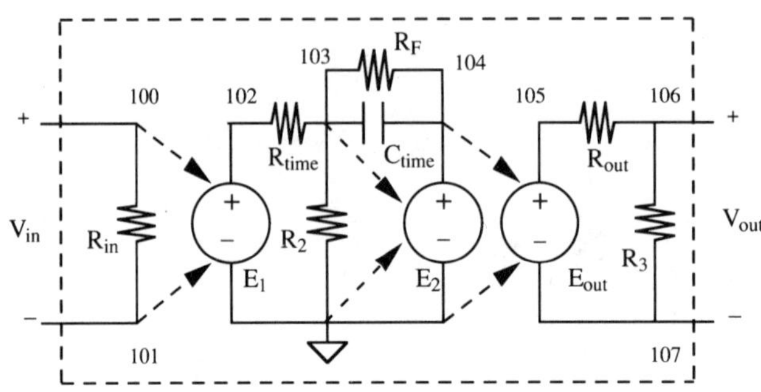

Figure 2.5 Integrator macromodel: block diagram and equivalent circuit.

Notes

$$[B] = \frac{\text{Amplitude of input signal}}{(2C_{time})(V_{p-p}\text{of output})(\text{Frequency of input signal})}$$

$$[C] = \text{Arbitrary Value}$$

$$[D] = 50(R_{time})$$

For the integrator macromodel, E_1 samples the input signal giving $E_1 = V_{in}$. E_2 and R_2 duplicate an op amp model with an open loop voltage gain of 1G and an input resistance of $R_2 = 1G\Omega$. The $R_{time}C_{time}$ combination produces the scaling time constant and integrates E_1. Feedback resistor R_F provides a DC stable circuit. The resulting equations are

$$V(104,0) = (-E1/R_{time})[R_F/(sR_FC_{time} + 1)] \tag{2.8}$$

By making $\omega R_F C_{time} \gg 1$ for all frequencies of interest, then

$$V(104,0) \approx [-1/(R_{time}C_{time})](V_{in}/s)$$
$$E_{out} = V(104,0) \approx -1/(R_{time}C_{time}) \int V_{in}dt \tag{2.9}$$

For a further discussion plus an example of this integrator macromodel, refer to Section 3.2.9. Also, many additional equivalent circuit drop-in models for advanced mathematical functions such as absolute value, series expansion, square, log and antilog functions, and the like are described in Chapter 3. All macromodels are primitive and cover only main transfer characteristics. They are directly applicable to SPICE or its extensions since all the elements are SPICE standard primitives.

2.4 PIECEWISE LINEAR APPROXIMATIONS

The nonlinear characteristics in analog functional modules[4] such as voltage/current saturation, hysteresis, and dead zone, cannot be described directly in a SPICE input file because SPICE does not include these nonlinear functional primitives. These functional limitations make it impossible to obtain complete solutions and also require long CPU times and large memory space if the nonlinear functional block is described by circuit element levels. However, a complete analysis of a circuit or a system containing nonlinear functional blocks or elements can be done if all the nonlinear parts are replaced by simple equivalent circuits.[5] These circuits are constructed with the standard elements recognized by the SPICE circuit simulators. Furthermore, these equivalent circuits enhance the flexibility and simulation capability of the existing simulators. Hence a generalized macromodeling methodology for nonlinear modules is presented to improve the accuracy and computational efficiency of macromodels.

The PWL function is one solution for modeling nonlinear characteristics, since a nonlinear function can be approximated by a sum of linear segments. These individual linear segments can be modeled easily by combining the following SPICE primitives:

- Modified diode model
- Dependent sources
- Independent voltage sources
- Resistors

The diode model in SPICE was used to formulate piecewise linear approximations. The diode acts as a voltage controlled switch, providing the capability to formulate PWL approximations. To develop the voltage controlled switch model from the SPICE diode model, consider the diode current equation given by

$$I_D = \text{Area} \cdot (I_n + I_b) \tag{2.10}$$

$$I_n = \text{Normal current} = I_S \cdot [e^{V/(N \cdot V_T)} - 1]$$

$$I_b = \text{Breakdown current} = I_{BV} \cdot e^{(-BV-V)/(N \cdot V_T)}$$

where

I_S = The saturation current of the diode (amps)
V = The voltage across the intrinsic diode (volts)
N = The emission coefficient
V_T = The thermal voltage (volts)
I_{BV} = The reserve current at the breakdown voltage (amps)
BV = The reverse breakdown voltage (volts)

If the default value of BV is used in the SPICE diode model, the breakdown current, I_b, will be automatically set to zero so that the normal current becomes the diode current.

Under this condition, the voltage across the diode can be expressed by

$$V = N \cdot V_T \cdot \ln\left(\frac{I_D}{I_S} - 1\right) \tag{2.11}$$

Equation (2.11) shows that the diode voltage is affected mainly by two diode model parameters: N and I_S. Between two model parameters, the emission coefficient is dominant and represents an efficient way to control the diode voltage because of the value compression of the natural logarithm. When the emission coefficient is less than the default value of 1, the voltage drop at the diode is reduced, while the diode current increases rapidly. This phenomenon is shown in Figure 2.6. Consequently, a very good voltage controlled switch model can be derived from the diode model by decreasing only the emission coefficient. The remaining diode model parameters use the default values.

To model the PWL functions, the voltage–voltage relationship is converted to the voltage–current domain because of the V-I characteristic of the modified diode model. As shown in Figure 2.7(a), the input voltage is converted to a current by the voltage–current block. Then the output current of the voltage–current block is converted to a voltage by current–voltage block. Thus the numerical values of breakpoints and slopes of the original function do not change. Figure 2.7(b) shows the basic structure of a generalized PWL macromodel, which consists of input, function, and output stages. The resistors R_{in} and R_{out} represent the input and output resistance of the actual functional modules

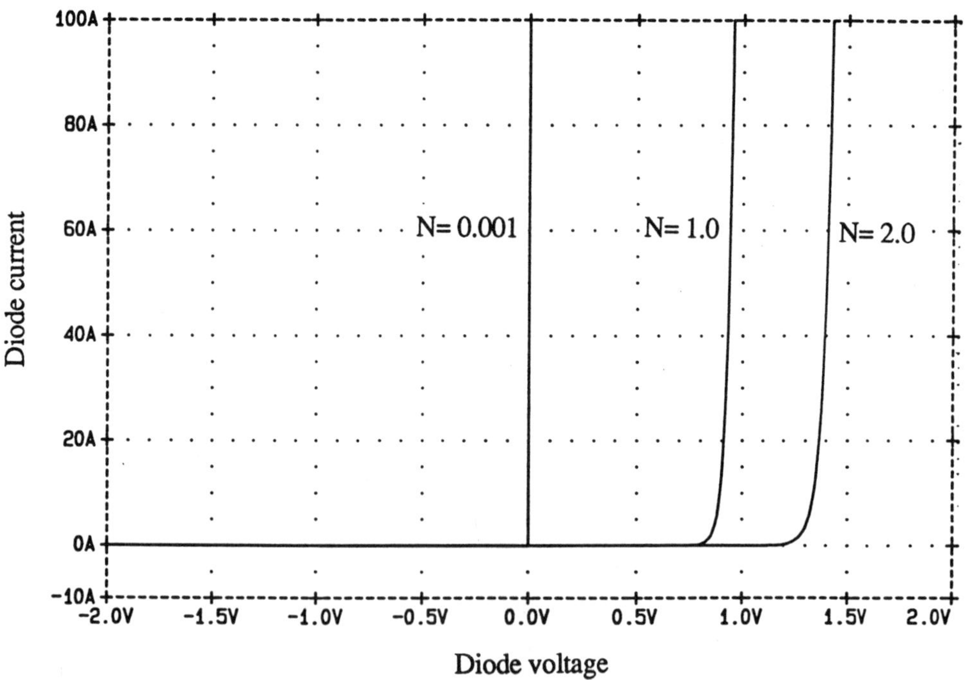

Figure 2.6 V-I characteristics of SPICE diode model with different emission coefficients.

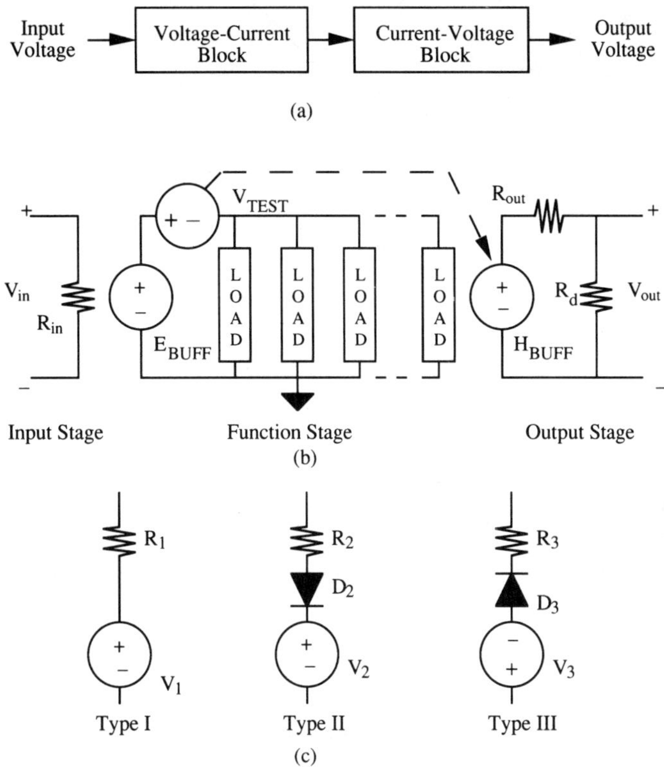

Figure 2.7 A general equivalent circuit model for PWL functions. (a) Functional block diagram. (b) Circuit schematic. (c) Types of parallel branch loads.

or systems. E_{BUFF} linearly reflects the input voltage to the function stage. If the gain of E_{BUFF} is unity, the function state voltage E_{BUFF} is identical to the input V_{in}. A dummy voltage source V_{TEST} (DC = 0 and AC = 0) measures the output current of voltage–current block (E_{BUFF}) since SPICE can measure only the current through independent voltage sources. Since there is no voltage drop at V_{TEST}, the parallel branches are directly affected by the V_{in}. H_{BUFF} in the output stage is a linear dependent CCVS whose controlling current is the current through V_{TEST} and whose gain is unity. This dependent source converts current to a voltage so that the input–output relationship is maintained in the voltage–voltage domain. The PWL characteristics are implemented by combinations of parallel branches that include three types of loads, as illustrated in Figure 2.7(c).

Now consider how the piecewise linear approximation function is modeled to the proposed equivalent circuit structure and how the element values are determined. Figure 2.8(a) and (b) illustrate simple PWL functions, which are composed of two linear segments. V_x is the input breakpoint voltage; M_0 and M_1 are slopes of the first and second linear segments, respectively. These PWL functions are implemented in the model

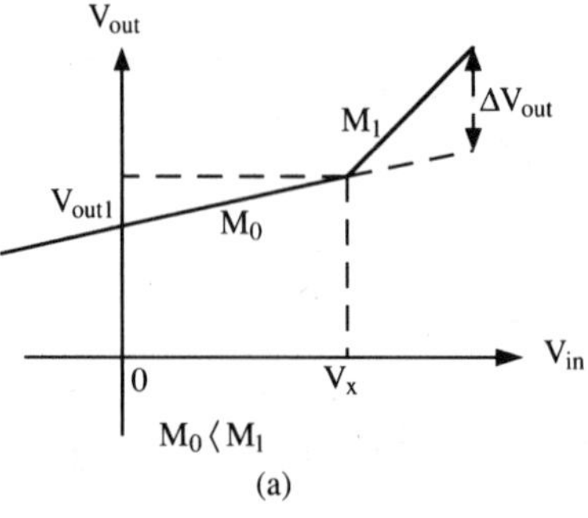

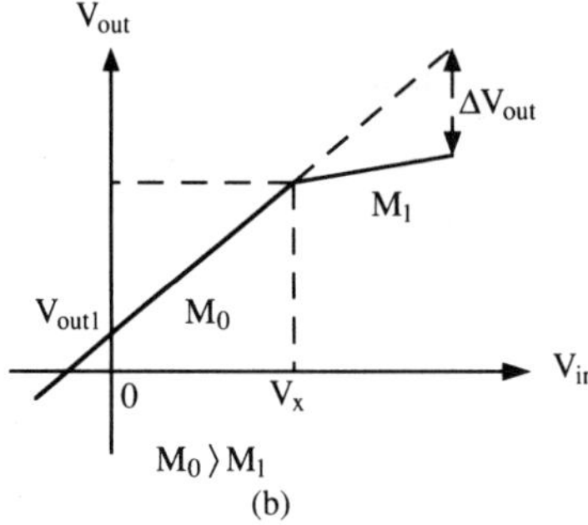

Figure 2.8 Simple PWL functions. (a) Increment. (b) Decrement.

by first assuming that V_{out} in the PWL function is identical to the current through V_{TEST} because $I(V_{TEST})$ is converted to V_{out} by H_{BUFF} in the output stage. The selection of the specific type of parallel branch load to use and the determination of the element values require examination of the given incremental PWL function like that presented in Figure 2.8(a). When the input voltage is less than the input breakpoint value V_x, the output voltage increases linearly with M_0 voltage gain. When V_{in} is equal to or larger than V_x, V_{out} increases with a larger slope M_1. In this case, a combination of Type I and II loads is suitable, since the current through V_{TEST} increases when the diode D_2 is "ON." In Figure 2.8(b), the combination of Type I and III loads is proper because the current through V_{TEST} decreases when the diode D_3 is "OFF." Thus the choice of parallel branch loads is made according to the following conditions:

$$
\begin{array}{ll}
M_0 < M_1 & \text{Type I and Type II} \\
 & V_2 = V_x \\
M_0 > M_1 & \text{Type I and Type III} \\
 & V_3 = V_x
\end{array}
\tag{2.12}
$$

In the incremental PWL macromodel, the current through V_{TEST} over the entire input voltage range is expressed by

$$I(V_{TEST}) = \begin{cases} \dfrac{1}{R_1} \cdot V_{in} - \dfrac{V_1}{R_1} \text{ when } V_{in} < V_2 \\ \dfrac{V_{in} - V_2}{R_2} + \dfrac{V_{in} - V_1}{R_1} \text{ when } V_{in} \geq V_2 \end{cases} \tag{2.13}$$

In Figure 2.8(a), the output voltage is described by

$$V_{out} = \begin{cases} M_0 \cdot V_{in} + V_{out1} \text{ when } V_{in} < V_x \\ M_1 \cdot (V_{in} - V_x) + (M_0 \cdot V_x + V_{out1}) \text{ when } V_{in} \geq V_x \end{cases} \tag{2.14}$$

From Equations (2.13) and (2.14), the incremental changes of the $I(V_{TEST})$ and V_{out}, when V_{in} is equal to or larger than $V_2 = V_x$, is given by

$$\Delta I(V_{TEST}) = \frac{V_{in}}{R_2} - \frac{V_2}{R_2}$$

$$\Delta V_{out} = (M_1 - M_0) \cdot V_{in} + (M_0 - M_1) \cdot V_x \tag{2.15}$$

Thus

$$M_0 - M_1 = -\frac{1}{R_2}$$

$$\therefore \quad R_2 = \frac{1}{M_1 - M_0} (\Omega) \tag{2.16}$$

In the other PWL function as shown in Figure 2.8(b), the parallel branch in the function stage consists of Type I and III loads. The circuit topology of this PWL function model is identical to that of the increasing PWL function macromodel except that a Type III load is used instead of a Type II. The current through V_{TEST} is expressed by

$$I(V_{TEST}) = \begin{cases} \dfrac{V_{in} - V_1}{R_1} - \dfrac{V_3 - V_{in}}{R_3} \text{ when } V_{in} < V_3 \\ \dfrac{V_{in} - V_1}{R_1} \text{ when } V_{in} \geq V_3 \end{cases} \tag{2.17}$$

The transfer function of the original decreasing PWL function is

$$V_{out} = \begin{cases} M_0 \cdot V_{in} + V_{out1} \text{ when } V_{in} < V_x \\ M_1 \cdot (V_{in} - V_x) + (M_0 \cdot V_x + V_{out1}) \text{ when } V_{in} \geq V_x \end{cases} \tag{2.18}$$

When the conditions of Equations (2.17) and (2.18) are compared, the value of V_3 can be easily determined as

$$V_3 = V_x \qquad \text{Input breakpoint value} \tag{2.19}$$

The changes of $I(V_{TEST})$ in Equation (2.17) and of the output voltage in Equation (2.18) when $V_{in} < V_3 = V_x$ are given by

$$\Delta I(V_{TEST}) = \frac{1}{R_3} \cdot V_{in} - \frac{V_3}{R_3}$$

$$\Delta V_{out} = (M_0 - M_1) \cdot V_{in} + (M_1 - M_0) \cdot V_x \tag{2.20}$$

Thus the value of R_3 is

$$R_3 = \frac{1}{M_0 - M_1}(\Omega) \tag{2.21}$$

For the Type I load elements, R_1 and V_1, two difference equations are required. First select two arbitrary data points in the given PWL function. After checking the status of the diodes in Type II and III at each data point, form two difference equations based on Kirchhoff's current law. Then solve the element values for the Type I load. Table 2.1 summarizes the generalized equations that define the element values. A more detailed procedure is described in the following example.

Consider a three-segment, piecewise linear function as shown in Figure 2.9.

Step 1 Determine the number and types of parallel branches required.
Number of parallel load in function stage = 3 because there are three segments.
Type I load One required
Type II load One required because of $M_1 < M_2$
Type III load One required because of $M_2 > M_3$

Step 2 On the basis of Figure 2.7, construct an equivalent circuit macromodel.

Step 3 Using Table 2.1, calculate the element values for the Type I, II, and III loads.

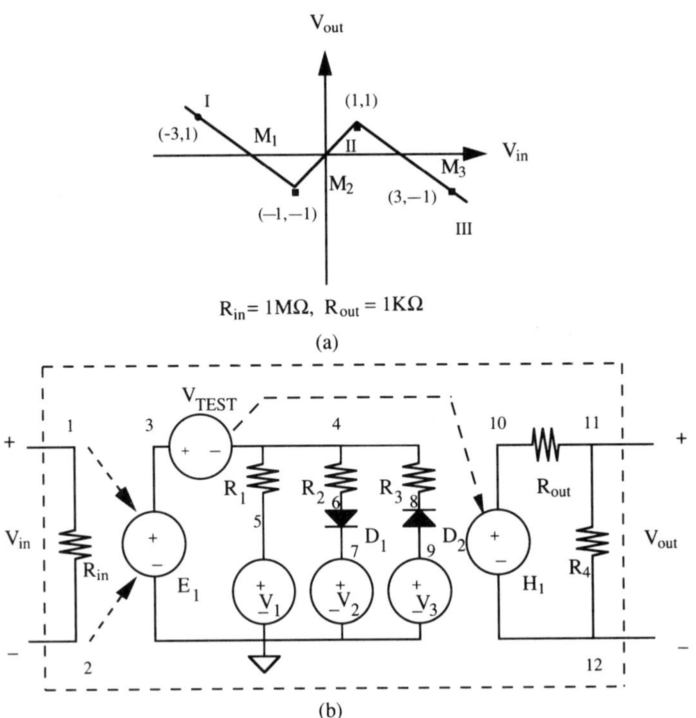

Figure 2.9 Example of PWL implementation. (a) Transfer characteristic. (b) Circuit implementation.

For the Type II load,

$$R_2 = \frac{1}{M_2 - M_1} = \frac{1}{2}(\Omega)$$

$$V_2 = \text{First breakpoint of input voltage} = -1(V)$$

For the Type III load,

$$R_3 = \frac{1}{M_2 - M_3} = \frac{1}{2}(\Omega)$$

$$V_3 = \text{Second breakpoint of input voltage} = 1(V)$$

For the element values of the Type I load, consider two data points such as (−3,1) and (3,−1) on the linear segments I and III, respectively.

TABLE 2.1 SUMMARY OF IMPLEMENTATION TECHNIQUE FOR PWL APPROXIMATIONS

PWL Function	Circuit	Element Value
	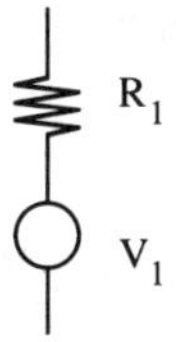 R_1, V_1 Type I	Required for every PWL macromodel
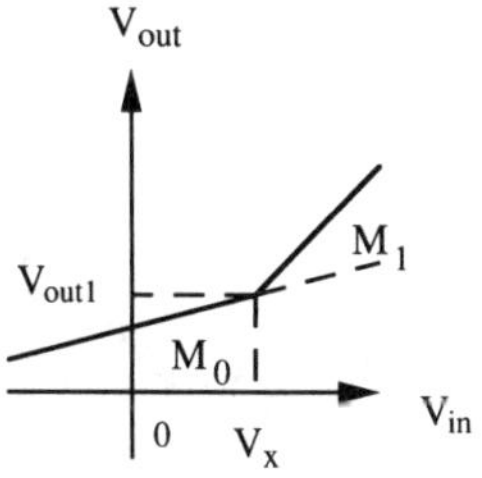 V_{out}, V_{out1}, M_1, M_0, V_{in}, 0, V_X $M_0 < M_1$	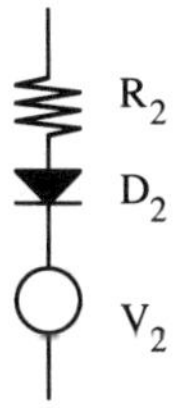 R_2, D_2, V_2 Type II	$R_2 = 1/(M_1 - M_0)$ (Ω) D_2= Modified diode model (N=0.001) $V_2 = V_X$ (V)
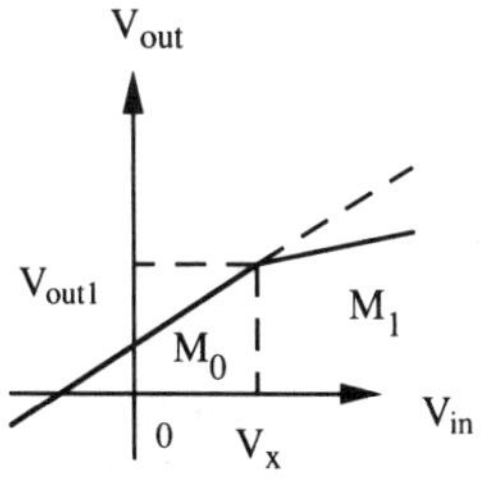 V_{out}, V_{out1}, M_1, M_0, V_{in}, 0, V_X $M_0 > M_1$	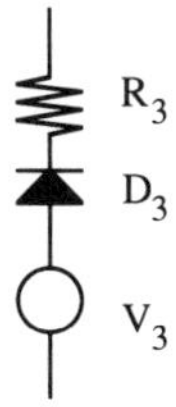 R_3, D_3, V_3 Type III	$R_3 = 1/(M_0 - M_1)$ (Ω) D_3= Modified diode model (N=0.001) $V_3 = V_x$ (V)

At (−3,1), diode D_1 is "ON" (short) while diode D_2 is "OFF" (open).

$$9R_1 + V_1 = -3$$

At (3,−1), the diode D_2 is "ON" while the diode D_3 is "OFF."

$$-9R_1 + V_1 = 3$$

So

$$R_1 = -\frac{1}{3}(\Omega) \text{ and } V_1 = 0(V)$$

The rest of the element values are

$R_{in} = 1\ M\Omega$, $R_{out} = 1\ K\Omega$, and $R_4 = 1T\Omega$

Gain of E_1 and H_1 = Unity D_1 and D_2 = Ideal diode model

Using this PWL macromodeling technique, many systems having highly nonlinear characteristics can be modeled. A complete program written in BASIC is developed for the automatic generation of PWL macromodels; it is listed in Chapter 6. The input data of the program are simple coordinate descriptions, while the output is an equivalent circuit structure macromodel in a format of a SPICE subcircuit. Both the simulation capability and the flexibility of the existing simulators can be enhanced by using this methodology without internal modifications of the source program.

2.5 BEHAVIORAL LEVEL FILTER MACROMODELS

Continuous-time domain transfer functions. The dynamic performance of a linear, time-invariant, continuous-time function is usually described in the complex frequency domain by applying the Laplace transform. The s-domain transfer function, H(s), can be represented graphically using a functional block diagram. In linear systems the functional block diagram consists of multiplication blocks, summing points, and branch points.[6] A branch point is defined as the point from which the output signal from a block goes concurrently to other blocks or summing points. This graphical expression of an s-domain transfer function can be reduced to simple equivalent circuit topologies composed of SPICE primitives such as resistors, capacitors, dependent control sources (including linear and polynomial), and transmission line elements.[7] Figure 2.10 shows a generalized procedure for modeling continuous and discrete transfer functions.

The analog filter is the most common continuous-time circuit. As an example, consider a third-order Chebyshev low-pass filter with 0.5 dB ripple width whose transfer function has been normalized to a cutoff frequency of 1 rad/sec and unity gain. The transfer function of this filter is given by

$$H(s) = \frac{0.7157}{s^3 + 1.2529 \cdot s^2 + 1.5349 \cdot s + 0.7157} \tag{2.22}$$

For a simplified graphical representation, Equation (2.22) is modified to define multiplication blocks, summing points, and branch points. The modification is produced by dividing each term of the numerator and the denominator by the highest power of s:

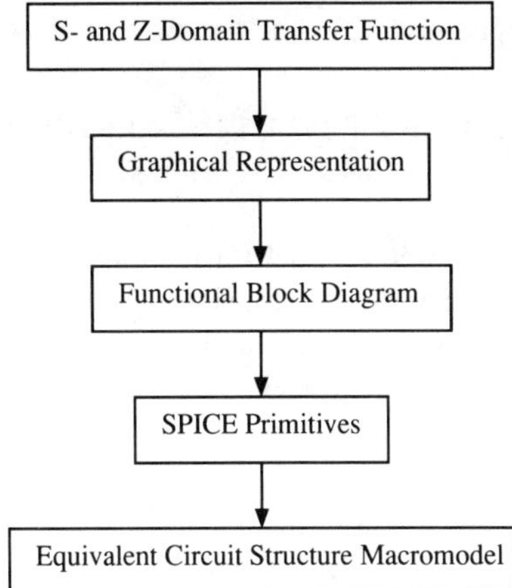

Figure 2.10 Macromodeling procedure for s- and z-domain transfer functions.

$$H_m(s) = \frac{0.7157 \cdot s^{-3}}{1 + 1.2529 \cdot s^{-1} + 1.5349 \cdot s^{-2} + 0.7157 \cdot s^{-3}} \quad (2.23)$$

Considering Equation (2.23) from a feedback theory viewpoint and using Mason's gain formula, each term of the denominator except the constant represents a feedback path whose coefficient is the feedback gain. The numerator of Equation (2.23) is the forward path, and its coefficient is the forward gain. The modified transfer function, $H_m(s)$ can be described by a functional block diagram, as shown in Figure 2.11.

Now consider how a equivalent circuit macromodel is constructed from the functional block diagram. The summing point can be realized by using the nonlinear polynomial dependent VCVS in SPICE. In this example, a four-dimensional polynomial dependent VCVS is required because there is one incoming and three feedback signals. The polynomial equation of the four-dimensional dependent VCVS in SPICE is given by

$$V_2 = P_0 + P_1 \cdot V_{in} + P_2 \cdot V_3 + P_3 \cdot V_4 + P_4 \cdot V_{out} + P_5 \cdot {V_{in}}^2 + \cdots \quad (2.24)$$

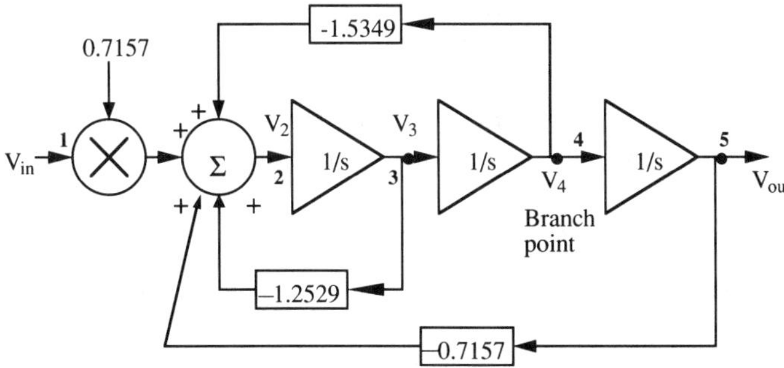

Figure 2.11 Functional block diagram of Equation (2.23).

Comparing Figure 2.11 and Equation (2.23), the coefficients of Equation (2.24) are easily determined as follows:

$$P_0 = 0,\ P_1 = 0.7157,\ P_2 = -1.2529,\ P_3 = -1.5349,\ P_4 = -0.7157$$

$$P_n = 0\ (n \geq 5)$$

An approximate equivalent circuit model for the integrator block is easily realized by using a linear VCCS, G_1, a resistor, and a capacitor. The circuit structure is shown in Figure 2.12. This is a simpler integrator macromodel than the one previously introduced in Section 2.3. This simpler integrator model is used here to illustrate the simulation methodology with a slight sacrifice in accuracy. However, the previous integrator could be substituted for this functional block.

G_1 = Voltage controlled current source
Controlling voltage = $V_{in}(s)$
Gain = Unity

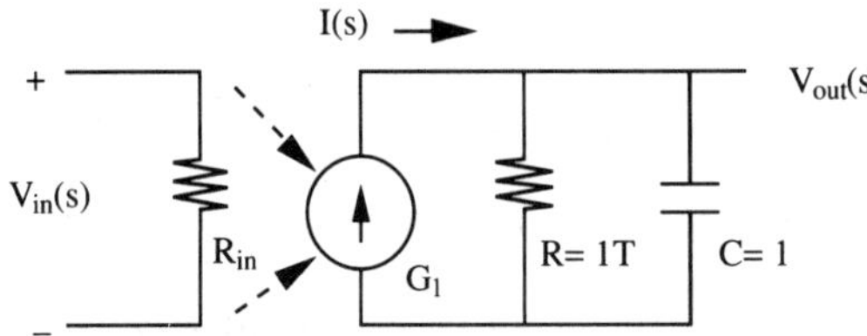

Figure 2.12 An equivalent circuit model for an integrator.

The controlling voltage of the G_1 is the voltage across R_{in}. The s-domain transfer function of the integrator model is

$$Z(s) = \frac{V_{out}(s)}{I(s)} = \frac{K \cdot \frac{1}{C}}{s + \frac{1}{R \cdot C}} \tag{2.25}$$

where K is the gain of the G_1. If C is a unit capacitance (= 1F) and the value of resistor, R, is very large (= 1TΩ), then the transimpedance, Z(s), of the integrator model becomes

$$Z(s) = \frac{V_{out}(s)}{I(s)} \approx \frac{K}{s} \tag{2.26}$$

The gain of the VCCS actually depends on the reciprocal of the time constant of the integrator, but in this example the gain is unity. Consequently, the equivalent circuit model of Figure 2.11 is constructed by cascading the submodels of the one summer and the three integrators. Figure 2.13(a) illustrates the equivalent circuit model and its SPICE code of Equation (2.22). A PSpice simulation result is shown in (b).

SPICE code for Figure 2.13

```
RIN 1 0 1MEG
ES 2 0 POLY(4) 1 0 3 0 4 0 5 0
+ 0 0.7157 -1.2529 -1.5349 -0.7157
RS 2 0 1
G1 0 3 2 0 1
```

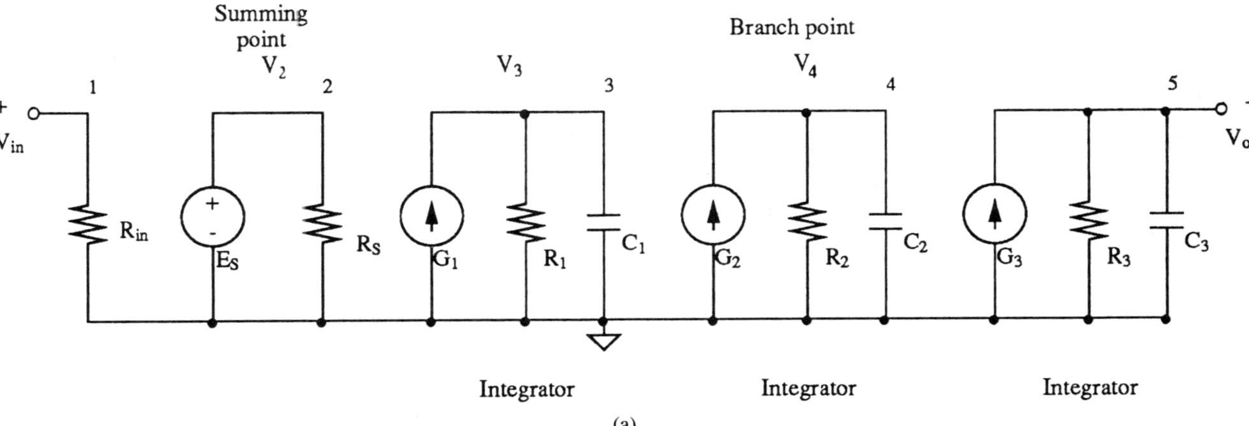

(a)

Figure 2.13 A macromodel for a third-order Chebyshev low-pass filter with 0.5 dB ripple width and unity gain. (a) Equivalent circuit model. (b) PSpice simulation result of normalized filter function. (c) PSpice simulation result of denormalized filter function.

```
R1 3 0 1T
C1 3 0 1
G2 0 4 3 0 1
R2 4 0 1T
C2 4 0 1
G3 0 5 4 0 1
R3 5 0 1T
C3 5 0 1
.OPTIONS LIMPTS=0
.AC DEC 20 0.001 10K
.PRINT AC VDB(5) VP(5)
.END
```

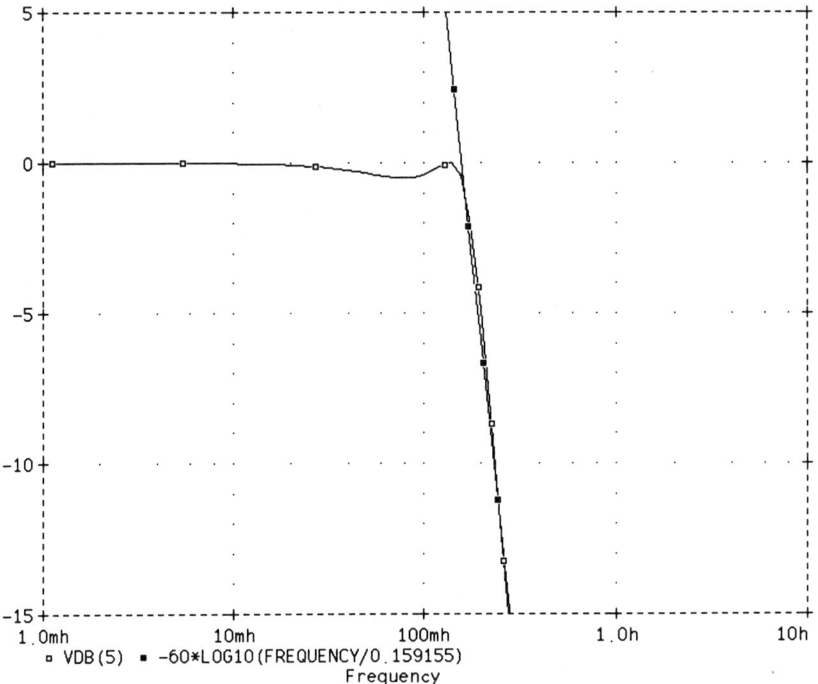

Figure 2.13(b)

Denormalizing this filter to a cutoff frequency of 1 KHz changes the macromodel constant as follows:

```
ES 2 0 POLY(4) 1 0 3 0 4 0 5 0
+ 0 0.7157 -1.2529 -1.5349 -0.7157              (Normalized)

ES 2 0 POLY(4) 1 0 3 0 4 0 5 0
+0 1.7753E11 -7.8722E3 -6.0595E7 -1.7753E11     (Denormalized)
```

The denormalized result is shown in Figure 2.13(c). Thus for the normalized filter, the E_5 polynomial VCVS in Figure 2.13(a) generates the following expression for the V_2 node voltage.

$$V_2 = 0.7157V_{in} - 1.2529s^{-1}V_3 - 1.53349s^{-2}V_4 - 0.7157s^{-3}V_{out} \tag{2.27}$$

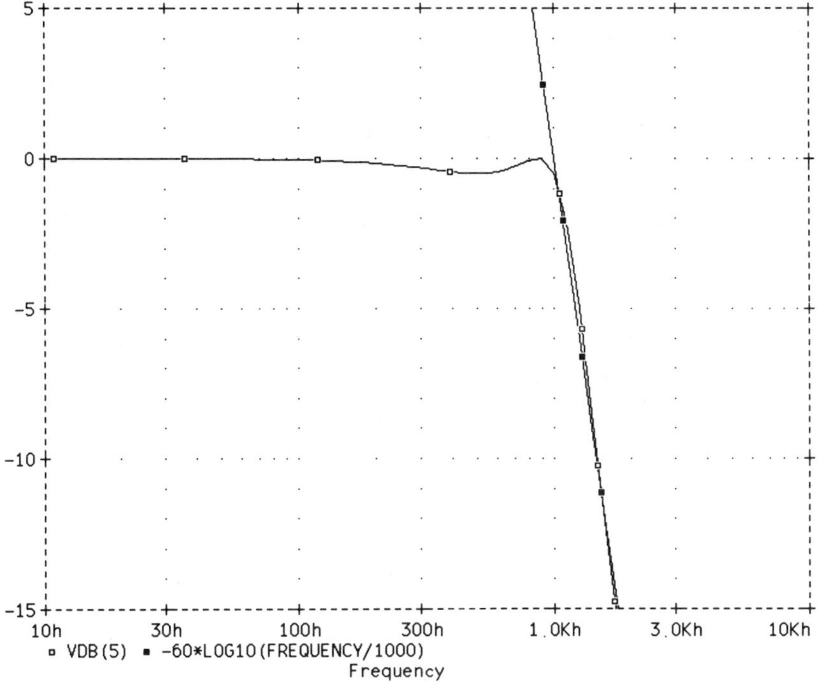

Figure 2.13(c)

Discrete-time sampled data filters (z-domain). This section shows how the macromodeling methodology for s-domain transfer functions is applied to z-domain transfer functions. This approach is valid for general-purpose circuit simulation programs such as SPICE. The z-domain transfer function is modeled in the continuous-time domain using a lossless transmission line element. The resulting continuous-time equivalent circuit for the discrete time transfer function retains sample data characteristics and is not restricted to frequencies below the Nyquist rate.

A linear, time-invariant, discrete time function is described by a z-transform expression. The basic algorithm for macromodeling a z-domain transfer function is almost the same as that for an s-domain transfer function. The only difference is the use of a discrete integrator instead of a continuous integrator. A usable equivalent circuit model for the discrete integrator is the *storistor*—a one-port storage element in which current is proportional to a delayed version of its applied voltage.[8] Therefore, the storistor represents a circuit element with memory (or pure delay) that retains the value of the

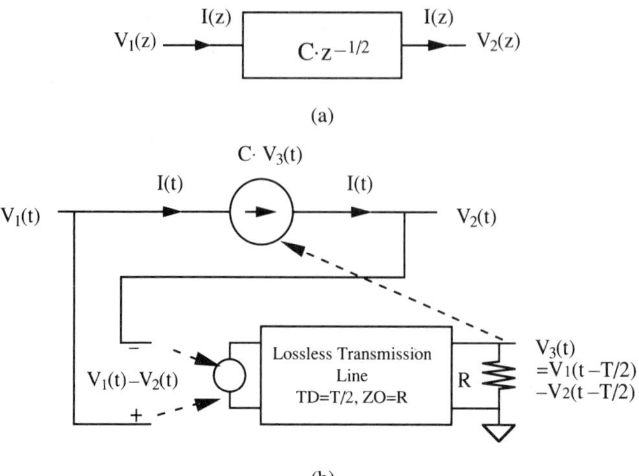

Figure 2.14 Storistor. (a) Symbol of storistor having a delay of one-half of switching period. (b) Continuous-time domain model of storistor in (a).

impressed voltage. The storistor preserves the sample data characteristics of the original discrete network, such as the periodic frequency characteristic.

It is possible to implement the storistor exactly in any general-purpose circuit simulation program that provides time delay elements. SPICE has such an element—the lossless transmission line. When properly terminated by its characteristic impedance, this element realizes a pure time delay necessary for the proper simulation of discrete systems. Figure 2.14(a) shows the schematic symbol (in the z-domain) of the storistor having a delay equal to T/2 (one-half of the switching period). Its continuous-time model is illustrated in Figure 2.14(b).

The symbols in Figure 2.14(a) correspond to the following equations:

$$I(z) = C \cdot z^{-1/2}[V_1(z) - V_2(z)] \text{ (z-domain)}$$

$$I(s) = C \cdot e^{-sT/2}[V_1(s) - V_2(s)] \text{ (s-domain)}$$

$$I(t) = C\left[V_1\left(t - \frac{T}{2}\right) - V_2\left(t - \frac{T}{2}\right)\right] \text{ (Continuous time-domain)}$$

If this storistor is connected to an ideal op amp, the storistor model can be simplified further. If the node voltage, $V_2(t)$, is connected to a virtual ground point, $V_2(t)$ is always zero. Thus,

$$V_3(t) = V_1\left(t - \frac{T}{2}\right)$$

$$I(t) = C \cdot V_3(t) = C \cdot V_1\left(t - \frac{T}{2}\right)$$

$$I(z) = C \cdot V_1(z) \cdot z^{-1/2} \tag{2.28}$$

Equation (2.28) is directly applied to Figure 2.15(a). If $V_1(z)$ is identical to the $V_{in}(z)$, the expression for $I_2(z)$ is given by

$$I_2(z) = V_{in}(z) \cdot C \cdot z^{-1/2} \tag{2.29}$$

where C is the capacitance of the storistor. Since the op amp is assumed ideal, its input current is zero. Thus the output voltage is expressed by

$$V_{out}(z) = -I_2(z) \cdot R = -V_{in}(z) \cdot R \cdot C \cdot z^{-1/2} \tag{2.30}$$

If C = 1 and R = 1, the z-domain transfer function of Figure 2.15(a) is

$$\frac{V_{out}(z)}{V_{in}(z)} = -R \cdot C \cdot z^{-1/2} = -z^{-1/2} \tag{2.31}$$

As shown in Equation (2.31), Figure 2.15(a) becomes a z-domain equivalent circuit for $-z^{-1/2}$. If these two blocks are cascaded, the z-domain transfer function is z^{-1}.

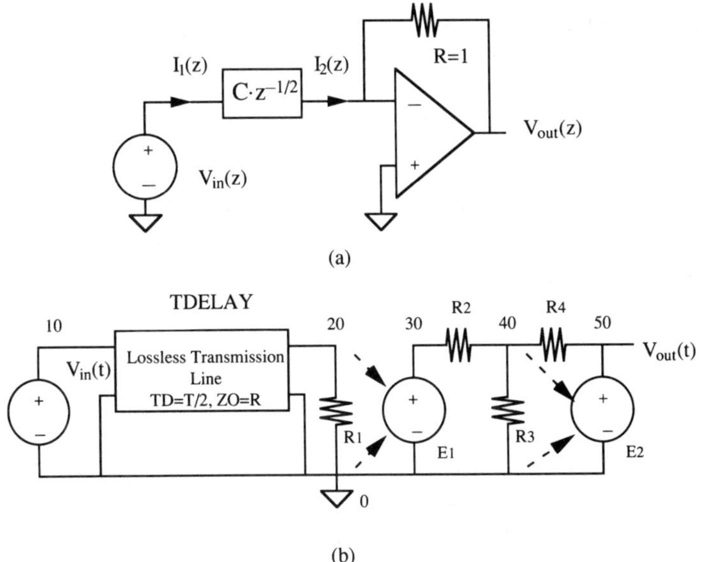

Figure 2.15 Simplified storistor model. (a) Storistor connected to an ideal op amp. (b) Continuous time domain model of (a).

SPICE code for Figure 2.15

```
.SUBCKT DINT 10 50
TDELAY 10 0 20 0 ZO=[A] TD=[B]        ; [A] = Characteristic
*                                               impedance
*                                     ; [B] = T/2
*                                     ; T = Sampling period
R1 20 0 [A]
E1 30 0 20 0 1
R2 30 40 1
R3 40 0 1T
R4 40 50 1
E2 50 0 40 0 -1T
.ENDS DINT
```

For the continuous-time model of Figure 2.15(a), Equation (2.31) corresponds to the s-domain form of

$$V_{out}(s) = -R \cdot C \cdot e^{-sT/2} \cdot V_{in}(s) \tag{2.32}$$

and as below for the continuous-time domain representation.

$$V_{out}(t) = -R \cdot C \cdot V_{in}\left(t - \frac{T}{2}\right) = -V_{in}\left(t - \frac{T}{2}\right) \tag{2.33}$$

The storistor model becomes unilateral and consists of a time-delayed voltage source in series with ideal op amp model. A continuous-time equivalent circuit model resulting from Equation (2.33) is shown in Figure 2.15(b). If two blocks of Figure 2.15(b) are cascaded, a continuous-time equivalent circuit model for z^{-1} is obtained.

Now consider an example z-domain transfer function such as

$$H(z) = \frac{1}{10142.4z^2 - 20141.4z + 10000} \tag{2.34}$$

The foregoing z-domain transfer function equation was derived from a normalized second-order Butterworth low-pass filter function using the backward transformation with the sampling frequency being 100 Hz.[9] Equation (2.35) represents the discrete integrator (z^{-1}) format.

$$H_m(z) = \frac{9.8596 \cdot 10^{-5} z^{-2}}{1 - 1.9859z^{-1} + 0.9860z^{-2}} \tag{2.35}$$

A block diagram representation of Equation (2.35) is shown in Figure 2.16. The summing point is modeled by using a three-dimensional nonlinear dependent VCVS. The discrete integrator (z^{-1}) is implemented by cascading two continuous-time models of $-z^{-1/2}$. The continuous-time macromodel for Equation (2.35) is shown in Figure 2.17(a). Figure 2.17(b) shows the macromodel simulation results used in PSpice version 4.0.[10] The SPICE code for this figure follows Figure 2.17(b). The continuous-time equivalent circuit model for the discrete-time transfer function retains sample data characteristics and is not restricted to frequencies below the Nyquist rate. Thus, the E_{sum} VCVS of Figure 2.17(a) generates the voltage V_3 as follows:

$$V_3 = 9.8586E - 5V_{in} + 1.9859Z^{-1}V_5 - 0.9860Z^{-2}V_{out} \tag{2.36}$$

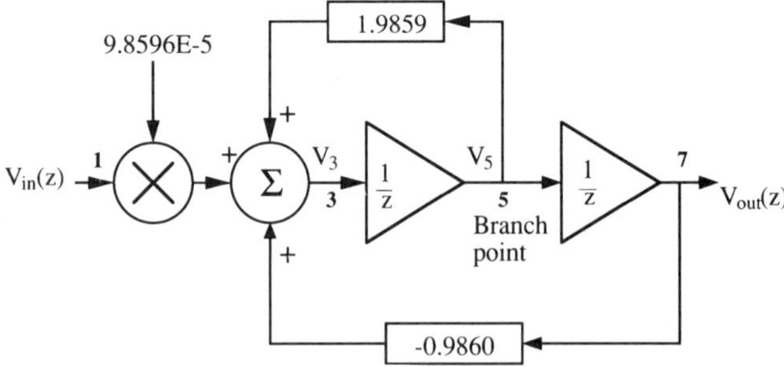

Figure 2.16 Functional block diagram of Equation (2.35).

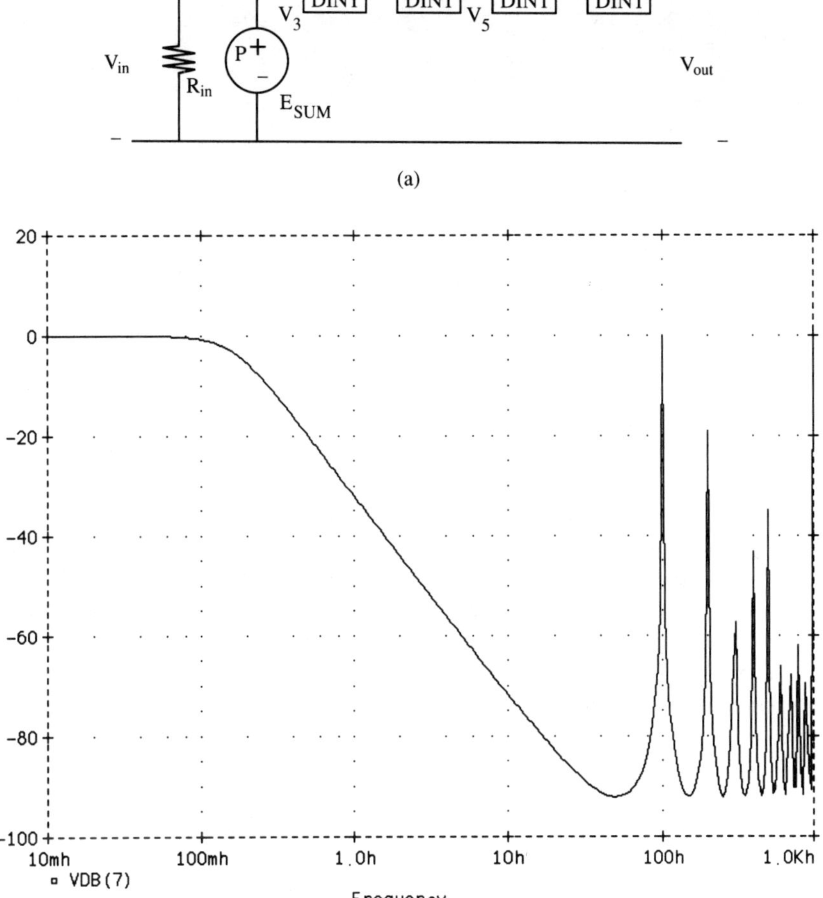

Figure 2.17 Continuous-time macromodel of Equation (2.35). (a) Circuit structure. (b) PSpice simulation results.

SPICE code for Figure 2.17(a)

```
VIN 1 0 AC=1
RIN 1 0 1MEG
ESUM 3 0 POLY(3) 1 0 5 0 7 0
+ 0 9.8596E-5  1.9859  -0.9860
XZ-0.5  3  4  DINT
XZ-1.0  4  5  DINT
XZ-1.5  5  6  DINT
XZ-2.0  6  7  DINT
.SUBCKT DINT 10 50
```

```
TDELAY 10 0 20 0 ZO=1K TD=0.005
R1 20 0 1K
E1 30 0 20 0 1
R2 30 40 1
R3 40 0 1T
R4 40 50 1
E2 50 0 40 0 -1T
.ENDS DINT
.OPTIONS LIMPTS=0
.AC DEC 60 0.01 1K
.PRINT AC VDB(7) VP(7)
.END
```

2.6 SUMMARY

The generalized macromodel approaches developed for arithmetic functions, piecewise linear approximations, and s- and z-domain transfer functions are easy and efficient ways to model linear and nonlinear blocks in complex analog circuits and systems. Using these approaches, both computational effort and simulation cost can be reduced. The macromodels are compatible with existing circuit simulators, which are enhanced and extended by utilizing these techniques. Furthermore, these modeling approaches permit mixed-mode simulation of analog and analog sampled data circuits using a SPICE-type simulation program.

REFERENCES

1. W. G. Magnuson, Jr., "Circuit Analysis and Simulation Programs," *IEEE Transactions on Nuclear Science*, Vol. NS-29, No. 1, February 1982, pp. 568–572; Donald O. Pederson, "A Historical Review of Circuit Simulation," *IEEE Transactions on Circuits and Systems*, Vol. CAS-31, No. 1, January 1984, pp. 103–111.
2. Robert Sparkes and Grame Boyle, "Evaluation of Macromodels for Mixed Analog/Digital Circuits," *Proceedings of the IEEE Custom Integrated Circuits Conference 1988*, Rochester, New York, pp. 3.4.1–3.4.6.
3. B. Epler, "SPICE 2 Application Notes for Dependent Sources," *IEEE Circuits and Devices Magazine*, September 1987, pp. 36–44; P. Braga and J. Fernandez, "Application of MicroCap-II's Polynomial Sources," *IEEE Circuits and Devices Magazine*, November 1989, pp. 30–31.
4. D. H. Sheingold, ed., *Nonlinear Circuits Handbook Designing with Analog Function Modules and ICs*, (Norwood, MA: Analog Devices, Inc., 1974).
5. J. Alvin Connelly and Pyung Choi, "A Generalized SPICE Macromodel for Piecewise Linear Circuits and Systems," *Extended Abstract of TECHCON '88*, Dallas, Texas, October 1988, pp. 366–369.

6. R. A. Gayakwad, *Op-Amps and Linear Integrated Circuits*, 2nd ed. (Englewood Cliffs, N.J.: Prentice Hall, 1988); R. J. Mayhan, *Discrete-Time and Continuous-Time Linear Systems* (Reading, Mass.: Addison-Wesley, 1984).

7. Pyung Choi and J. Alvin Connelly, "Macromodeling Methodology for Continuous and Discrete Time Transfer Functions," *Proceedings of the Thirty-second Midwest Symposium on Circuits and Systems*, Urbana, Illinois, August 1989, pp. 997–1000.

8. B. D. Nelin, "Analysis of Switched-Capacitor Networks Using General-Purpose Circuit Simulation Program," *IEEE Transactions on Circuits and Systems*, Vol. CAS-30, No. 1, January 1983, pp. 43–48.

9. H. S. Malvar, "Transform Analog Filters into Digital Equivalents," *Electronic Design*, April 30, 1981, pp. 145–148; A. M. Bush and D. C. Fielder, "Simplified Algebra for the Bilinear and Related Transformations," *IEEE Transactions on Audio and Electroacoustics*, April 1973, pp. 127–128.

10. PSpice: Electrical Circuit Simulator, Microsim Corporation, Irvine, California.

3

Drop-In Macromodels

3.1 INTRODUCTION

Although SPICE has proved to be a powerful analysis tool for circuit designers, the simulation of large-scale integrated circuits (ICs) and complex systems has been impeded by certain functional program limitations. Moreover, the analysis of large-scale ICs or complex IC systems often surpasses the computer's memory capability, the simulator circuit size capability, or the inherent numerical accuracy of the computer. Even with adequate computer memory, the required simulation time can make the analysis impractical.

The use of higher level models such as macromodels and behavioral models for an IC or for a portion of an IC system represents a viable solution to these problems. Furthermore, higher level models have become a necessity because of the evolution of modern, complex integrated circuits. Many higher level models, especially macromodels, for operational amplifiers and comparators have been developed and widely used.[1] These macromodels have simple network topologies that significantly reduce the complexity of the original circuits. Macromodels provide shorter and less costly simulation times or permit the simulation of more complex IC systems for the same time and cost.

In this chapter a large collection of behavioral, functional, and commercial macromodels are introduced, explained, and illustrated. These macromodels can be used much as functional blocks or modules of laboratory equipment are used to simulate and evaluate the performance of circuits and systems. Using these drop-in macromodels with various test signal sources allows a circuit and/or system designer to determine the performance of a design rapidly and easily. Furthermore, the designer can simulate a large system using the macromodels to identify which block or stage is the critical one for limiting the overall performance. Then a more complete device-level model can be inserted for this critical block while continuing to use the other drop-in macromodel functions.

3.2 BEHAVIORAL MACROMODELS

3.2.1 Voltage Summer

Voltage summing is one basic application of an op amp. Summation is macromodeled by using only one dependent, nonlinear polynomial voltage controlled voltage source (VCVS). The summer does not include an op amp macromodel. The transfer function for the voltage summer is:

$$V_{out} = K_1 \cdot V_{in1} + K_2 \cdot V_{in2} + K_3 \cdot V_{in3} + \ldots$$

$$V(106, 108) = K_1 V(100, 101) + K_2 V(102, 103) + K_3 V(104, 105) + \ldots$$

SPICE code for Figure 3.2

```
RIN1 100 101 [A]                ; [A] = Input resistance I
RIN2 102 103 [B]                ; [B] = Input resistance II
RIN3 104 105 [C]                ; [C] = Input resistance N
EOUT 106 108 POLY(N)            ; (N) = Number of variables
+ 100 101 102 103 104 105
+ 0 [K1] [K2] [K3]              ; [KN] = Coefficient of variable
RD 107 108 1G
ROUT 106 107 [D]                ; [D] = Output resistance
```

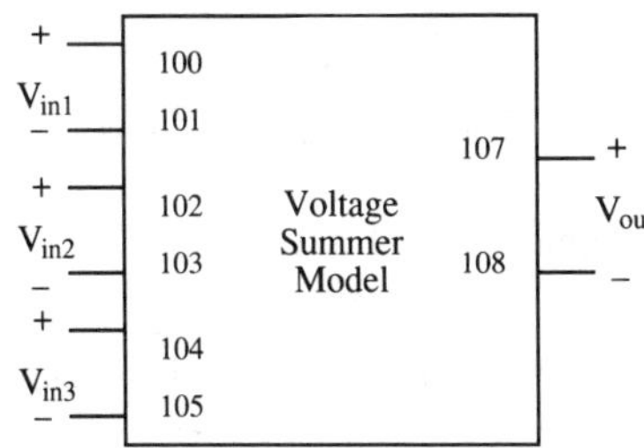

Figure 3.1 Functional block for the voltage summer.

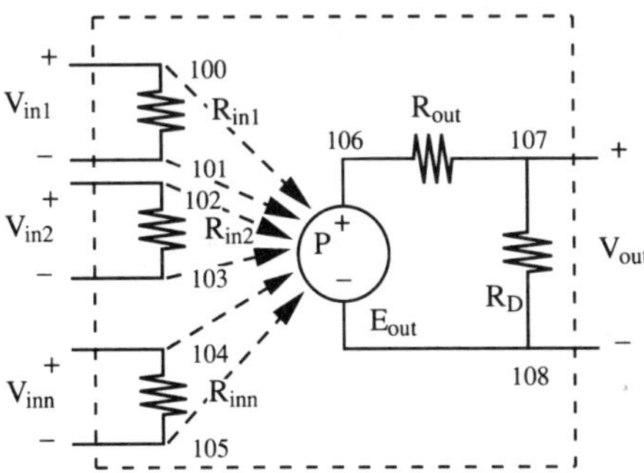

Figure 3.2 Equivalent circuit for the voltage summer.

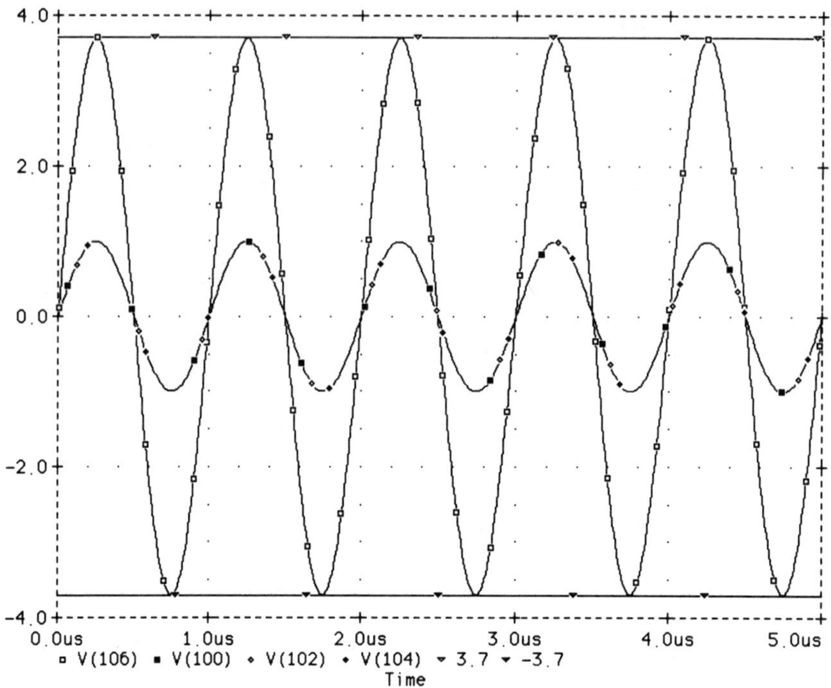

Figure 3.3 Simulation results for the voltage summer.

Example

$$V_{out} = V_{in1} + 2V_{in2} + 0.7V_{in3}$$

Input resistance of each input port = 1 MΩ
Output resistance = 1 KΩ
Nodes 101, 103, 105, and 108 are grounded in this example.

```
RIN1 100 0 1MEG              ; Input resistance I = 1 MΩ
RIN2 102 0 1MEG              ; Input resistance II = 1 MΩ
RIN3 104 0 1MEG              ; Input resistance III = 1 MΩ
EOUT 106 0 POLY(3)           ; Number of variables = 3
+ 100 0 102 0 104 0
+ 0 1 2 0.7                  ; K1 = 1, K2 = 2, K3 = 0.7
RD 107 0 1G
ROUT 106 107 1K              ; Output resistance= 1 KΩ

VIN1 100 0 SIN(0 1 1MEG)
VIN2 102 0 SIN(0 1 1MEG)
VIN3 104 0 SIN(0 1 1MEG)
.TRAN 10N 5U
```

```
.PRINT TRAN V(106)
.PROBE                               ; PSpice
.END
```

3.2.2 Voltage Multiplier

The behavioral characteristics of an analog multiplier is modeled using a nonlinear polynomial VCVS. The voltage gain of the multiplier is determined by properly specifying the coefficient, K, of the nonlinear polynomial VCVS. The transfer function for the voltage multiplier, shown in Figure 3.4, is:

$$V_{out} = K \times \{V_{in1} \times V_{in2}\}$$

The equation for the polynomial source in Figure 3.5 is:

$$V(104,106) = K \cdot V(100,101) \cdot V(102,103)$$

SPICE code for Figure 3.5

```
RIN1 100 101 [A]                    ; [A] = Input resistance I
RIN2 102 103 [B]                    ; [B] = Input resistance II
EOUT 104 106 POLY(2)
+ 100 101 102 103
+ 0 0 0 0 [K]                       ; [K] = Multiplication factor
RD 105 106 1G
ROUT 104 105 [C]                    ; [C] = Output resistance
```

Example

$$V_{out} = 10 \times \{V_{in1} \times V_{in2}\}$$

Input resistance of each input port = 1 MΩ
Output resistance = 1 KΩ
Nodes 101, 103, and 106 are grounded in this example.

```
RIN1 100 0 1MEG              ; Input resistance I = 1 MΩ
RIN2 102 0 1MEG              ; Input resistance II = 1 MΩ
EOUT 104 0 POLY(2)           ; Number of variables = 2
+ 100 0 102 0
+ 0 0 0 0 10                 ; Multiplication factor = 10
RD 105 0 1G
ROUT 104 105 1K              ; Output resistance = 1 KΩ
```

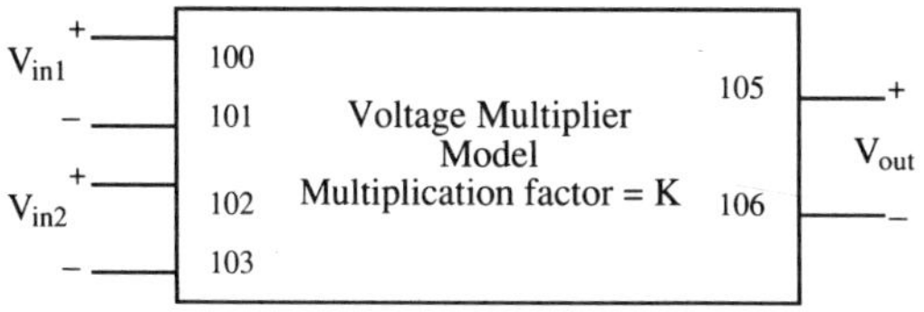

Figure 3.4 Functional block for the voltage multiplier.

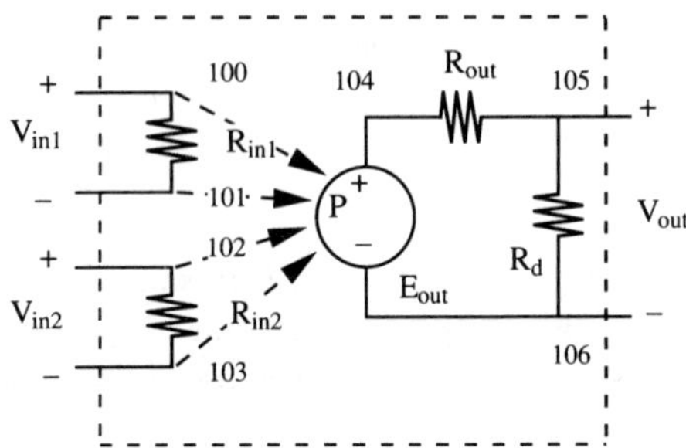

Figure 3.5 Equivalent circuit for the voltage multiplier.

```
VIN1 100 0
VIN2 102 0
.DC  VIN1 -1 1 0.25 VIN2 -1 1 0.25
.PRINT DC V(105)
.PROBE                              ; PSpice
.END
```

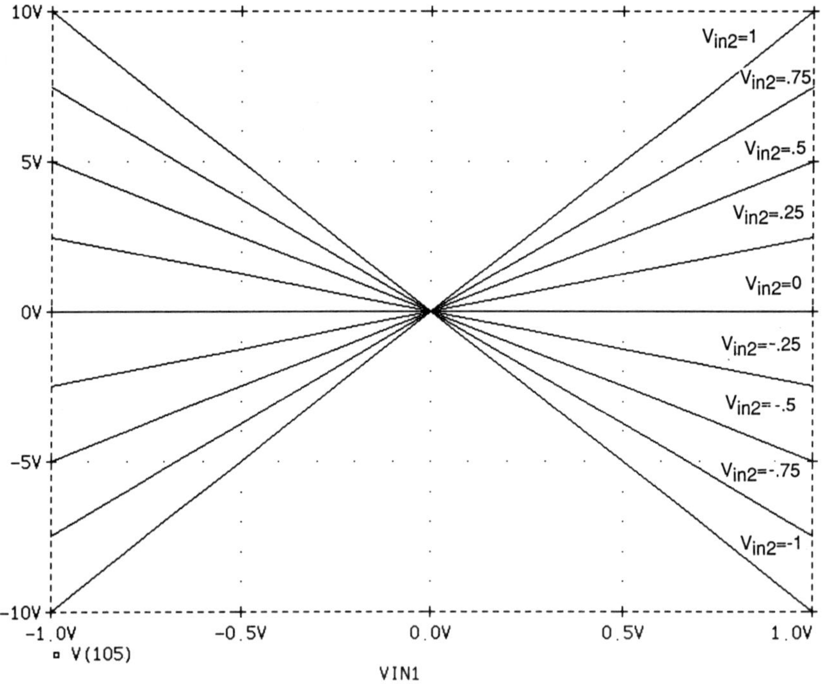

Figure 3.6 Simulation results for the voltage multiplier.

3.2.3 Voltage Divider

Analog voltage division is modeled by using a linear dependent VCCS and a nonlinear polynomial VCCS. These two dependent sources are connected at the same node with a large resistance. The large resistance accommodates leakage current so that Kirchhoff's current law (KCL) is satisfied. The transfer function for the voltage divider is:

$$V_{out} = \frac{C \cdot V_1}{D \cdot V_2}$$

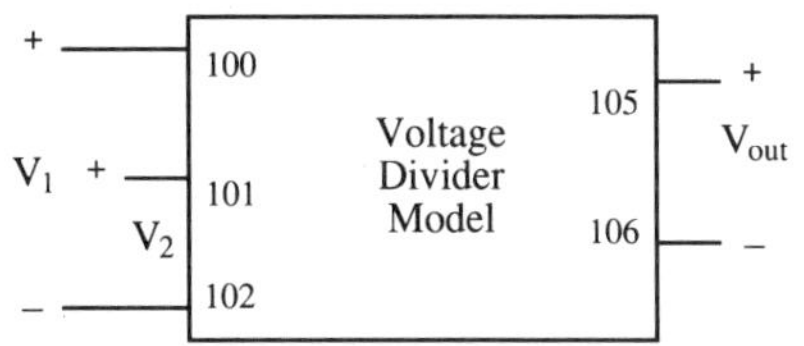

Figure 3.7 Functional block for the voltage divider.

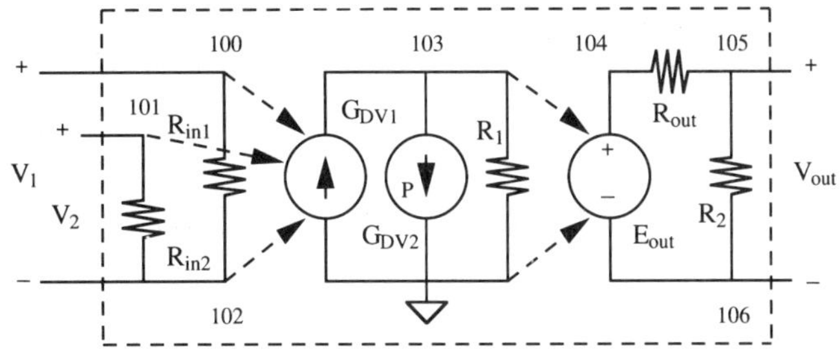

Figure 3.8 Equivalent circuit for the voltage divider.

The equations for Figure 3.8 are:

$I(G_{DV1}) = C \cdot V(100{,}102)$

$I(G_{DV2}) = D \cdot V(101{,}102) \cdot V(103{,}0)$

$V(103{,}0) = R_1[I(G_{DV1}) - I(G_{DV2})]$

$V(103{,}0) = R_1[C \cdot V(100{,}102) - D \cdot V(101{,}102) \cdot V(103{,}0)]$

$V(103{,}0)[1 + R_1 \cdot D \cdot V(101{,}102)] = R_1 \cdot C \cdot V(100{,}102)$

$V(103{,}0) = [R_1 \cdot C \cdot V(100{,}102)]/[1 + R_1 \cdot D \cdot V(101{,}102)]$

Since the current through R_1 is negligible because of large value of R_1,

$$I(G_{DV1}) = I(G_{DV2})$$

Taking the limit for a large value of R_1 gives the desired relationship:

$$V(103{,}0) = [C \cdot V(100{,}102)]/[D \cdot V(101{,}102)]$$

SPICE code for Figure 3.8

```
RIN1 100 102 [A]                    ; [A] = Input resistance I
RIN2 101 102 [B]                    ; [B] = Input resistance II
GDV1 0 103 100 102 [C]              ; [C] = Gain of GDV1
GDV2 103 0 POLY(2) 101 102 103 0
+ 0 0 0 0 [D]                       ; [D] = Gain of GDV2
R1 103 0 1G
EOUT 104 106 103 0 1
R2 105 106 1G
ROUT 104 105 [E]                    ; [E] = Output resistance
```

Example

$V_{out} = V_1 / (2 \cdot V_2)$
Input resistances = 1 MΩ
Output resistances = 1 KΩ
Nodes 102 and 106 are grounded in this example.

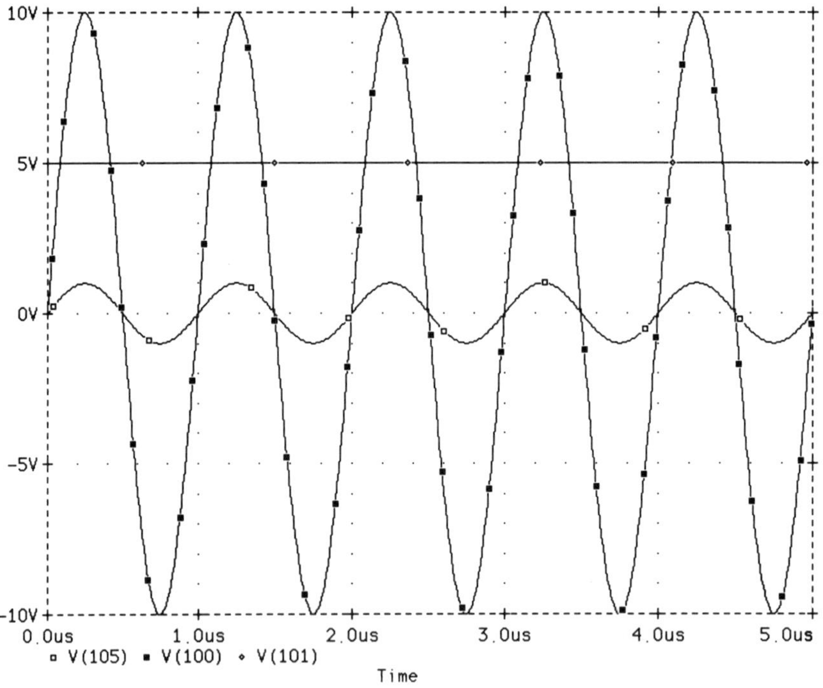

Figure 3.9 Simulation results for the voltage divider.

```
RIN1 100 0 1MEG                    ; Input resistance I = 1 MΩ
RIN2 101 0 1MEG                    ; Input resistance II = 1 MΩ
GDV1 0 103 100 0 1                 ; Gain of GDV1 = 1
GDV2 103 0 POLY(2) 101 0 103 0
+ 0 0 0 0 2                        ; Gain of GDV2 = 2
R1 103 0 1G
EOUT 104 0 103 0 1
R2 105 0 1G
ROUT 104 105 1K                    ; Output resistance = 1 KΩ
V1 100 0 SIN(0 10 1MEG)
V2 101 0 DC=5
.TRAN 10N 5U
.PRINT DC V(105)
.PROBE                             ; PSpice
.END
```

3.2.4 Squarer

The squaring function is modeled by using a one-dimensional nonlinear polynomial VCVS. The third coefficient of the polynomial source sets K, the conversion gain of the squaring circuit. The transfer function for the squarer is:

$$V_{out} = K \cdot V_{in}^2$$

The equation for Figure 3.11 is:

$$V(102,104) = K \cdot [V(100,101)]^2$$

SPICE code for Figure 3.11

```
RIN1 100 101 [A]                   ; [A] = Input resistance
EOUT 102 104 POLY(1) 100 101
```

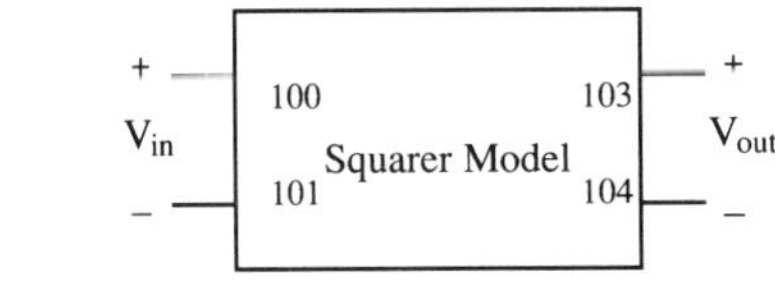

Figure 3.10 Functional block for the squarer.

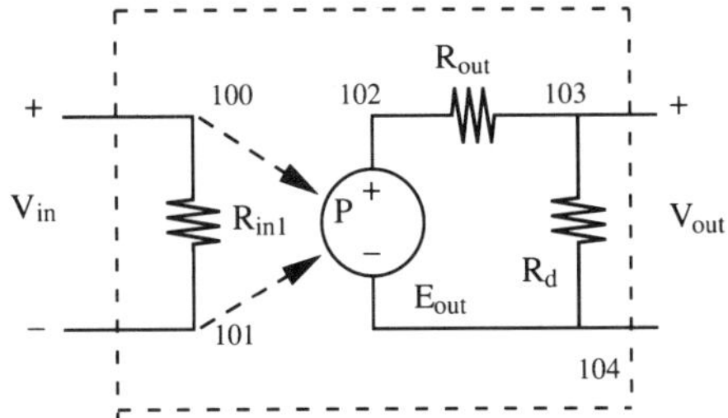

Figure 3.11 Equivalent circuit for the squarer.

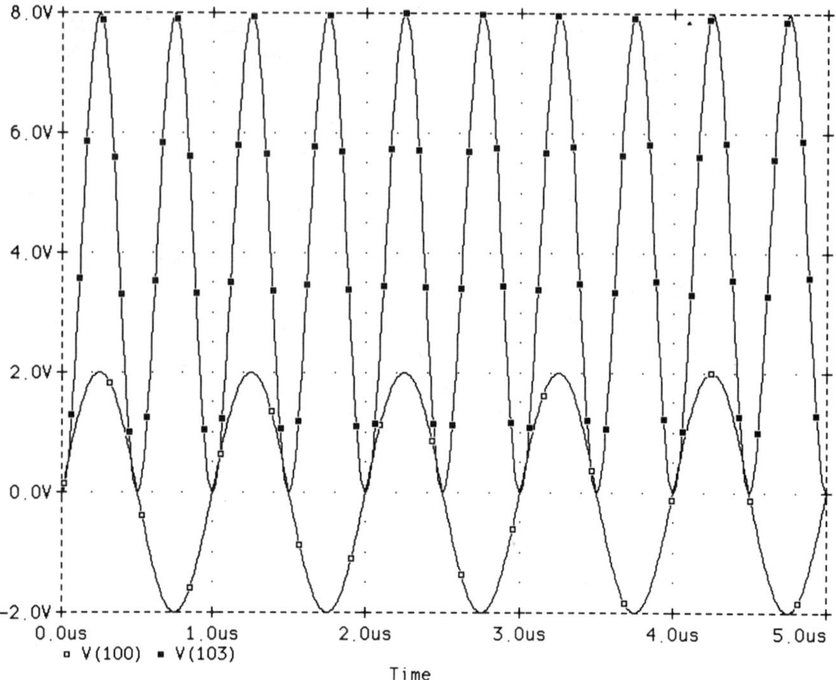

Figure 3.12 Simulation results for the squarer.

```
+ 0 0 [K]                          ; [K] = Scaling coefficient
RD 103 104 1G
ROUT 102 103 [B]                   ; [B] = Output resistance
```

Example

$V_{in} = 2\sin(2\pi \cdot 10^6 \cdot t)$
$V_{out} = 2(V_{in})^2$
Input resistance = 1 MΩ
Output resistance = 1KΩ
Nodes 101 and 104 are grounded in this example.

```
RIN1 100 0 1MEG                ; Input resistance = 1 MΩ
EOUT 102 0 POLY(1) 100 0
+ 0 0 2                        ; Scaling coefficient K = 2
RD 103 0 1G
ROUT 102 103 1K                ; Output resistance = 1 KΩ

VIN1 100 0 SIN(0 2 1MEG)
.TRAN 10N 5U
.PRINT TRAN V(100) V(103)
```

```
.PROBE
.END
```

3.2.5 Ideal Transformer

This macromodel for an ideal transformer is valid for DC, AC, and transient analyses. The current and voltage relationships between the primary and secondary windings are matched to the theoretical turns ratios (see Figure 3.13).

The equations for Figure 3.14 are:

$$I(F_1) = (N_2/N_1) \cdot I(V_{SENSE2})$$

$$I(F_2) = (N_1/N_2) \cdot I(V_{SENSE1})$$

where N_1 and N_2 are the number of turns on the primary and secondary coils, respectively.

SPICE code for Figure 3.14

```
R1 100 101 1
VSENSE1 101 102 DC=0 AC=0
F1 102 100 VSENSE2 [A]              ; [A] = (N2/N1)
F2 105 103 VSENSE1 [B]              ; [B] = (N1/N2)
R2 103 104 [C]                      ; [C] = (N2/N1)^2
VSENSE2 104 105 DC=0 AC=0
```

Example

$N_1 = 1$, $N_2 = 2$

Nodes 102 and 105 in the equivalent circuit are grounded.

```
R1 100 101 1
VSENSE1 101 0 DC=0 AC=0
F1 0 100 VSENSE2 2                  ; [A] = (N2/N1) = 2
F2 0 103 VSENSE1 0.5                ; [B] = (N1/N2) = 0.5
R2 103 104 4                        ; [C] = (N2/N1)^2 = 4
VSENSE2 104 0 DC=0 AC=0

VIN 100 0 SIN(0 1 1K)
.TRAN 10U 4M
.PRINT TRAN V(100) V(103)
```

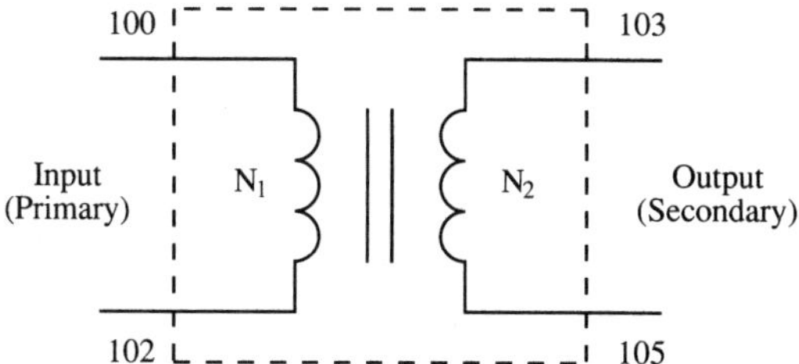

Figure 3.13 Functional block for the ideal transformer.

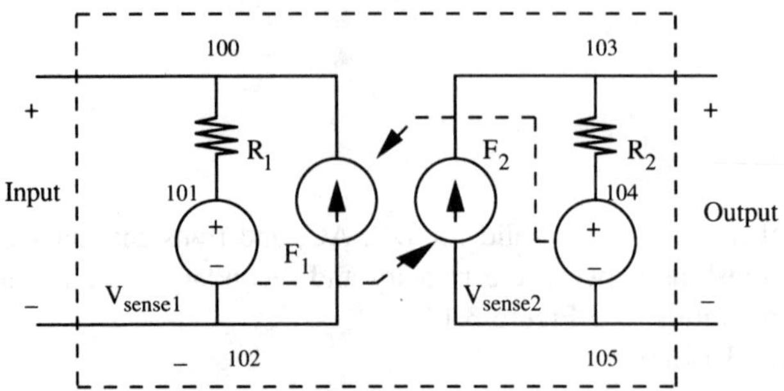

Figure 3.14 Equivalent circuit for the ideal transformer.

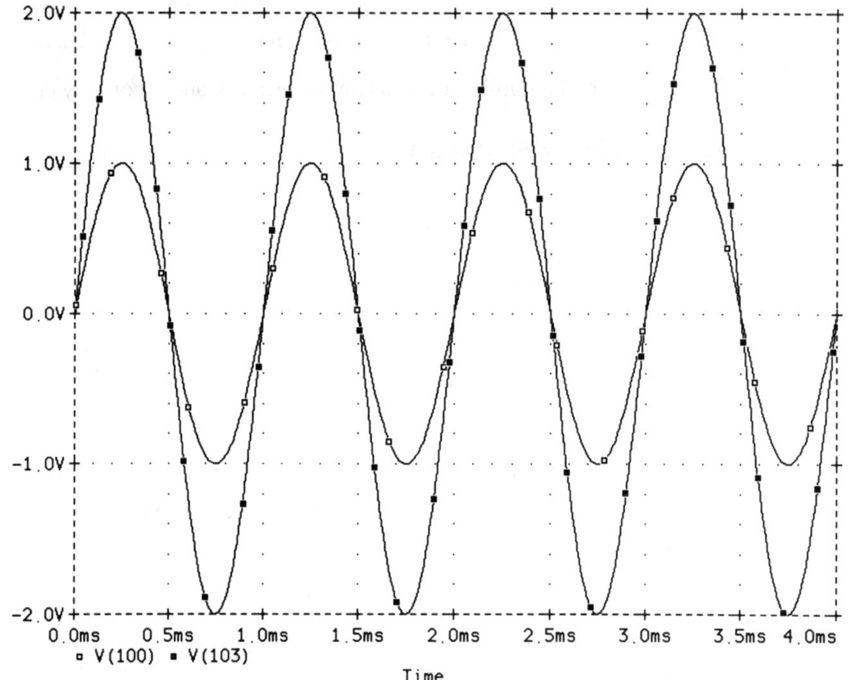

Figure 3.15 Simulation results for the ideal transformer.

```
.PROBE                                ; PSpice case
.END
```

3.2.6 Square Root Extractor

This macromodel for taking the square root of an input voltage has many applications in which rms quantities are required. The input voltage must be positive. Here K is a constant scaling factor. The transfer function for the square root extractor is:

$$V_{out} = K \cdot \sqrt{V_{in}}$$

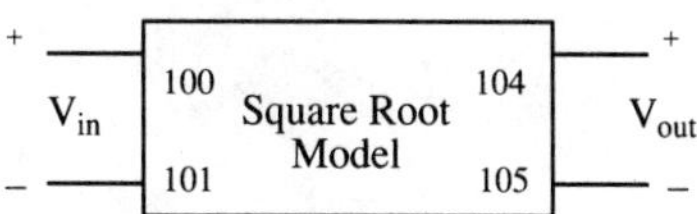

Figure 3.16 Functional block for the square root extractor.

The equations for Figure 3.17 are:

$$I(G_1) = V(100,101) = V_{in}$$

$$I(G_2) = V(102,0)^2$$

Since the current through R_1 is negligible because of the large value of R_1,

$$I(G_1) = I(G_2)$$

and

$$V(102,0) = V(100,101)^{1/2} = V_{in}^{1/2}$$

$$V(103,0) = K \cdot V(100,101)^{1/2} = K \cdot V_{in}^{1/2}$$

SPICE code for Figure 3.17

```
RIN 100 101 [A]                     ; [A] = Input resistance
G1 0 102 100 101 1
G2 102 0 POLY(1) 102 0 0 0 1
R1 102 0 1T
EOUT 103 105 102 0 [K]              ; [K] = Voltage gain
R2 104 105 1G
ROUT 103 104 [B]                    ; [B] = Output resistance
```

Example

$V_{out} = 2 \cdot (V_{in})^{1/2}$
Input resistance = 1MΩ
Output resistance = 1KΩ
Nodes 101 and 105 are grounded in this example.

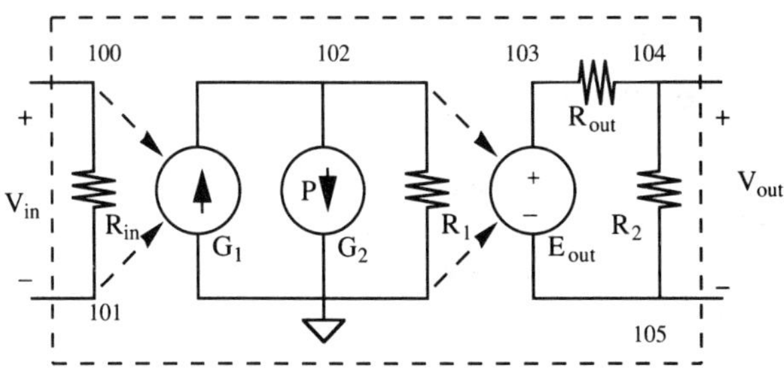

Figure 3.17 Equivalent circuit for the square root extractor.

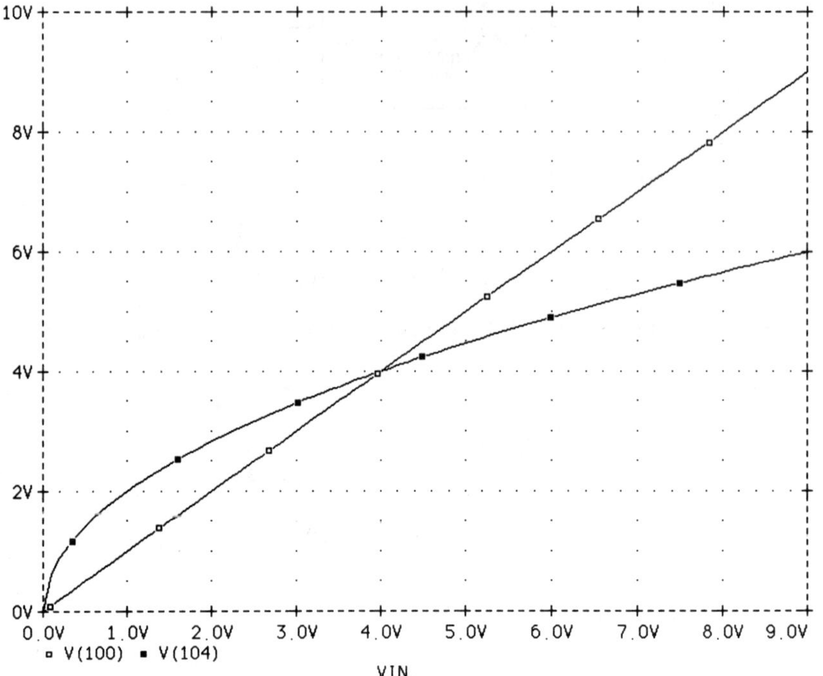

Figure 3.18 Simulation results for the square root extractor.

```
RIN 100 0 1MEG                    ; Input resistance = 1 MΩ
G1 0 102 100 0 1
G2 102 0 POLY(1) 102 0 0 0 1
R1 102 0 1T
EOUT 103 0 102 0 2                ; Voltage gain = 2
R2 104 0 1G
ROUT 103 104 1K                   ; Output resistance = 1 KΩ
VIN 100 0
.DC VIN 0 9 0.1
.PRINT DC V(104)
.PROBE                            ; PSpice
.END
```

3.2.7 Triangle-to-Sine-Wave Converter

A nonlinear polynomial dependent VCVS can produce a triangle-to-sine-wave conversion function. The series expansion of a sine function whose peak magnitude is unity is the approach used for modeling this converter. A conversion constant is required to scale the radian value of the sine wave output to the real value of the triangle wave. The transfer function for the triangle-to-sine-wave converter, shown in Figure 3.19, is:

$$V_{out} = K \cdot \sin(C \cdot V_{in})t \qquad C = \text{ Conversion constant } = \frac{\pi}{2}$$

$$\approx K \cdot \left\{C \cdot V_{in} - \frac{1}{3!}(C \cdot V_{in})^3 + \frac{1}{5!}(C \cdot V_{in})^5 - \frac{1}{7!}(C \cdot V_{in})^7 + - \cdots\right\}$$

$$\approx K \cdot \{1.5708 \cdot V_{in} - 0.6460 V_{in}{}^3 + 7.9693 \times 10^{-2} V_{in}{}^5$$

$$- 4.6818 \times 10^{-3} V_{in}^7 + - \cdots\}$$

The equation for Figure 3.20 is:

$$V(103,0) \approx K \cdot \left\{C \cdot V_{in} - \frac{1}{3!}(C \cdot V_{in})^3 + \frac{1}{5!}(C \cdot V_{in})^5 - \frac{1}{7!}(C \cdot V_{in})^7 + - \cdots\right\}$$

$$= K \cdot V(102,0)$$

where C is the conversion constant defined previously and V_{in} is V(100, 101).

SPICE code for Figure 3.20

```
RIN 100 101 [A]                    ; [A] = Input resistance
E1 102 0 POLY(1) 100 101
+0 1.5708 0 -0.6460 0 7.9693E-2
+0 -4.6818E-3
R1 102 0 1
EOUT 103 105 102 0 [K]             ; [K] = Voltage gain
R2 104 105 1G
ROUT 103 104 [B]                   ; [B] = Output resistance
```

Notes

1. $V_{peak\text{-}to\text{-}peak}$ of triangle wave input must be 2 V with a DC level of zero volts. Otherwise, the sine wave output is distorted.
2. The accuracy of the output wave depends on the number of terms included as coefficients of the polynomial dependent VCVS, E_1.

Figure 3.19 Functional block for the triangle-to-sine-wave converter.

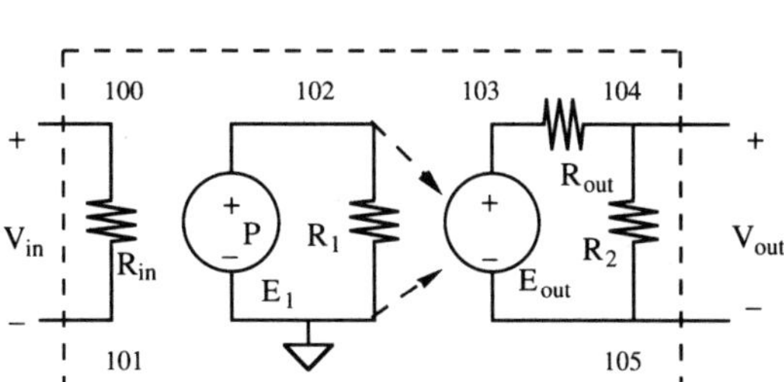

Figure 3.20 Equivalent circuit for the triangle-to-sine-wave converter.

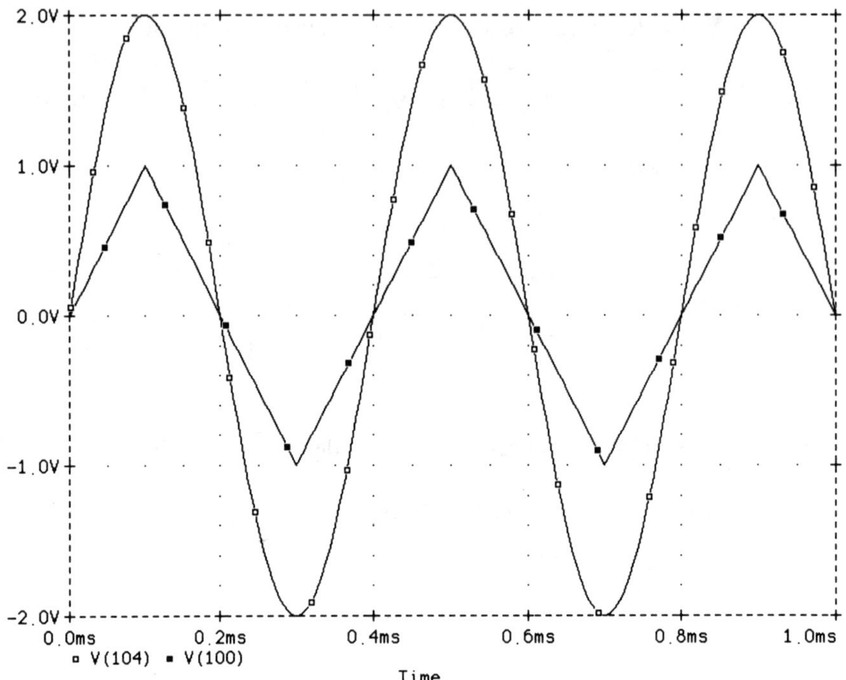

Figure 3.21 Simulation results for the triangle-to-sine-wave converter.

Example

Input signal = Triangle wave with 1 V peak magnitude.
Expected output signal = Sine wave with 2 V magnitude.
Input resistance = 1 MΩ
Output resistance = 1 KΩ
Nodes 101 and 105 are grounded in this example.

```
RIN 100 0 1MEG                       ; Input resistance = 1 MΩ
E1 102 0 POLY(1) 100 0
+0 1.5708 0 -0.6460 0 7.9693E-2
+0 -4.6818E-3
R1 102 0 1
EOUT 103 0 102 0 2                   ; Voltage gain = 2
R2 104 0 1G
ROUT 103 104 1K                      ; Output resistance = 1 KΩ
.OPTIONS LIMPTS=0
.VIN 100 0 PWL(0 0 0.1M 1 0.3M -1 0.5M 1
+ 0.7M -1 0.9M 1 1.0M 0)
.TRAN 5U 1.0M
.PRINT TRAN V(104)
```

```
.PROBE                          ; PSpice
.END
```

3.2.8 Phase Shifter

Phase shift or delay is an important test signal in many applications such as phase modulation, Doppler radar, and phase locked loops. This macromodel delays a sinusoidal input signal by any desired phase angle from 0 to 360 degrees as specified by the user. The output signal has the same magnitude as the input (see Figure 3.22).

In Figure 3.23, E_1 samples the input signal and generates the current $I(V_{test})$ through the capacitor C_1, where

$$V_{in} = V_{max} \sin(\omega t + \beta)$$

$$E_1 = V_{in}$$

$$I(V_{test}) = I(C_1) = (C_1)[d(V_{in})/dt] = C_1[\omega V_{max} \cos(\omega t + \beta)]$$

$$H_1 = BI(V_{test})$$

$$B = 1/[(2\pi)(f_{in})] = 1/\omega = T/2\pi$$

$$C_1 = 1$$

Hence

$$V(104) = V_{max} \cos(\omega t + \beta)$$

Let T denote the (known) period of the input signal. Since the input is a sine (or cosine) signal, V(104) will be a cosine (or sine) signal. The sine and cosine signals are used to generate the shifted output signal as follows:

$$E_{out} = C(V_{in}) + D[V(104)]$$

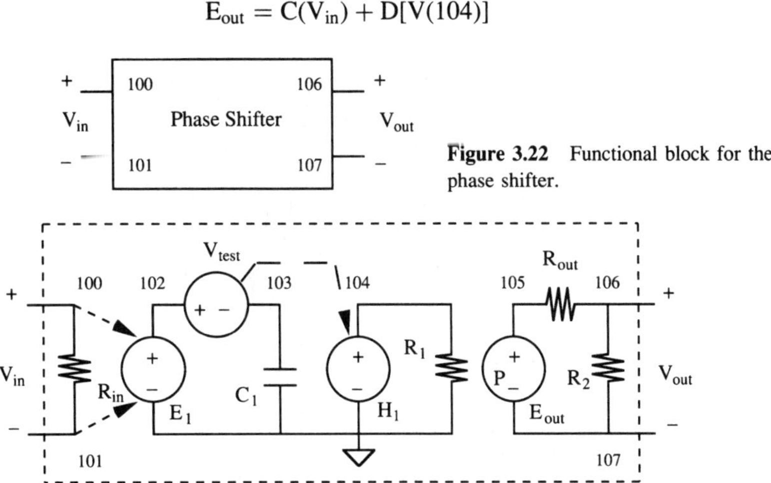

Figure 3.22 Functional block for the phase shifter.

Figure 3.23 Equivalent circuit for the phase shifter.

where C = cos ϕ and D = sin ϕ with ϕ in radians. With R_2 very large, $V_{out} \approx E_{out}$, so that

$$E_{out} = V_{max}[(\cos\phi)(\sin(\omega t + \beta)) + (\sin\phi)(\cos(\omega t + \beta))].$$

Note that V_{out} will lead the V_{in} by ϕ radians. In other words, V_{in} will be shifted by $-\phi$ radians. Therefore in order to shift the input by $+\phi$ radians, [B] = $-T/2\pi$.

SPICE code for Figure 3.23

```
RIN 100 101 [A]                    ; [A] = Input resistance
E1 102 0 100 101 1
VTEST 102 103 DC=0 AC=0
C1 103 0 1
H1 104 0 VTEST [B]                 ; [B] = -T/2π for positive shift
*                                          +T/2π for negative shift
R1 104 0 1
EOUT 105 107 POLY(2)  100 101 104 0
+ 0 [C] [D]                        ; [C] = cos φ, [D] = sin φ
R2 106 107 1G
ROUT 105 106 [B]                   ; [D] = Output resistance
```

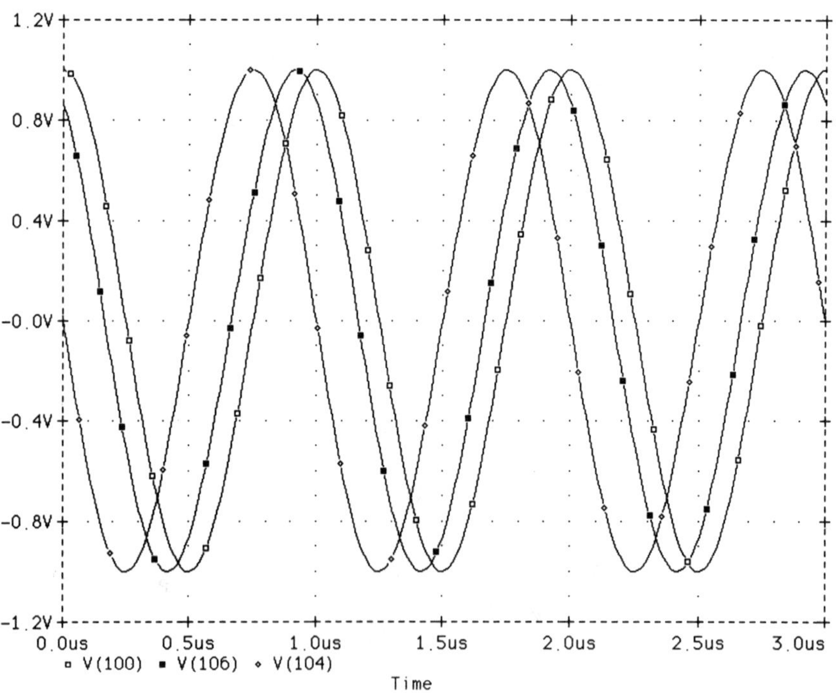

Figure 3.24 Simulation results for the phase shifter.

Notes:

1. The waveform of the input signal must be sinusoidal.
2. The frequency of the input signal must be known.

Example

Input = sin (2π x 10^6t + $\pi/2$)
Input resistance = 1 MΩ
Output resistance = 1 KΩ
$\phi = -30$ degrees
Nodes 101 and 107 are grounded in this example.

```
RIN 100 0 1MitEG                        ; Input = MΩ
E1 102 0 100 0 1
VTEST 102 103 DC=0 AC=0
C1 103 0 1
H1 104 0 VTEST 0.1591549E-6             ; T/2π = 0.1591549E − 6
R1 104 0 1
EOUT 105 0 POLY(2)  100 0 104 0
+ 0 0.8660254 0.5                       ; cos φ = 0.8660254, sin φ = 0.5
R2 106 0 1G
ROUT 105 106 1K                         ; Output resistance = 1 K

VIN 100 0 SIN(0 1 1MEG 0 0 90)
.OP
.TRAN 10N 3U 0 10N
.PRINT TRAN V(104)
.PROBE                                  ; PSpice case
.END
```

3.2.9 Integrator

The analog integrator macromodel is formed by taking the voltage across a capacitor whose current is a linear function of the input voltage. Three VCVS elements are used to form the macromodel. The scaling factor K is the reciprocal of the integrator's time constant. The transfer function for the integrator is:

$$V_{out} = K \cdot \int V_{in} dt$$

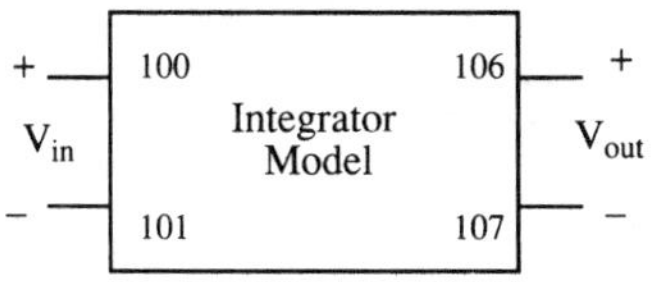

+ 100 106 +
V_{in} Integrator Model V_{out}
− 101 107 −

Figure 3.25 Functional block for the integrator.

As shown in Figure 3.26, E_1 samples the input signal, giving $E_1 = V_{in}$. E_2 and R_2 duplicate an op amp model with an open loop gain of 1G and an input resistance of $R_2 = 1G\Omega$. The $R_{time} - C_{time}$ combination integrates E_1. Feedback resistor R_f provides a DC stable circuit. The resulting equations are

$$V(104,0) = (-E_1/R_{time})[R_f/(sR_fC_{time} + 1)]$$

By making $R_f >> 1$, then

$$V(104,0) \approx [-1/(R_{time}C_{time})](V_{in}/s)$$

$$E_{out} = V(104,0) \approx -1/(R_{time}C_{time}) \int V_{in}dt$$

SPICE code for Figure 3.26

```
RIN 100 101 [A]              ; [A] = Input resistance
E1 102 0 100 101 1
RTIME 102 103 [B]            ; [B] = See note
R2 103 0 1G
CTIME 103 104 [C]            ; [C] = See note
RF 103 104 [D]               ; [D] = See note
E2 0 104 103 0 1G
EOUT 105 107 104 0 1
ROUT 105 106 [E]             ; [E] = Output resistance
R3 106 107 1G
```

Notes

$$[B] = \frac{\text{Amplitude of input signal}}{(2C_{time})(V_{p-p}\text{of output})(\text{Frequency of input signal})}$$

[C] = Arbitrary value

[D] = $50(R_{time})$

Example

Input resistance = 1 MΩ

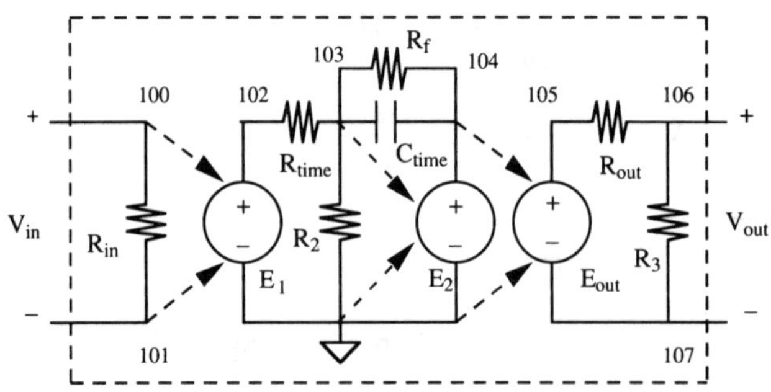

Figure 3.26 Equivalent circuit for the integrator.

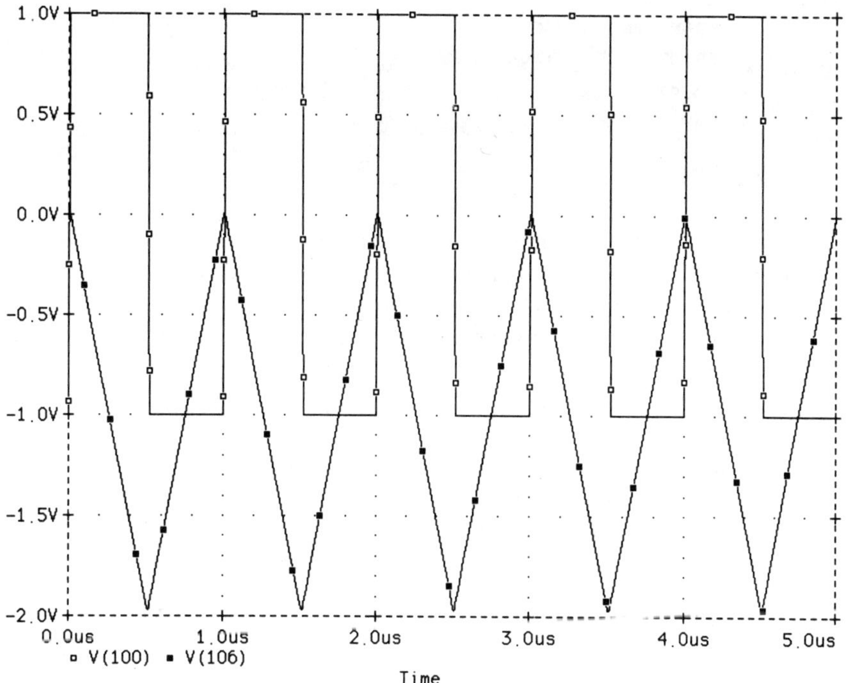

Figure 3.27 Simulation results for the integrator.

Output resistance = 1 KΩ

Input signal specifications:

- Amplitude = 1 V
- Frequency = 1 MHz
- Waveform = Pulse train

Nodes 101 and 107 are grounded in this example.

```
RIN 100 0 1MEG                  ; Input resistance = 1 MΩ
E1 102 0 100 0 1
RTIME  102 103  0.25            ; [B] = 0.25 Ω
R2 103 0 1G
CTIME 103 104 1U                ; [C] = 1 μF
RF 103 104 12.5                 ; [D] = 12.5 Ω
E2 0 104 103 0 1G
EOUT 105 0 104 0 1
ROUT 105 106 1K                 ; Output resistance = 1 KΩ
R3 106 0 1G

VIN 100 0 PULSE (-1 1 0 0 0 0.5U 1U)
.IC V(104)=0
```

```
.TRAN 10N 5U UIC
.PRINT TRAN  V(106)
.PROBE                                  ; PSpice
.END
```

3.2.10 Differentiator

The analog differentiator macromodel is formed by applying the input voltage across a capacitor. The current through this capacitor is converted to a voltage by forcing it through a resistance. Three VCVS elements are used to form the macromodel. The scaling factor K is the differentiator's time constant. The transfer function for the differentiator, shown in Figure 3.28, is:

$$V_{out} = -K \cdot \frac{d}{dt}(V_{in})$$

In Figure 3.29, E_1 samples the input signal, giving $E_1 = V_{in}$. E_2 and R_2 form an op amp model with the gain of 1 G and input resistance of $R_2 = 1\ G\Omega$. The $R_{time} - C_{time}$ combination sets the time constant scaling factor. With R_2 and E_2 very large, the following equations describe the differentiator:

$$V(104, 0) = (-sC_{time}R_{time})(E_1) \Rightarrow -K\ dV_{in}/dt$$

$$E_{out} = V(104, 0) = -K\ dV_{in}/dt$$

SPICE code for Figure 3.29

```
RIN 100 101 [A]              ; [A] = Input resistance
E1 102 0 100 101 1
CTIME 102 103 [B]            ; [B] = See note
RTIME 103 104 [C]                ; [C] = See note
R2 103 0 1G
E2 104 0 103 0 -1G
EOUT 105 107 104 0 1
```

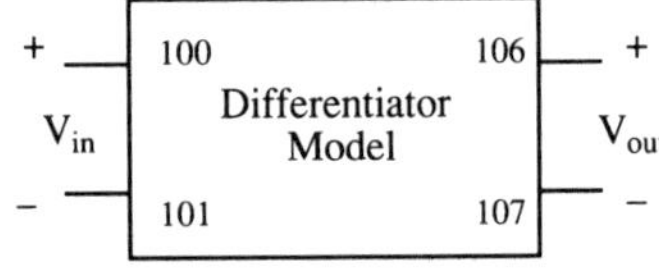

Figure 3.28 Functional block for the differentiator.

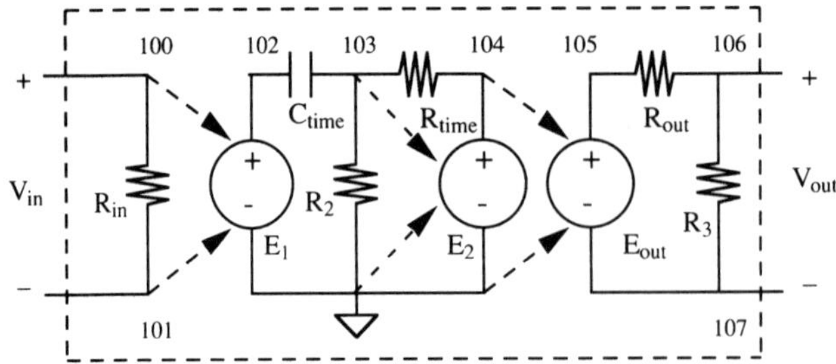

Figure 3.29 Equivalent circuit for the differentiator.

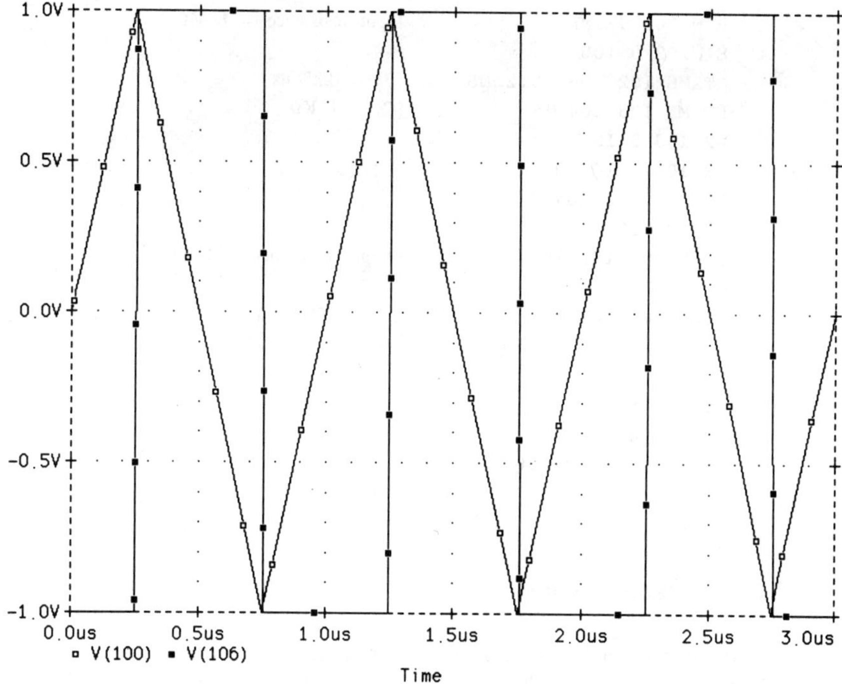

Figure 3.30 Simulation results for the differentiator.

```
R3 106 107 1G
ROUT 105 106 [D]          ; [D] = Output resistance
```

Notes To determine the values of [B] and [C] use the following equations:

[B] = [1E-3 * T] / [4 * Amplitude of the input signal] where T is the period of the input signal.
[C] = [1E3 * (Desired amplitude of output)]

Example

Input resistance = 1 MΩ

Input signal specifications:

 Amplitude = 1 V

 Frequency = 1 MHz

 Waveform = Pulse train

Desired amplitude of output voltage = 1 V

Output resistance = 1 KΩ

Nodes 101 and 107 are grounded in this example.

```
RIN 100 0 1MEG              ; Input resistance = 1 MΩ
E1 102 0 100 0 1
CTIME 102 103  0.25NS       ; [B] = 0.25 ns
RTIME 103 104 1K            ; [C] = 1 KΩ
R2 103 0 1G
E2 104 0 103 0 -1G
EOUT 105 0 104 0 1
R3 106 107 1G
ROUT 105 106 1K             ; Output resistance = 1 KΩ

VIN 100 0 PWL(0  0  0.25  1  0.75U  -1 1.25U 1 1.75U  -1  2.25U  1
+ 2.75U  -1  3.25U  1  3.75U  -1  4.25U  1  4.75U  -1  5.0U  0)
.IC V(106)=0
.OPTIONS PIVTOL=1E-30
.TRAN 10N 5U UIC
.PRINT TRAN V(106)
.PROBE                      ; PSpice
.END
```

3.2.11 Absolute Value

The main transfer function of this macromodel, shown in Figure 3.31, is identical to that of a full wave rectifier. The nonlinear transfer characteristic is modeled by the piecewise linear approximation described in Chapter 2. This macromodel is valid in both DC sweep and transient analyses.

In Figure 3.32, the independent voltage source V_{TEST} is used to measure the current from E_1. Since the conversion gains of both controlled sources are unity, the following equations are valid:

$$V(102,0) = V(100,101) = V(103) = V_{in}$$

$$V(105,107) = I(V_{TEST})$$

SPICE code for Figure 3.32

```
RIN 100 101 [A]                 ; [A] = Input resistance
E1 102 0 100 101 1
```

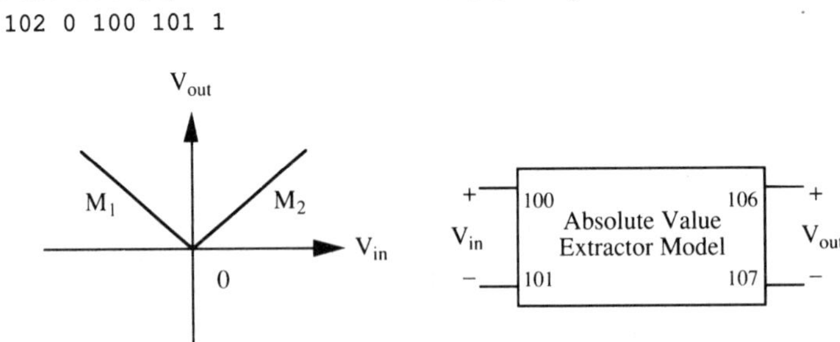

Figure 3.31 Transfer characteristic and functional block for absolute value.

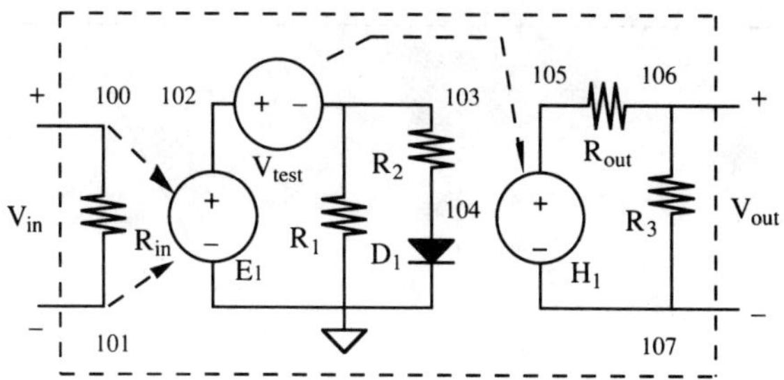

Figure 3.32 Equivalent circuit for absolute value.

```
VTEST 102 103 DC=0.0 AC=0.0
R1 103 0 [B]                    ; [B] = 1/M1
R2 103 104 [C]                  ; [C] = 1/(M2 - M1)
D1 104 0 DIODE
H1 105 107 VTEST 1
R3 106 107 1G
ROUT 105 106 [D]                ; [D] = Output resistance
.MODEL DIODE D(N=0.001)
```

Notes

1. If the value of [B] or [C] is infinite, let [B] or [C] be 1 T.
2. In the transient analysis, the maximum internal time step should be less than 1/200 of the period of input signal (or half of the printing step) to obtain the correct output waveform.

Example

Input resistance = 1 MΩ

Output resistnce = 1 KΩ

$M_1 = -1.0$

$M_2 = 1.0$

Nodes 101 and 107 are grounded in this example.

```
RIN 100 0 1MEG                  ; Input resistance = 1 MΩ
E1 102 0 100 0 1
VTEST 102 103 DC=0.0 AC=0.0
R1 103 0 -1                     ; 1/M1 = -1
R2 103 104 0.5                  ; 1/(M2 - M1) = 0.5
D1 104 0 DIODE
H1 105 0 VTEST 1
R3 106 0 1G
ROUT 105 106 1K                 ; Output resistance = 1 KΩ
.MODEL DIODE D(N=0.001)
VIN 100 0
.DC VIN -5 5 0.1
.PRINT DC V(106)
```

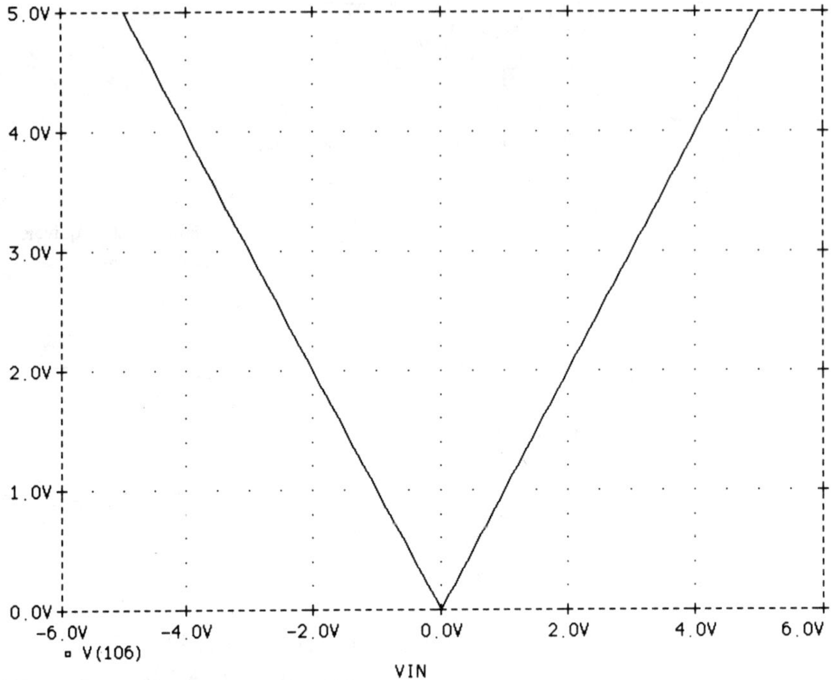

Figure 3.33 Simulation results for absolute value.

```
.PROBE                                  ; PSpice case
.END
```

3.2.12 Peak Detector

A peak detector network measures the positive peak voltage of an input signal. This macromodel, shown in Figure 3.34, can be generated automatically using the diskettes and instructions in Chapter 6.

In Figure 3.35, the diode D_1 is an ideal type with the emission coefficient N = 0.001. It acts as a voltage controlled switch.

SPICE code for Figure 3.35

```
RIN 100 101 [A]                         ; [A] = Input resistance
E1 102 0 100 101 1
D1 102 103 DIODE
```

+ +
V_{in} 100 Peak Detector Model 105 V_{out}
– 101 106 –

Figure 3.34 Functional block for the peak detector.

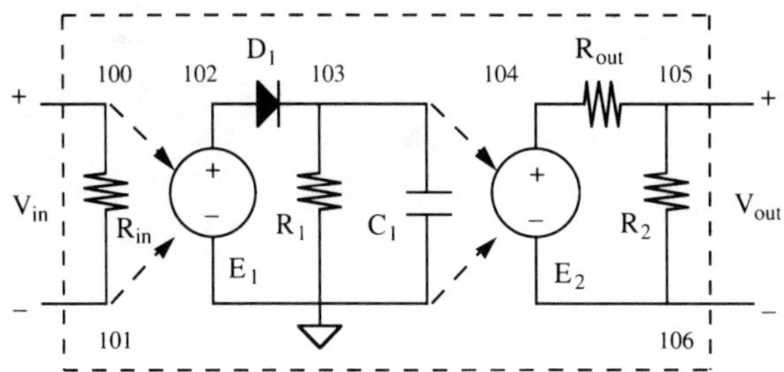

Figure 3.35 Equivalent circuit for the peak detector.

```
R1 103 0 1MEG
C1 103 0 [B]                        ; [B] = See note
E2 104 106 103 0 1
R2 105 106 1G
ROUT 104 105 [C]                    ; [C] = Output resistance
.MODEL DIODE D(N=0.001)
```

Note

1. $[B] = 10^{-4} \times T_{in}$, where T_{in} is the period of the input signal.

Example

Input resistance = 1 MΩ
Output resistance = 1 KΩ
Input signal = Arbitrary piecewise linear source
Nodes 101 and 106 are grounded in this example.

```
RIN 100 0 1MEG                      ; Input resistance = 1 MΩ
E1 102 0 100 0 1
D1 102 103 DIODE
R1 103 0 1MEG
C1 103 0 10N                        ; [B] = 10 nF
E2 104 0 103 0 1
R2 105 0 1G
ROUT 104 105 1K                     ; Output resistance = 1 KΩ
.MODEL DIODE D(N=0.001)

VIN 100 0 PWL(0 0 0.2U 1 0.35U -1
+  0.5U 3 0.7U 1 0.95U 2 1.3U -3
+  1.5U 4 1.75U -1 2U 0)
.TRAN 10N 2U
.PRINT TRAN V(105)
.PROBE                              ; PSpice case
.END
```

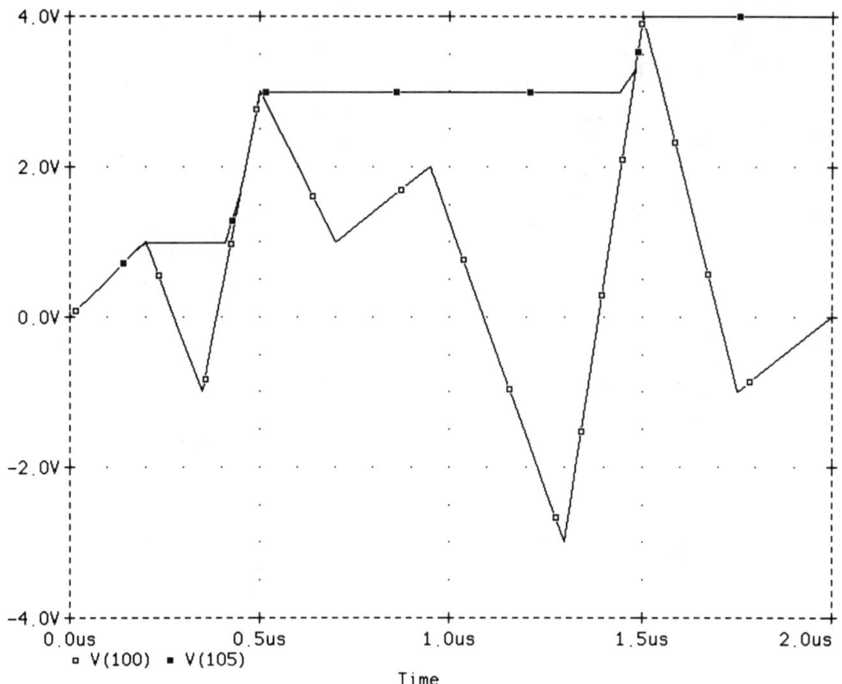

Figure 3.36 Simulation results for the peak detector.

3.2.13 Frequency Multiplier

The frequency multiplier function is widely used in communication systems and in frequency synthesizers. The magnitude of the input and output sinusoidal signals must be the same. The output frequency can be set to any desired integer multiple of the input frequency. The frequency multiplier macromodel is based on trigonometric relationships. The transfer function for the frequency multiplier, shown in Figure 3.37, is:

$$\text{Output frequency, } f_{out} = K * \text{ Input frequency, } f_{in}$$

The branch currents in the second stage of Figure 3.38 are

$$I(G_1) = 1 - V(100,101)^2 = 1 - V_{in}^2 = 1 - \sin^2 \omega t \text{ when the input is a sine wave}$$

$$I(G_2) = V(102,0)^2$$

Since the R_1 value is very large, the branch current in R_1 is very small. Hence

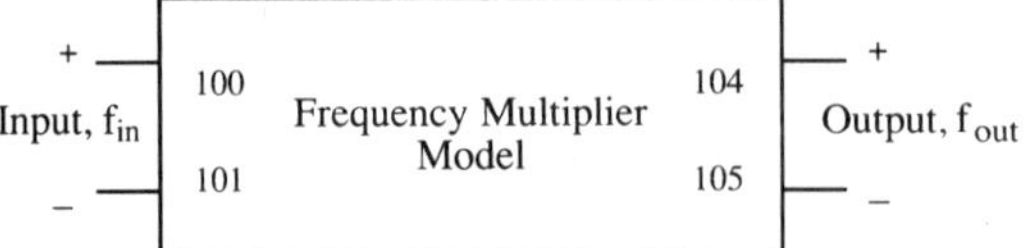

Figure 3.37 Functional block for the frequency multiplier.

$$I(G_1) = 1 - V_{in}^2 = V(102,0)^2 = I(G_2)$$

$$V(102,0) = (1 - V_{in}^2)^{1/2} = (1 - \sin^2 \omega t)^{1/2} = \cos \omega t$$

If the desired output frequency is double the input frequency, the voltage between nodes 103 and 105 is:

$$V(103,0) = 2 \cdot V(100,101) \cdot V(102,0) = 2V_{in}(1 - V_{in}^2)^{1/2}$$
$$= 2 \sin \omega t \cdot \cos \omega t = \sin 2\omega t$$

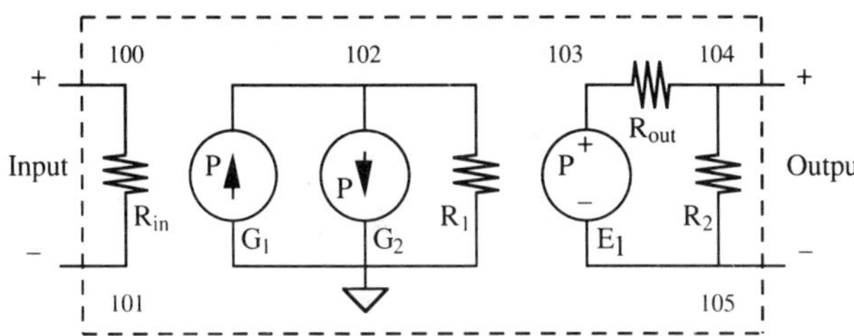

Figure 3.38 Equivalent circuit for the frequency multiplier.

SPICE code for Figure 3.38

```
RIN 100 101 [A]                                ; [A] = Input resistance
G1 0 102 POLY(1) 100 101 1 0 -1
G2 102 0 POLY(1) 102 0 0 0 1
R1 102 0 1G
E1 103 105 POLY(2) 100 101 102 0
+ P0 P1 P2 P3 P4 P5 P6 P7 P8 P9 P10 P11
R2 104 105 1G
ROUT 103 104 [B]                               ; [B] = Output resistance
```

Notes:

1. The waveform of input signal must be sinusoidal.
2. The peak-to-peak voltage of the input signal must be 2 V, and its average voltage (DC value) must be zero.
3. Set $P_4 = 2$ when $f_{out} = 2 \cdot f_{in}$ and set the rest of the coefficients to zero. These coefficients are obtained from the trigonometric identity $\sin 2x = 2(\sin x)(\cos x)$.
4. Set $P_1 = 3$ and $P_6 = -4$ when $f_{out} = 3 \cdot f_{in}$. The rest of the coefficients are set to zero. Here the trigonometric identity $\sin 3x = 3 \sin x - 4\sin^3 x$ is used.
5. Set $P_4 = 4$ and $P_{11} = -8$ when $f_{out} = 4 \cdot f_{in}$. The rest of the coefficients are set to zero. Again use the trigonometric identity, $\sin 4x = 4(\sin x)(\cos x) - 8(\sin^3 x)(\cos x)$.
6. If the macromodel does not work properly during the transient analysis, reduce the maximum internal time step.

Example

$f_{out} = 2 \cdot f_{in}$ ($f_{in} = 1$ MHz)
Input resistance = 1 MΩ
Output resistance = 1 KΩ
Nodes 101 and 105 are grounded in this example.

```
RIN 100 0 1MEG                        ; Input resistance = 1 MΩ
G1 0 102 POLY(1) 100 0 1 0 -1
G2 102 0 POLY(1) 102 0 0 0 1
R1 102 0 1G
E1 103 0 POLY(2) 100 0 102 0
+ 0 0 0 0 2                           ; P4 = 2
R2 104 0 1G
ROUT 103 104 1K                       ; Output resistance = 1 KΩ

VIN 100 0 SIN(0 1 1MEG)
.TRAN 10N 4U 0 10N
.PRINT TRAN V(104)
.PROBE                                ; PSpice case
.END
```

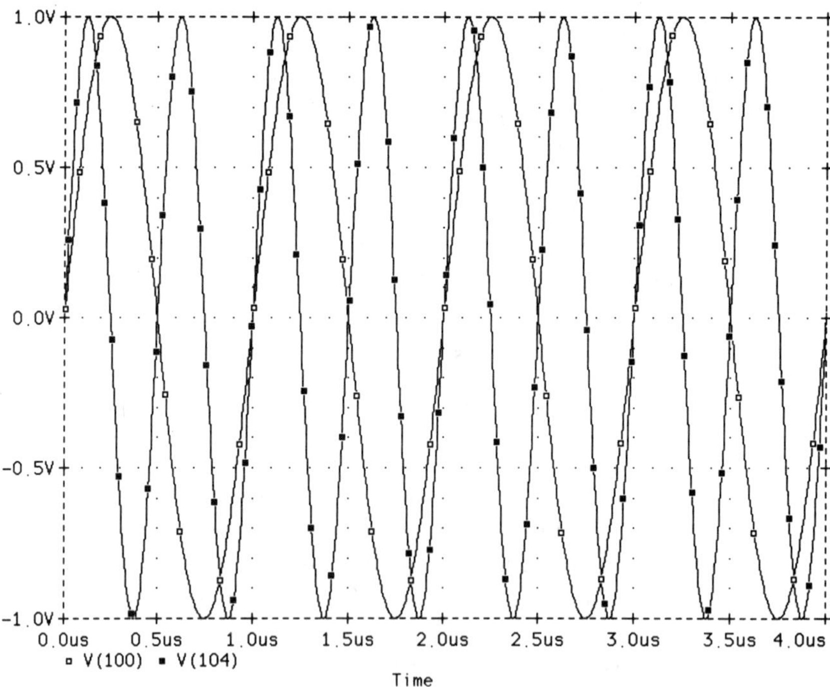

Figure 3.39 Simulation results for the frequency multiplier.

3.2.14 Frequency Divider (by 2)

Frequency division is also widely used in frequency synthesis. The magnitude of the input and output signals must be the same. The output frequency is half of the input frequency. This function is also modeled using trigonometric relationships. The transfer function for the frequency divider, shown in Figure 3.40, is:

$$\text{Output frequency, } f_{out} = \text{ Input frequency, } f_{in}/2$$

The branch currents in the second stage, shown in Figure 3.41, are

$$I(G_1) = 1 - V(100{,}101)^2 = 1 - V_{in}^2 = 1 - \sin^2 \omega t \text{ when the } V_{in} \text{ is a sine wave}$$

$$I(G_2) = V(102{,}0)^2$$

Since the R_2 value is made very large, the branch current in R_2 is very small. Hence,

$$I(G_1) = 1 - V_{in}^2 = V(102{,}0)^2 = I(G_2)$$

$$V(102{,}0) = (1 - V_{in}^2)^{1/2} = (1 - \sin^2 \omega t)^{1/2} = \cos \omega t$$

The branch currents in the third stage are

$$I(G_3) = [1 - V(102{,}0)]/2 = [1 - \cos \omega t]/2$$

$$I(G_4) = V(103{,}0)^2$$

If R_3 is made very large, then

$$I(G_3) = [1 - \cos \omega t]/2 = V(103{,}0)^2 = I(G_4)$$

$$V(103{,}0) = [(1 - \cos \omega t)/2]^{1/2} = \sin(\omega t/2)$$

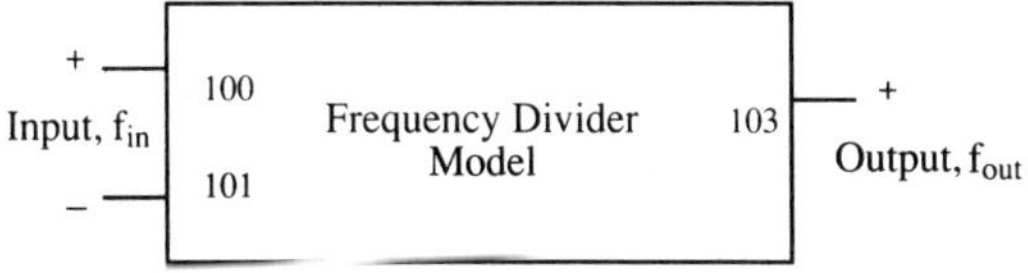

Figure 3.40 Functional block for the frequency divider (by 2).

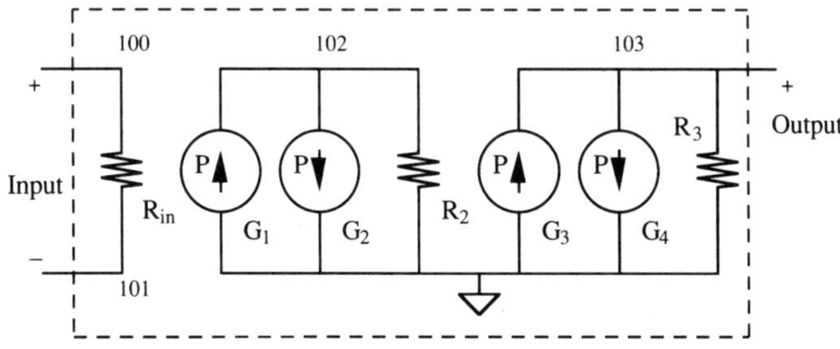

Figure 3.41 Equivalent circuit for the frequency divider (by 2).

SPICE code for Figure 3.41

```
RIN 100 101 [A]                                ; [A] = Input resistance
G1 0 102 POLY(1) 100 101 1 0  -1
G2 102 0 POLY(1) 102 0 0 0 1
R2 102 0 1G
G3  0 103 POLY(1) 102 0 0.5   -0.5
G4 103 0 POLY(1) 103 0 0 0 1
R3 103 0 1G
```

Notes

1. The waveform of the input must be sinusoidal.
2. The peak-to-peak voltage of the input signal must be 2 V.
3. If the macromodel does not work properly during the transient analysis, reduce the maximum internal time step.

Example

$f_{out} = f_{in} / 2$ (f_{in}= 1MHz)

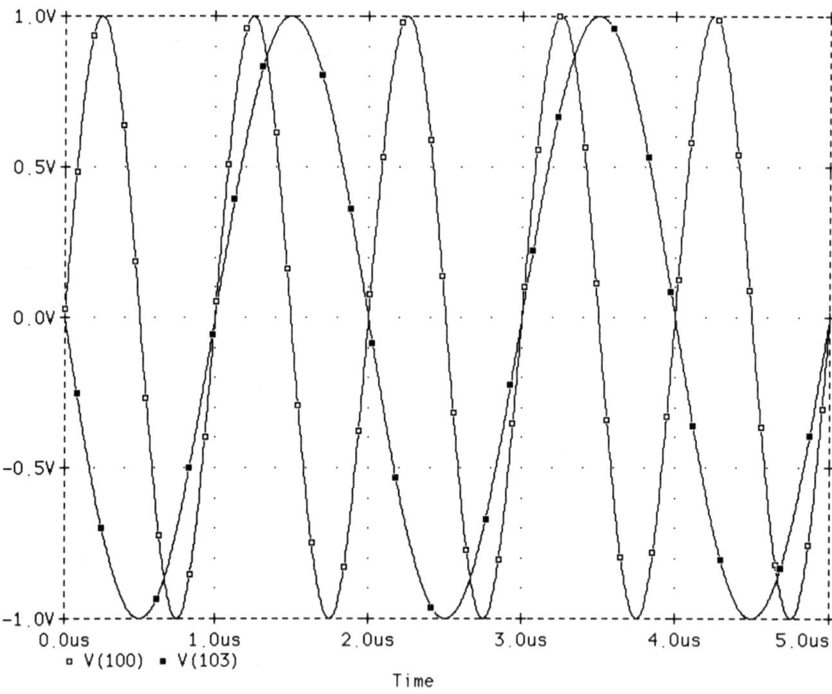

Figure 3.42 Simulation results for the frequency divider (by 2).

Input frequency = 1 MHz

Input resistance = 1 MΩ

Node 101 is grounded in this example.

```
RIN 100 0 1MEG                          ; Input resistance = 1 MΩ
G1 0 102 POLY(1) 100 0 1 0 -1
G2 102 0 POLY(1) 102 0 0 0 1
R2 102 0 1G
G3 0 103  POLY(1) 102 0 0.5   -0.5
G4 103 0 POLY(1) 103 0 0 0 1
R3 103 0 1G

VIN 100 0 SIN(0 1 1MEG)
.OPTIONS ITL1=200 ITL5=0 RELTOL=1E-8
+ ITL4=200
.TRAN 10N 5U 0 5N
.PROBE                                  ; PSpice case
.END
```

3.2.15 Frequency Adder/Subtractor

Frequency addition and subtraction can be modeled by using the sum of angles and difference of angles relationships in trigonometry. The modeling procedure is similar to that used in the frequency multiplication and division macromodels. The transfer function for the frequency adder-subtractor, shown in Figure 3.43, is:

$$\text{Output frequency} = \text{Frequency of input 1} \pm \text{Frequency of input 2}$$

The middle stage in the macromodel of Figure 3.44 performs a 90 degree phase shift with respect to the inputs 1 and 2.

$$V(104) = [1 - V(100{,}101)^2]^{1/2} = (1 - V_{\text{in}1}^2)^{1/2} = (1 - \sin^2 \omega_1 t)^{1/2} = \cos \omega_1 t$$

$$V(105) = [1 - V(102{,}103)^2]^{1/2} = (1 - V_{\text{in}2}^2)^{1/2} = (1 - \sin^2 \omega_2 t)^{1/2} = \cos \omega_2 t$$

By multiplying V(100, 101) by V(105, 0) and also by multiplying V(102, 103) by V(104), E_{out} will model the sum and difference of angles as follows:

$$\begin{aligned} V(106{,}108) &= [V(100{,}101) \cdot V(105{,}0)] \pm [V(102{,}103) \cdot V(104{,}0)] \\ &= [\sin \omega_1 t \cdot \cos \omega_2 t] \pm [\sin \omega_2 t \cdot \cos \omega_1 t] \\ &= \sin(\omega_1 \pm \omega_2)t \end{aligned}$$

SPICE code for Figure 3.44

```
RIN1 100 101 [A]                        ; [A] = Input resistance 1
RIN2 102 103 [B]                        ; [B] = Input resistance 2
G1 0 104 POLY(1) 100 101 1 0 -1
```

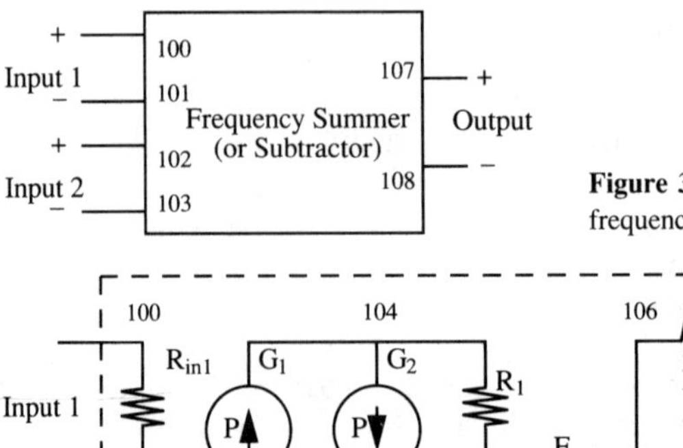

Figure 3.43 Functional block for the frequency adder/subtractor.

Figure 3.44 Equivalent circuit for the frequency adder/subtractor.

```
G2 104 0 POLY(1) 104 0 0 0 1
G3 0 105 POLY(1) 102 103 1 0 -1
G4 105 0 POLY(1) 105 0 0 0 1

R1 104 0 1G
R2 105 0 1G
EOUT 106 108 POLY(4)
+ 100 101 104 0 102 103 105 0
+ 0 0 0 0 0 0 0 0 1 0 [C]          ; [C] = 1 for adder
                                   ; [C] = -1 for subtractor

RD 107 108 1G
ROUT 106 107 [D]                   ; [D] = Output resistance
```

Notes

1. The waveform of the inputs 1 and 2 must be sinusoidal.
2. The peak-to-peak voltage of the inputs must be 2 V.
3. During transient analysis, reduce the internal time step if the macromodel does not work properly.

Example

$f_{out} = f_{in1}(1\ \text{MHz}) + f_{in2}(1\ \text{MHz})$
Input 1 $= \sin(2\pi \times 10^6 t)$
Input 2 $= \sin(2\pi \times 10^6 t)$
Input resistance I = 1 MΩ

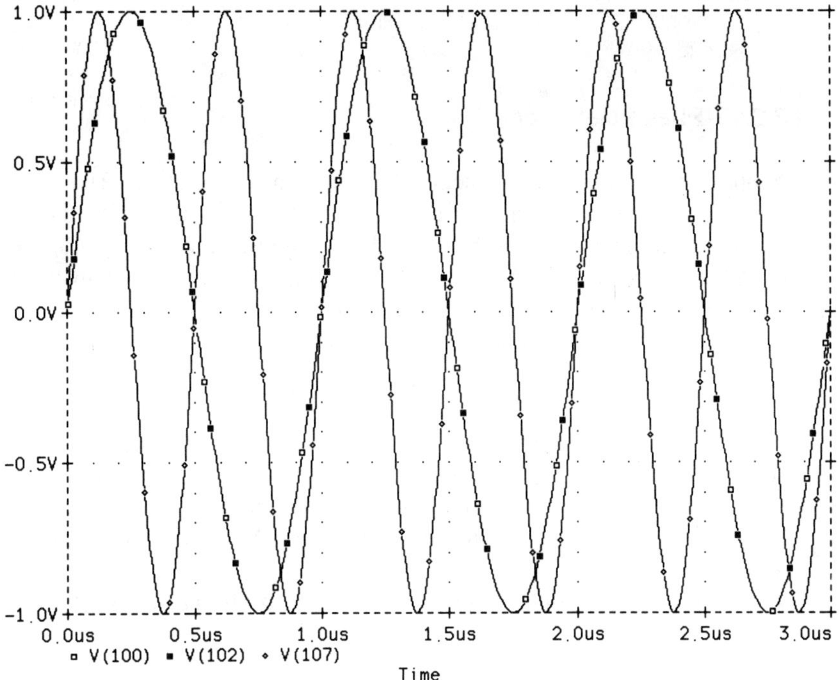

Figure 3.45 Simulation results for the frequency adder/subtractor.

Input resistance II = 1 MΩ

Output resistance = 1 KΩ

Nodes 101, 103, and 108 are grounded in this example.

```
RIN1 100 0 1MEG            ; Input resistance 1 = 1 MΩ
RIN2 102 0 1MEG            ; Input resistance 2 = 1 MΩ
G1 0 104 POLY(1) 100 0 1 0 -1
G2 104 0 POLY(1) 104 0 0 0 1
G3 0 105 POLY(1) 102 0 1 0 -1
G4 105 0 POLY(1) 105 0 0 0 1
R1 104 0 1G
R2 105 0 1G
EOUT 106 0 POLY(4)
+ 100 0 104 0 102 0 105 0
+ 0 0 0 0 0 0 0 0 1 0 1 ; [C] = 1 when adder case
RD 107 0 1G
ROUT 106 107 1K            ; Output resistance = 1 KΩ

VIN1 100 0 SIN(0 1 1MEG)
VIN2 102 0 SIN(0 1 1MEG)
.OPTIONS ITL1=200
.TRAN 10N 3U 0 5N
```

```
.PROBE                          ; Pspice case
.END
```

3.2.16 Phase Detector

A phase detector is a key functional block in a PLL system. The purpose of a phase detector is to produce an output voltage that is proportional to the phase difference between two input signals. Often this is accomplished by multiplying two input signals of the same frequency. A generalized macromodel in Figure 3.46 is developed where the user can set the desired conversion gain.

The voltage at node 103 in Figure 3.47 is

$$V(103{,}0) = V(100{,}102) \cdot 10^{7} - V(104{,}0) \cdot 10^{-11}$$

The maximum and minimum voltages at node 104 are set by the independent voltage sources V_1 and V_2. Hence, the waveform at V(104, 0) is a square wave with a fixed peak-to-peak amplitude regardless of the input signal.

The conversion gain of the phase detector is K_d in units of volts/radian. The voltage at node 107 is

$$V(107{,}109) = K_a \cdot C_f \cdot [V(101{,}102) \cdot V(104{,}0)]$$

where $K_a \cdot C_f = (K_d \cdot \pi)/(2V_1V_2)$

SPICE code for Figure 3.47

```
RIN1 100 102 [A]                ; [A] = Input resistance 1
RIN2 101 102 [B]                ; [B] = Input resistance 2
E1 103 0 POLY(2) 100 102 104 0
```

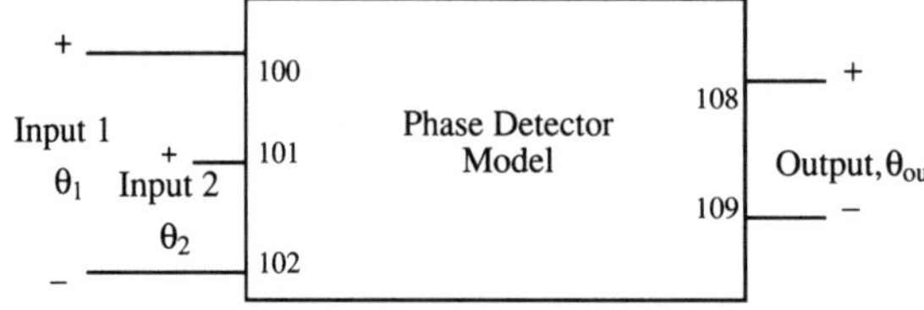

Figure 3.46 Functional block for the phase detector.

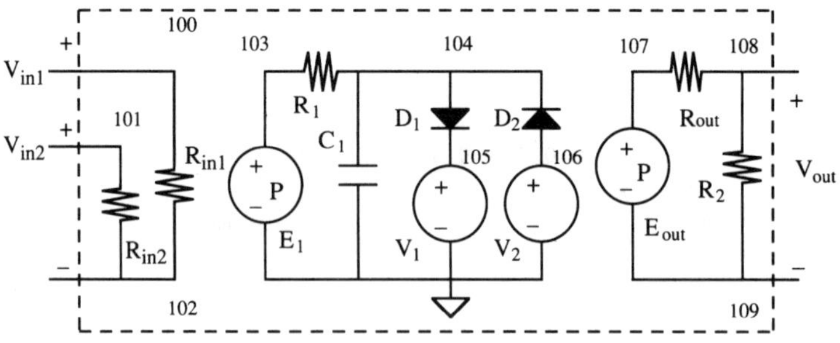

Figure 3.47 Equivalent circuit for the phase detector.

```
+ 0 1E7 -1E-11
R1 103 104 1T
C1 104 0 1E-15
D1 104 105 SWITCH
V1 105 0 DC=10V
D2 106 104 SWITCH
V2 106 0 DC=-10V
.MODEL SWITCH D(N=0.001)
EOUT 107 109 POLY(2) 104 0 101 102
+ 0 0 0 0 [C]                    ; [C] = Ka * Cf/10
*                                ; Cf = Correction factor
R2 108 109 1G
ROUT 107 108 [D]                 ; [D] = Output resistance
```

Notes

1. To determine the correction factor C_f:
   (a) Develop a macromodel for the phase detector with $C_f = 1$.
   (b) Simulate the phase detector model with 1 radian phase difference between the two signals.
   (c) Find the expected average output voltage and the simulated average output voltage.
   (d) The correction factor is defined as:

$$C_f = \frac{\text{Expected average value}}{\text{Simulated average value}}$$

Example

Input resistance 1 = 1 MΩ
Input resistance 2 = 1 MΩ
Output resistance = 1 KΩ
Conversion gain K_d = 0.66 V/rad
Two input signals are sinc waves
Frequency of two input signals is 1 MHz
Phase difference of two signals = 32.7 degrees
Relative phase error of two signals = 1 radian
Nodes 102 and 109 are grounded in this example.

```
RIN1 100 0 1MEG          ; Input resistance 1 = 1 MΩ
RIN2 101 0 1MEG          ; Input resistance 2 = 1 MΩ
E1 103 0 POLY(2) 100 0 104 0
+ 0 1E7 -1E-11
R1 103 104 1T
C1 104 0 1E-15
D1 104 105 SWITCH
```

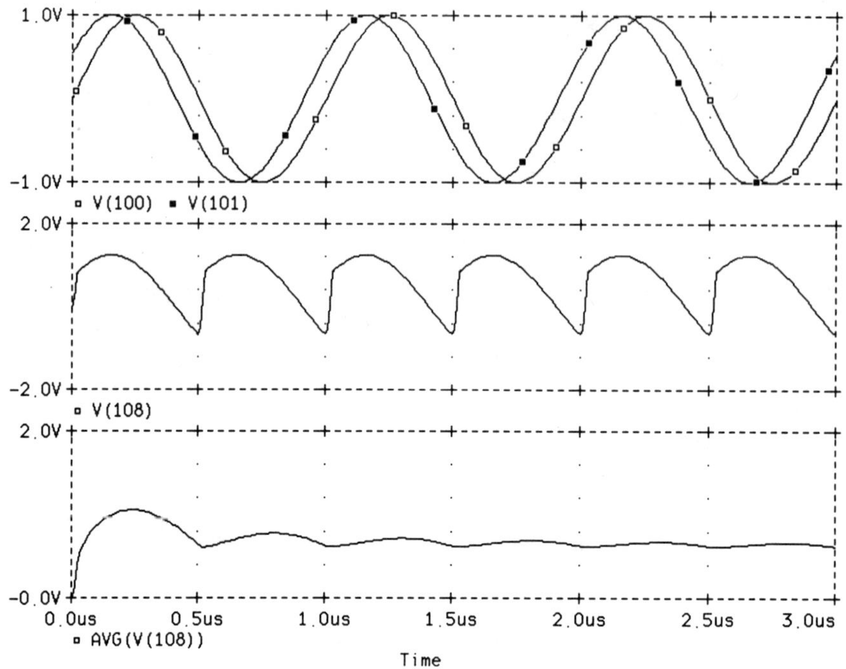

Figure 3.48 Simulation results for the phase detector.

```
V1 105 0 DC=10
D2 106 104 SWITCH
V2 106 0 DC=-10
.MODEL SWITCH D(N=0.001)
EOUT 107 0 POLY(2) 104 0 101 0
+ 0 0 0 0 0.12408          ; [C] = (0.66*1.88)/10
R2 108 0 1G
ROUT 107 108 1K            ; Output resistance = 1 KΩ

VIN1 100 0 SIN(0 1 1MEG)
VIN2 101 0 SIN(0 1 1MEG 0 0 32.7)
.TRAN 10N  5U 0 10N
.OPTIONS LIMPTS=0
.PRINT TRAN V(100)  V(101)  V(108)
.PROBE                     ; PSpice case
.END
```

Notes. To get the average value of the output:

1. Run the PROBE function.
2. Plot the output V(108) using the ADD TRACE option.
3. Select the ADD TRACE option again.

4. Find the average by typing AVG(V(108)).
5. Read the average value from the graph or use the cursor to display this value .
6. Print a hard copy of this display.

3.2.17 Delay Line

The transmission line model in SPICE is used for modeling signal delay (see Figure 3.49). By matching the characteristic impedance, ZO, of the transmission line to the load resistance, any arbitrary time delay can be realized by properly defining the TD value (see Figure 3.50).

SPICE code for Figure 3.50

```
R1 100 101 1G
E1 102 0 100 101 1
TDELAY 102 0 103 0 ZO=1K TD=[A]            ; [A] = Delay time
R2 103 0 1K
```

Notes

1. Resistance of R_2 must be identical to the characteristic impedance of the transmission line in order to prevent standing wave reflections.
2. TD is delay time in seconds.

Example

Delay time = 0.5 μsec
Frequency of input signal = 1 MHz
Node 101 is grounded in this example.

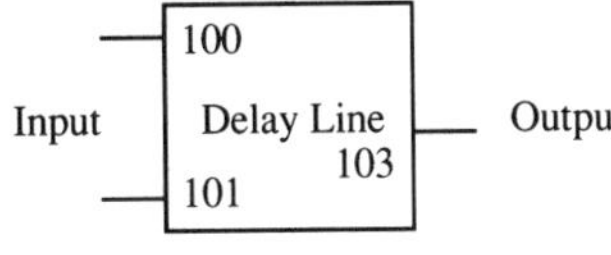

Figure 3.49 Functional block for the delay line.

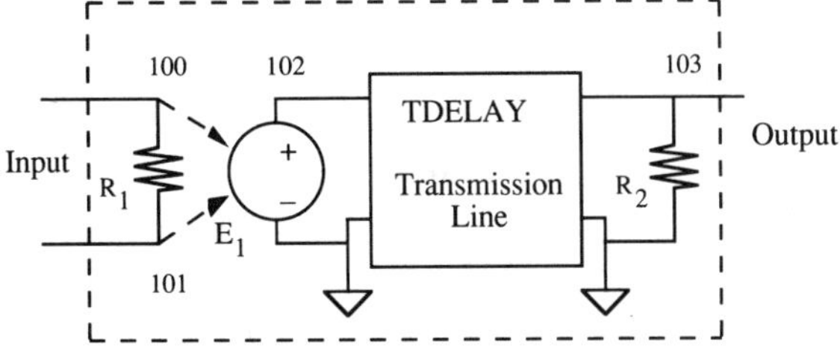

Figure 3.50 Equivalent circuit for the delay line.

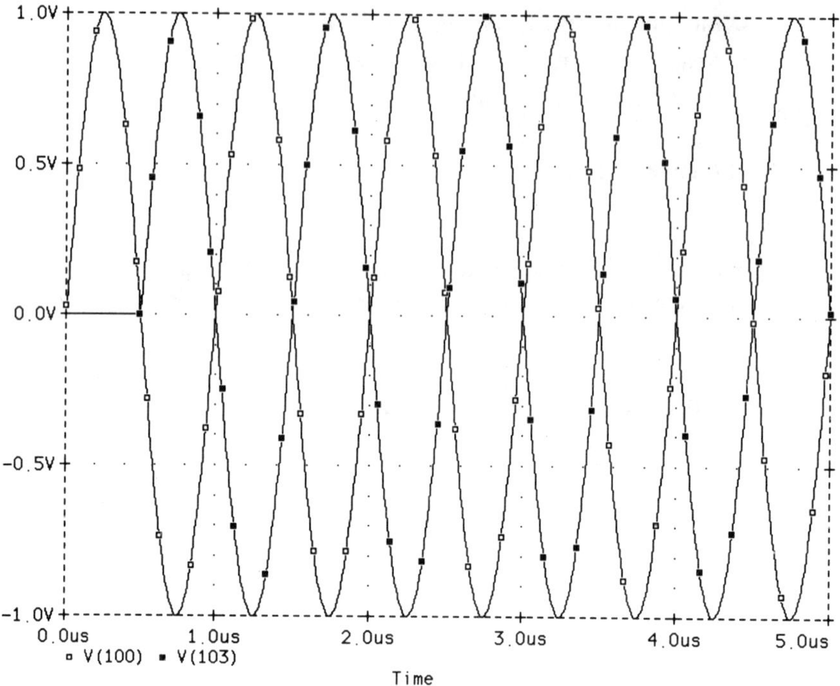

Figure 3.51 Simulation results for the delay line.

```
R1 100 0 1G
E1 102 0 100 0 1
TDELAY 102 0 103 0 ZO=1K TD=0.5U      ; Delay time = 0.5 μsec
R2 103 0 1K

VIN 100 0 SIN(0 1 1MEG)
.TRAN 10N 5U
.PRINT TRAN V(103)
.PROBE                                ; PSpice
.END
```

3.2.18 Schmitt Trigger Simulator

The Schmitt trigger is widely used in many electronic systems to reduce jitter and noise. The function is difficult to simulate because the positive feedback produces double-valued output levels that cause convergence problems in SPICE. A simple equivalent circuit structure macromodel solves the simulation difficulties. Furthermore, this macromodel reduces the overall simulation time. This macromodel, shown in Figure 3.52, is available for automatic computer generation using the techniques of Chapter 6.

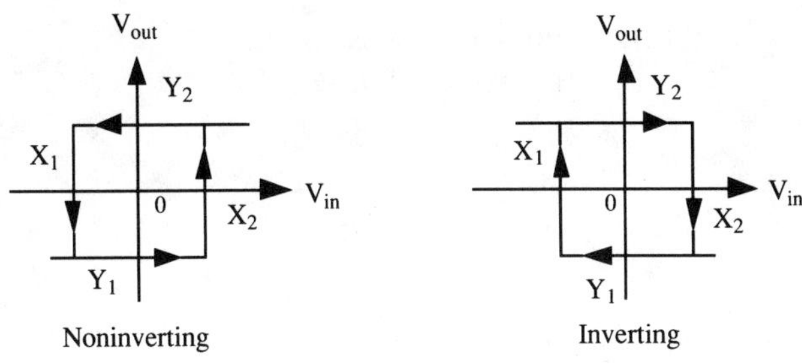

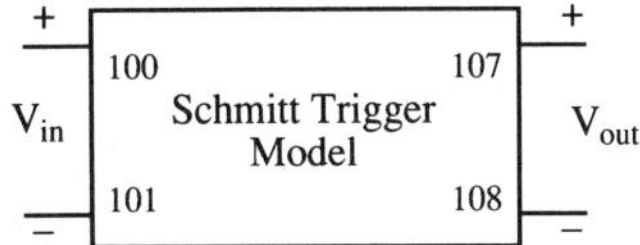

Figure 3.52 Transfer characteristics and functional block for the Schmitt trigger simulator.

The voltage at node 102, shown in Figure 3.53, is:

$$V(102,0) = [V(100,101) \cdot (X_2 + X_1)/2 + V(103,0) \cdot (X_2 - X_1)/(Y_2 - Y_1)] \cdot 10^7$$

where X_1 and X_2 are input break voltages and Y_1 and Y_2 are the corresponding output voltages. The independent voltage sources V_1 and V_2 are defined by

$$V_1 = (Y_2 - Y_1)/2$$

$$V_2 = -(Y_2 - Y_1)/2$$

The voltage at node 106 is

$$V(106,108) = (Y_2 + Y_1)/2 \pm V(103,0)$$

where the + sign is used when an inverting transfer characteristic is desired. Use the − sign for a noninverting Schmitt trigger.

SPICE code for Figure 3.53

```
RIN 100 101 [A]                    ; [A] = Input resistance
E1 102 0 POLY(2)
```

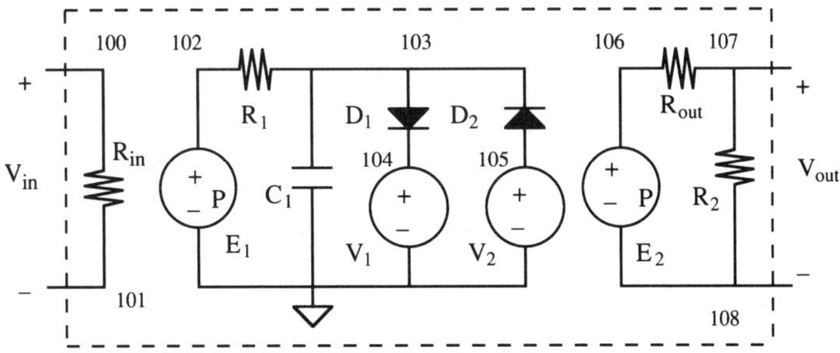

Figure 3.53 Equivalent circuit for the Schmitt trigger simulator.

```
+ 100 101 103 0 [B]E7 -1E7 [X]E7; [B] = (X2 + X1)/2
                                  ; [X] = (X2 - X1)/(Y2 - Y1)
R1 102 103 1T
C1 103 0 1E-15
D1 103 104 DIODE
V1 104 0 DC=[Y]                   ; [Y] = (Y2 - Y1)/2
D2 105 103 DIODE
V2 105 0 DC=[-Y]
E2 106 108 POLY(1) 103 0 [C] [D]; [C] = (Y2 + Y1)/2
                                  ; [D] = 1 when inverting
R2 107 108 1G                     ; [D] = -1 when noninverting
ROUT 106 107 [E]                  ; [E] = Output resistance
.MODEL DIODE D(N=0.001)
```

Notes

1. If convergence problems occur in the transient analysis, increase the C_1 value from 1E-15 to 1E-14 or larger.
2. Do not use a DC-sweep analysis because the model may not converge. Instead, use a transient analysis with a piecewise linear (PWL) source to generate a very slow ramp.

Example

Noninverting type
Input resistance = 1 MΩ
Output resistance = 1 KΩ
Nodes 101 and 108 are grounded in this example.

Transfer function:

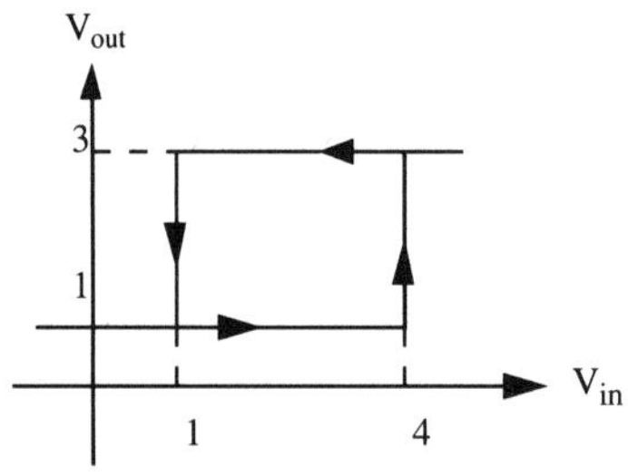


```
RIN 100 0 1MEG                    ; Input resistance = 1 MΩ
E1 102 0 POLY(2)
+ 100 0 103 0 2.5E7 -1E 1.5E7     ; (X2 + X1)/2 = 2.5
R1 102 103 1T                     ; (X2 - X1)/(Y2 - Y1) = 1.5
C1 103 0 1E-15
D1 103 104 DIODE
V1 104 0 DC= 1                    ; (Y2 - Y1)/2 = 1
```

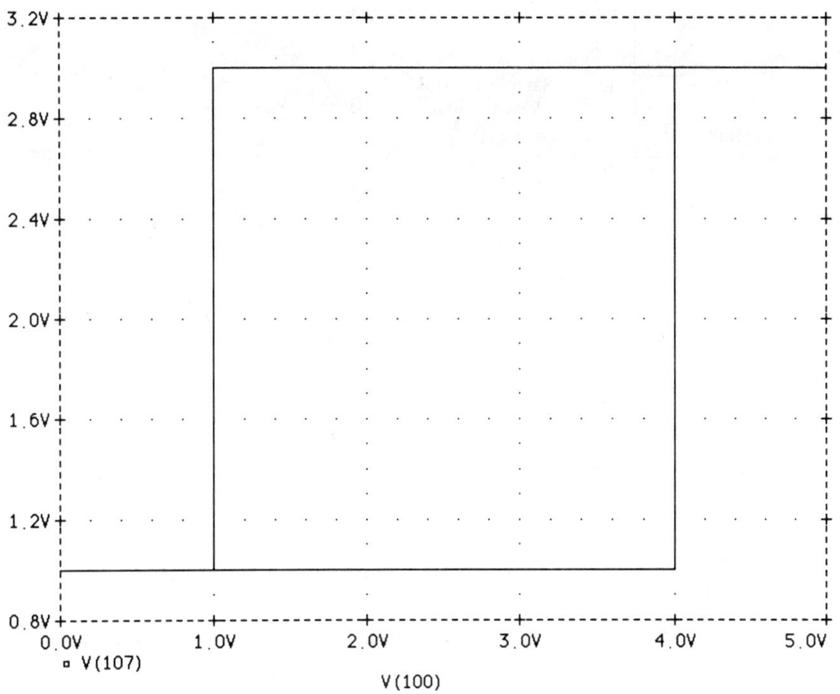

Figure 3.54 Simulation results for the Schmitt trigger simulator.

```
D2 105 103 DIODE
V2 105 0 DC=-1                    ; -(Y2 - Y1)/2 = -1
E2 106 0 POLY(1) 103 0 2 -1       ; (Y2 + Y1)/2 = 2, [D] = -1
R2 107 0 1G
ROUT 106 107 1K                   ; Output resistance = 1 KΩ
.MODEL DIODE D(N=0.001)

VIN 100 0 PWL(0 -5 2U 5 4U -5)
.TRAN 10N 4U
.OPTIONS LIMPTS=0
.PRINT TRAN V(107)
.PROBE                            ; PSpice case
.END
```

3.2.19 Sample-Hold

The sample-and-hold circuit, as its name implies, samples an input signal and holds the last sampled value until the input is sampled again. The sampling function can be modeled by a nonlinear polynomial voltage controlled current source (VCCS). When sampling, the current source is "ON." When "OFF," the sampled data are stored as a voltage

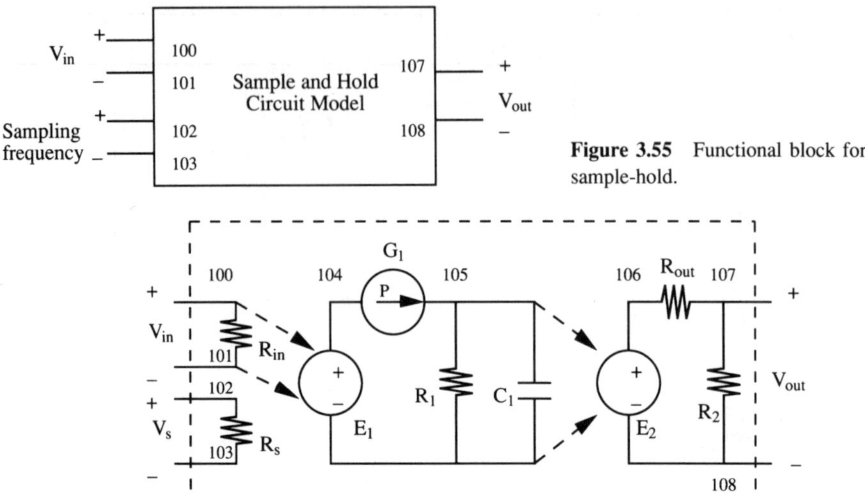

Figure 3.55 Functional block for the sample-hold.

Figure 3.56 Equivalent circuit for the sample-hold.

on an ideal capacitor. This macromodel, shown in Figure 3.55, can be generated automatically using the procedures available in Chapter 6.

The current of the VCCS, shown in Figure 3.56, G_1, is

$$I(G_1) = V(104,105) \cdot V(102,103)$$

When the input signal is sampled, $I(G_1)$ is large. This means that the capacitor C_1 is charging during this period. While holding, $I(G_1)$ is negligibly small because of the "OFF" state of G_1.

SPICE code for Figure 3.56

```
RIN 100 101 [A]                     ; [A] = Input resistance
RS 102 103 1
E1 104 0 100 101 1
G1 104 105 POLY(2)   104 105 102 103
+ 0 0 0 0 1
R1 105 0 [B]                        ; [B] See notes
C1 105 0 [C]                        ; [C] See notes
VS 102 103 PULSE(0 1E3 0 0 0 [D] [E])     ; [D] = Ts/100, [E] = Ts
E2 106 108 105 0 1
R2 107 108 1G
ROUT 106 107 [F]                    ; [F] = Output resistance
```

Notes

1. [B] · [C] = 100 · T_{in} where T_{in} is the period of the input waveform.
2. T_s is the sampling period.

Example

Input signal = 5 sin($2\pi \times 10^6$t)

Sampling period = 100 ns

Input resistance = 1 MΩ

Output resistance = 1 KΩ

Nodes 101, 103, and 108 are grounded.

```
RIN 100 0 1MEG                       ; Input resistance = 1 MΩ
RS 102 0 1
E1 104 0 100 0 1
G1 104 105 POLY(2)  104 105 102 0 0 0 0 0 1
R1 105 0 1K                          ; [B] = 1 KΩ
C1 105 0 100N                        ; [C] = 100 nF
VS 102 0 PULSE(0 1E3 0 0 0 1N 100N)  ; Ts/100 = 1 nsec, Ts = 100 nsec
E2 106 0 105 0 1
R2 107 0 1G
ROUT 106 107 1K                      ; Output resistance = 1 KΩ

VIN 100 0 SIN(0 5 1MEG)
.TRAN 10N 2U
```

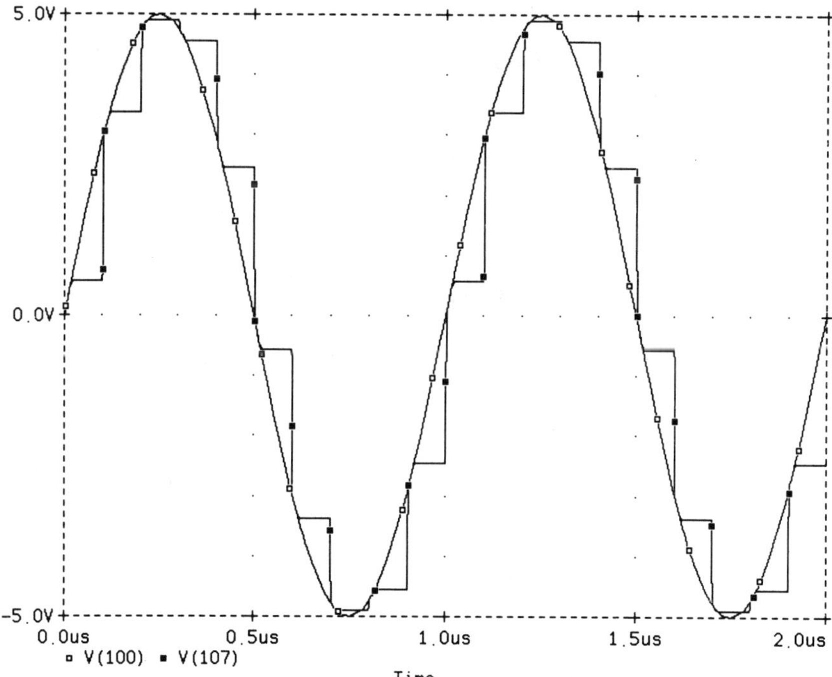

Figure 3.57 Simulation results for the sample-hold.

```
.OPTIONS LIMPTS=0 ITL5=0
.PRINT TRAN V(107)
.PROBE                          ; PSpice case
.END
```

3.2.20 Pulse Width Modulator

Pulse width modulation (PWM), as shown in Figure 3.58, is a very popular transmission technique in telecommunication systems. A summer and a Schmitt trigger model are used to model the PWM function. First, two input signals of V_{in} and a voltage ramp are merged. Then the combined signal is compared to a reference voltage. If the amplitude of the merged signal is larger than the reference, the output, V_{out}, will become the maximum voltage, K_1. The output is at the minimum voltage, K_2, when the merged signal is less than the reference. The defining equations for Figure 3.59 are

$$\begin{aligned} V(102,0) = {} & \text{Voltage gain} \cdot [V_{in} - V(\text{Reference}) - V(\text{Input offset}) \\ & + V(\text{Rampwave})] + 10^{-15} \cdot V(104) \\ = {} & \text{Voltage gain} \cdot [V(100,0) - V(101,0) - P_0 \text{ in } E_1 \\ & + V(107)] + 10^{-15} \cdot V(104) \\ V(104) = {} & 10^6 \cdot V(102,0) \end{aligned}$$

SPICE code for Figure 3.59

```
RIN1 100 101 [A]          ; [A] = Input resistance I
RIN2 107 0 [B]            ; [B] = Input resistance II
E1 102 0 POLY(4) 100 0 101 0 104 0 107 0
```

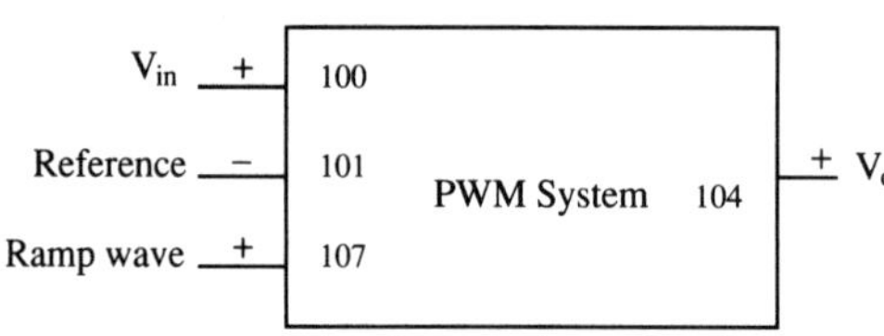

Figure 3.58 Functional block for the pulse width modulator.

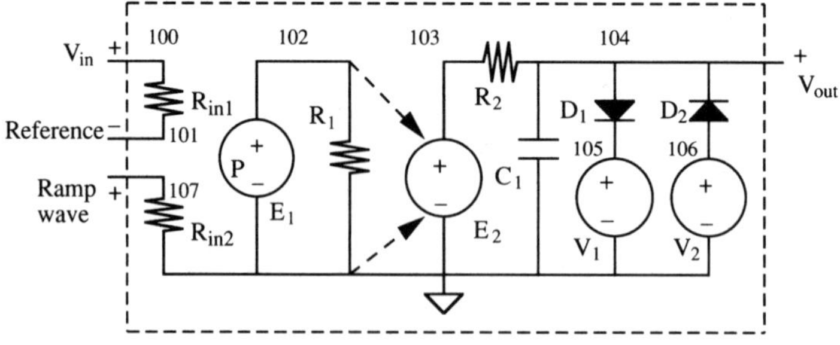

Figure 3.59 Equivalent circuit for the pulse width modulator.

```
+ [P0] [P1] [P2] [P3] [P4]   ; [P0] = −(Voltage gain · Input offset)
                             ; [P1] = Voltage gain of PWM system
                             ; [P2] = −(Voltage gain)
                             ; [P3] = 1E−15
                             ; [P4] = Voltage gain of PWM system
R1 102 0 1
E2 103 0 102 0 1MEG
R2 103 104 1T
C1 104 0 1E-15
D1 104 105 DIODE
V1 105 0 DC=[B]              ; [B] = Maximum output voltage = K1
D2 106 104 DIODE
V2 106 0 DC=[C]              ; [C] = Minimum output volatge = −K2
.MODEL DIODE D(N=0.001)
```

Example

Voltage gain of PWM system = 100
Reference voltage = 0.5 V
Input resistance I = 1 MΩ
Input resistance II = 1 MΩ

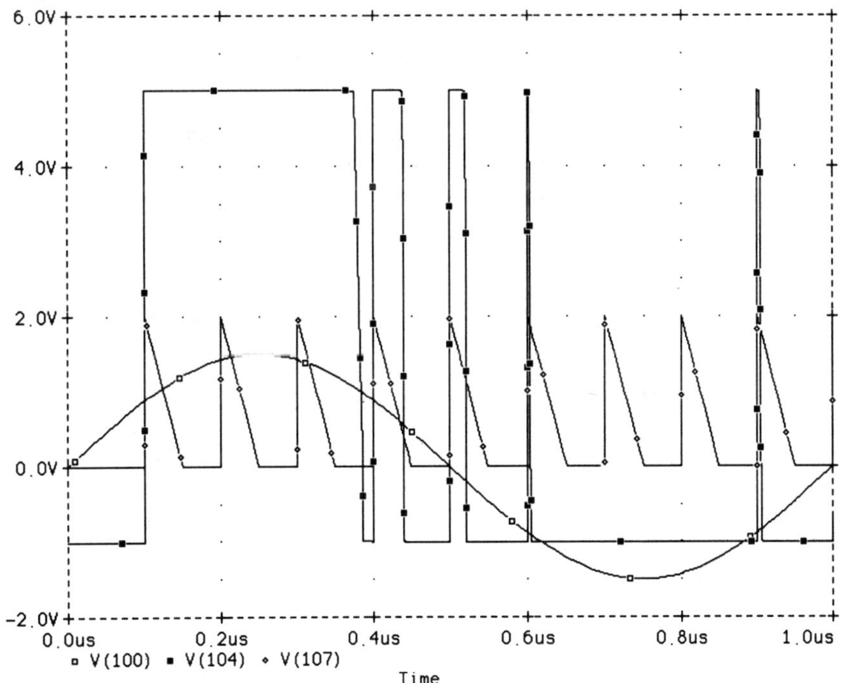

Figure 3.60 Simulation results for the pulse width modulator.

Maximum voltage of output wave = 5 V

Minimum voltage of output wave = −1 V

```
RIN1 100 101 1MEG            ; Input resistance I = 1 MΩ
RIN2 107 0 1MEG              ; Input resistance II = 1 MΩ
E1 102 0 POLY(4) 100 0 101 0
+ 104 0 107 0
+ 0 100 -100 1E-15 100       ; Voltage gain of PWM system = 100
                             ; −(Voltage gain ) = −100
                             ; [P3] = 1E-15
                             ; [P4] = 100
R1 102 0 1
E2 103 0 102 0 1MEG
R2 103 104 1T
C1 104 0 1E-15
D1 104 105 DIODE
V1 105 0 DC=5                ; Maximum output voltage = 5
D2 106 104 DIODE
V2 106 0 DC=-1               ; Minimum output voltage = −1
.MODEL DIODE D(N=0.001)

VIN 100 0 SIN(0 1.5 1MEG)
VREF 101 0 DC=0.5
VCARRIER 107 0 PWL (
+ 0 0   99.99N 0 100N 2 150N 0 199.99N 0 200N 2 250N 0
+      299.99N 0 300N 2 350N 0 399.99N 0 400N 2 450N 0
+      499.99N 0 500N 2 550N 0 599.99N 0 600N 2 650N 0
+      699.99N 0 700N 2 750N 0 799.99N 0 800N 2 850N 0
+      399.99N 0 900N 2 950N 0 999.99N 0 1000N 2)
.IC V(104)=0
.TRAN 10N 1U 0 10N UIC
.PRINT TRAN V(104)
.PROBE
.END
```

3.2.21 Amplitude Modulator

Amplitude modulation (AM) is widely used in communication systems. The output signal is composed of a carrier frequency whose amplitude is controlled by an input modulating signal. The output is produced by multiplication of the input modulation

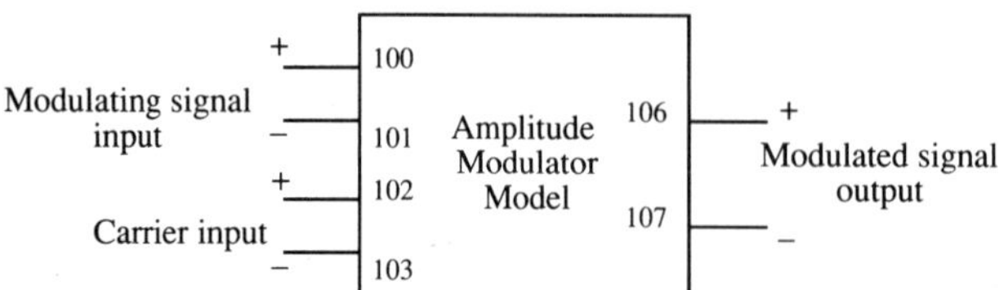

Figure 3.61 Functional block for the amplitude modulator.

signal and the carrier signal, as shown in Figure 3.61. The equation for Figure 3.62 is:

$$V(104,0) = V(100,101) \cdot V(102,103) = \text{Modulating signal} \cdot \text{Carrier}$$

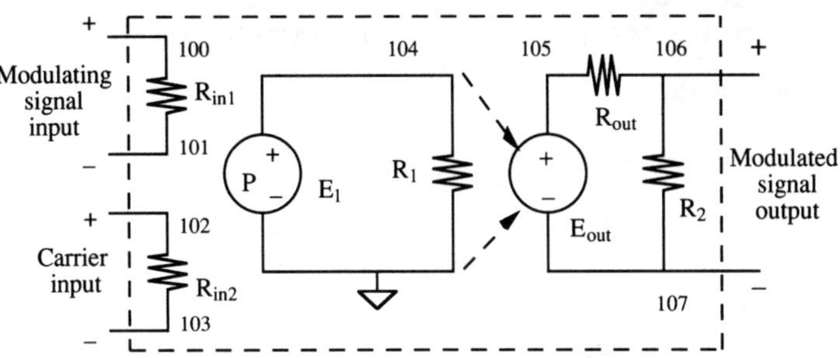

Figure 3.62 Equivalent circuit for the amplitude modulator.

SPICE code for Figure 3.62

```
RIN1 100 101 [A]          ; [A] = Input resistance of modulating
*                                 signal port
RIN2 102 103 [B]          ; [B] = Input resistance of carrier port
E1 104 0 POLY(2) 100 101 102 103
```

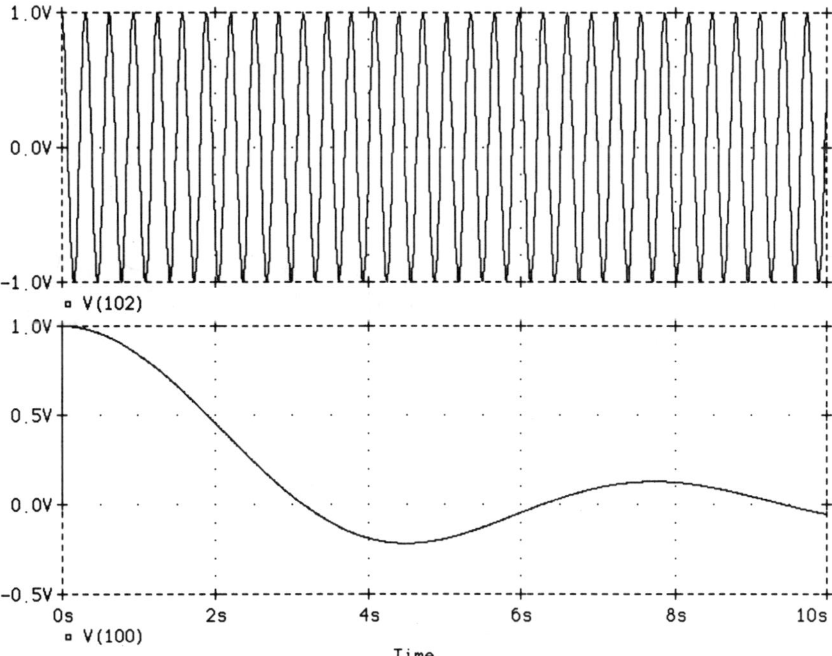

Figure 3.63 Carrier and modulation simulations for the amplitude modulator.

```
+ 0 0 0 0 1
R1 104 0 1
EOUT 105 107 104 0 1
R2 106 107 1G
ROUT 105 106 [C]            ; [C] = Output resistance
```

Example

Modulating signal = $S_m(t) = \frac{\sin t}{t}$

Carrier = $\cos\left(2\pi \cdot \frac{10}{\pi}\right)t = \cos 20t$

Input resistance of modulation signal port = 1 MΩ

Input resistance of carrier port = 1 MΩ

Output resistance = 1 KΩ

Nodes 101, 103, and 107 are grounded in this example.

```
RIN1 100 0 1MEG             ; Input resistance I = 1 MΩ
RIN2 102 0 1MEG             ; Input resistance II = 1 MΩ
E1 104 0 POLY(2) 100 0 102 0 0 0 0 0 1
R1 104 0 1
EOUT 105 0 104 0 1
R2 106 0 1G
```

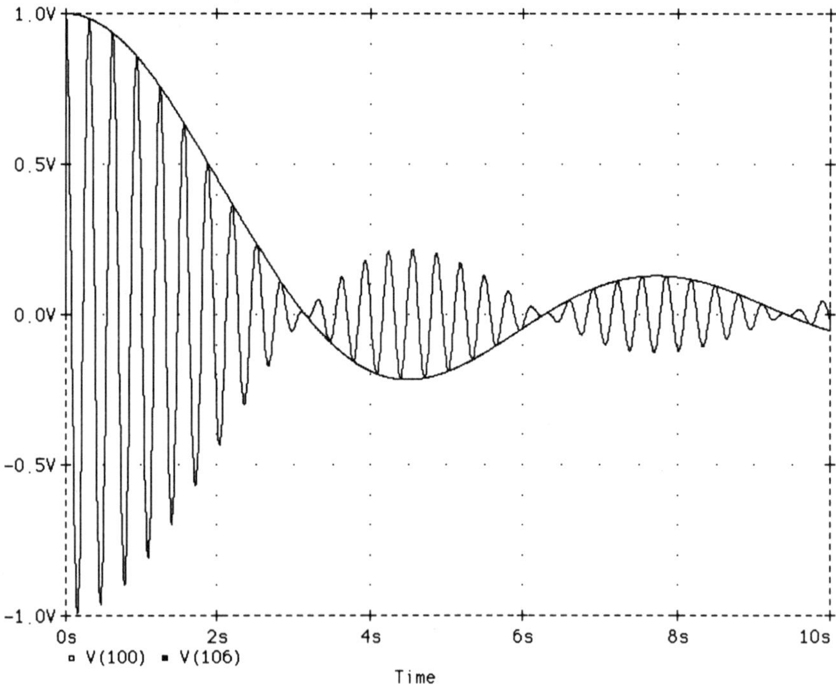

Figure 3.64 Carrier and output simulations for the amplitude modulator.

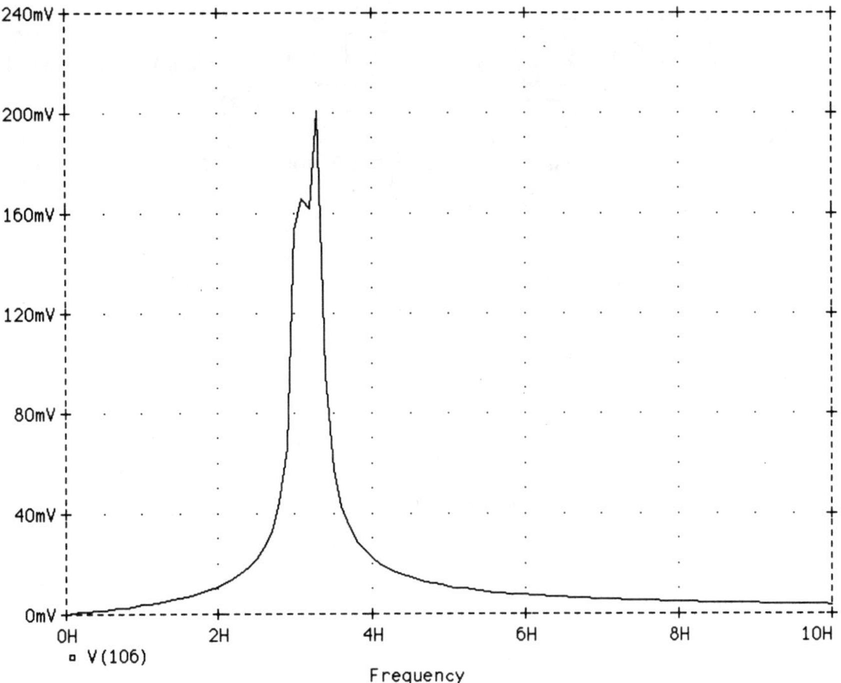

Figure 3.65 Fourier frequency spectrum simulation results for the amplitude modulator.

```
ROUT 105 106 1K             ; Output resistance = 1 KΩ

VINS 150 0 SIN(0 1 0.15915)      ;―――――――――
RINS 150 0 1                     ;
VINP 151 0 PWL(0 0 10 10)        ;

RINP 151 0 1                     ; Sm(t) = sin t / t
G1 0 10 150 0 1                  ;
G2 100 0 POLY(2) 151 0 100 0     ;
+0 0 0 0 1
                                 ; ―――――――――
VCARRIER 102 0 SIN(0 1 3.1831 0 0 90)   ; Carrier
.OPTIONS LIMPTS=0
.TRAN 10M 10 0 10M
.PRINT TRAN V(100) V(106)
.PROBE                           ; PSpice case
.END
```

3.2.22 Log Amplifier

The log function is approximated by using the series expansion of the natural logarithm (see Figure 3.66). The accuracy of the macromodel depends on the number of terms

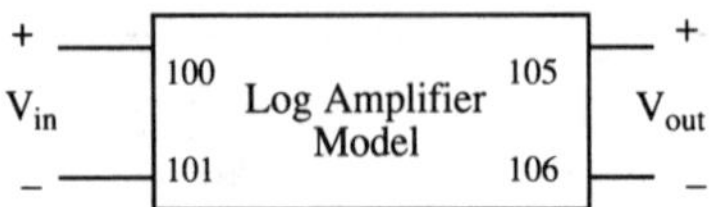

Figure 3.66 Functional block for the log amplifier.

included in the expansion. The transfer function for the log amplifier of Figure 3.67 is:

$$V_{out} = K \cdot \ln(V_{in})$$
$$\approx K \cdot 2\left[\frac{V_{in}-1}{V_{in}+1} + \frac{1}{3}\left(\frac{V_{in}-1}{V_{in}+1}\right)^3 + \frac{1}{5}\left(\frac{V_{in}-1}{V_{in}+1}\right)^5 + \cdots\right]$$

If $K = 1/2.3026$, $V_{out} = \log_{10}(V_{in})$

The voltages at nodes 102 and 103 are:

$$V(102{,}0) = 1 + V(100{,}101) = V_{in} + 1$$
$$I(G_1) = -1 + V(100{,}101) = V_{in} - 1$$
$$I(G_2) = V(102{,}0) \cdot V(103{,}0) = (V_{in} + 1) \cdot V(103{,}0)$$

$I(G_1)$ is approximately equal to $I(G_2)$ since R_2 is very large. Therefore,

$$V(103{,}0) = (V_{in} - 1)/(V_{in} + 1)$$

The voltage at node 104 is

$$V(104{,}106) = K \cdot \ln(V_{in}) = 2 \cdot K\{[(V_{in}-1)/(V_{in}+1)] + [(V_{in}-1)/(V_{in}+1)]^3/3 + [(V_{in}-1)/(V_{in}+1)]^5/5 + [(V_{in}-1)/(V_{in}+1)]^7/7 + \cdots\}$$

SPICE code for Figure 3.67

```
RIN 100 101 [A]          ; [A] = Input resistance
E1 102 0 POLY(1) 100 101 1 1
R1 102 0 1
G1 0 103 POLY(1) 100 101 -1 1
G2 103 0 POLY(2) 102 0 103 0
+ 0 0 0 0 1
```

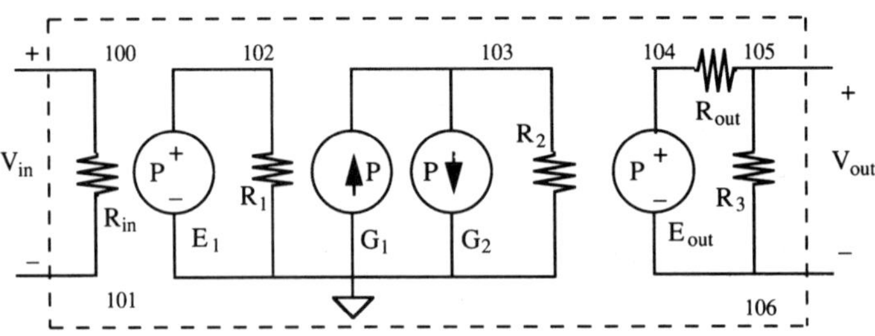

Figure 3.67 Equivalent circuit for the log amplifier.

```
R2 103 0 1G
EOUT 104 106 POLY(1) 103 0
+ 0 2*K 0 0.6667*K 0 0.4*K        ; [K] = Voltage gain
+0 0.2857*K 0 0.2222*K 0 0.1818*K
R3 105 106 1G
ROUT 104 105 [B]                  ; [B] = Output resistance
```

Example

$V_{out} = 2\ln(V_{in})$

Input resistance = 1 MΩ

Output resistance = 1 KΩ

Nodes 101 and 106 are grounded in this example.

```
.OPTIONS LIMPTS=0
RIN 100 0  1MEG            ; Input resistance = 1 MΩ
E1 102 0 POLY(1) 100 0 1 1
R1 102 0 1
G1 0 103 POLY(1) 100 0 -1 1
G2 103 0 POLY(2) 102 0 103 0
+ 0 0 0 0 1
```

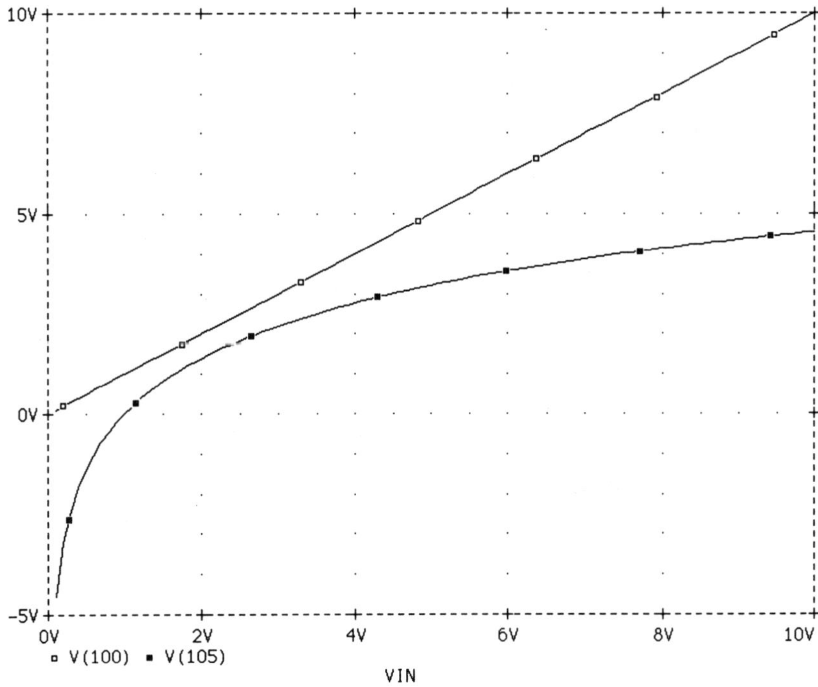

Figure 3.68 Simulation results for the log amplifier.

```
R2 103 0 1G
EOUT 104 0 POLY(1) 103 0
+ 0 4 0 1.3333 0 0.8 0 0.5714      ; Voltage gain = 2
+0 0.4444 0 0.3636
R3 105 0 1G
ROUT 104 105 1K                    ; Output resistance = 1 KΩ
VIN 100 0
.DC VIN 0.1 10 0.1
.PRINT DC V(105)
.PROBE                             ; PSpice
.END
```

Note. This macromodel is accurate for input voltages from 0.1 V through 10 V.

3.2.23 The Nth Root Extractor

This macromodel takes the Nth (2nd, 3rd, 4th, etc.) root of the input signal. N can be an integer or any real number. The macromodel of Figure 3.69 is accurate for input voltages in the range from 0.1 V to 10 V. The transfer function for the Nth root extractor is:

$$V_{out} = (V_{in})^{1/N}$$

The equations for the log amplifier macromodel of Figure 3.70 are:

$$V(104) = \ln(V_{in})$$

$$V(105) = (1/N) \cdot [V(104)] = (1/N) \cdot [\ln(V_{in})]$$

The series approximation of ln (V_{in}) is used:

$$\ln(V_{in}) = 2\left[\left(\frac{V_{in}-1}{V_{in}+1}\right) + \frac{1}{3}\left(\frac{V_{in}-1}{V_{in}+1}\right)^3 + \frac{1}{5}\left(\frac{V_{in}-1}{V_{in}+1}\right)^5 + \frac{1}{7}\left(\frac{V_{in}-1}{V_{in}+1}\right)^7 + \cdots\right]$$

and

$$E_{out} = \text{Exp}[V(105)] = (V_{in})^{1/N}$$

The series approximation for Exp(x) is

$$e^{(x)} = 1 + x + \frac{x^2}{2!} + \frac{x^3}{3!} + \frac{x^4}{4!} + \frac{x^5}{5!} + \cdots$$

SPICE code for Figure 3.70

```
RIN 100 101 [A]           ; [A] = Input resistance
E1 102 0 POLY(1) 100 101 1 1
```

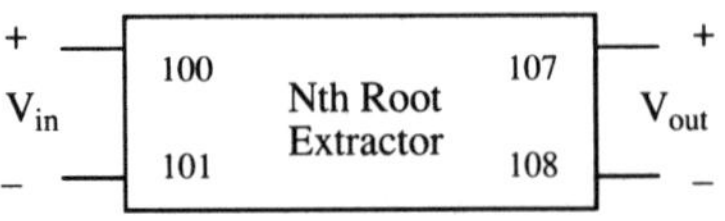

Figure 3.69 Functional block for the Nth root extractor.

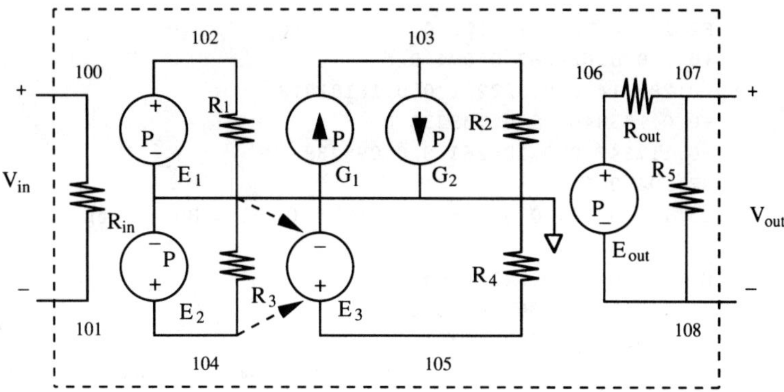

Figure 3.70 Equivalent circuit for the Nth root extractor.

```
R1 102 0 1
G1 0 103 POLY(1) 100 101 -1 1
G2 103 0 POLY(2) 102 0 103 0
+0 0 0 0 1
R2 103 0 1G
E2 104 0 POLY(1) 103 0
+0 2 0 0.666667 0 0.4 0
+0.285714 0 0.222222 0 0.18181818
+0 0.1538461 0 0.133333 0
+0.117647 0 0.105263 0 0.095238
R3 104 0 1
E3 105 0 104 0 [B]          ; [B] = 1/N (N = Root value)
R4 105 0 1
EOUT 106 108 POLY(1) 105 0
+1 1 0.5 0.16667 41.6667E-3
+8.3333E-3 1.38889E-3 198.412E-6
ROUT 106 107 [C]            ; [C] = Output resistance
R5 107 108 1T
```

Example A square root macromodel with the following specifications:

Input resistance = 100 KΩ
Output resistance = 1 KΩ

```
RIN 100 0 100K              ; Input resistance = 100 KΩ
E1 102 0 POLY(1) 100 0 1 1
R1 102 0 1
G1 0 103 POLY(1) 100 0 -1 1
G2 103 0 POLY(2) 102 0 103 0
+0 0 0 0 1
R2 103 0 1G
```

```
E2 104 0 POLY(1) 103 0
+0 2 0 0.666667 0 0.4 0
+0.285714 0 0.222222 0 0.18181818
+0 0.1538461 0 0.133333 0
+0.117647 0 0.105263 0 0.095238
R3 104 0 1
E3 105 0 104 0 0.5          ; 1/N = 0.5 (2 = Root value)
R4 105 0 1
EOUT 106 0 POLY(1) 105 0
+1 1 0.5 0.16667 41.6667E-3
+8.3333E-3 1.3889E-3 198.412E-6
ROUT 106 107 1K             ; Output resistance = 1 KΩ
R5 107 0 1T

VIN 100 0 PWL(0 0 100U 10)
.DC VIN 0.1 10 0.1
.OP
.TRAN 1U 110U 0 1U
.PRINT  TRAN V(107)
.PROBE                      ; PSpice case
.END
```


Figure 3.71 Simulation results for the Nth root extractor.

Notes

1. Nodes 101 and 108 are grounded.
2. The macromodel is accurate for input voltages in the range from 0.1 V to 10 V.

3.2.24 Noise Sources: Voltage and Current

Noise analysis (.NOISE) in SPICE is performed in conjunction with the .AC option. SPICE will calculate a noise spectral density for every resistor in the circuit. It also calculates shot and flicker noise spectral densities for every semiconductor device in the circuit. For for modeling purposes it is often advantageous to lump all the noise produced by an amplifier or other functional block into two noise sources, E_n and I_n, as shown in Figure 3.72.

SPICE calculates the spectral density of the thermal noise in a resistor according to

$$I_T^2 = \frac{4kT}{R} \qquad \left(\text{(units of)}\ \frac{A^2}{Hz}\right) \tag{3.1}$$

where k is Boltzmann's constant $= 1.38 x 10^{-23}$ W-s/K and T is temperature in Kelvin. The default temperature is 27 C or 300K. This resistor noise is modeled in SPICE as a current source in parallel with a noiseless resistor, as shown in Figure 3.73.

Suppose we desire to create E_n and I_n noise voltage and current sources of $3nV/\sqrt{Hz}$ and $10pA/\sqrt{Hz}$, respectively. First we create two thermal noise current standards of $1pA/\sqrt{Hz}$ by using two resistors of value

$$R_n = \frac{4(1.38 x 10^{-23})(300)}{10^{-24}} = 16.56\ K\Omega \tag{3.2}$$

Two separate sources are necessary since it is normally assumed that the noise sources are uncorrelated. If E_n and I_n are 100% correlated, only one standard source is needed. Next we sense the noise currents produced in these resistors by using independent (dummy) voltage sources set to zero volts. Finally, we reflect each standard noise current to the appropriate nodes by using current controlled dependent sources, as shown in Figure 3.74.

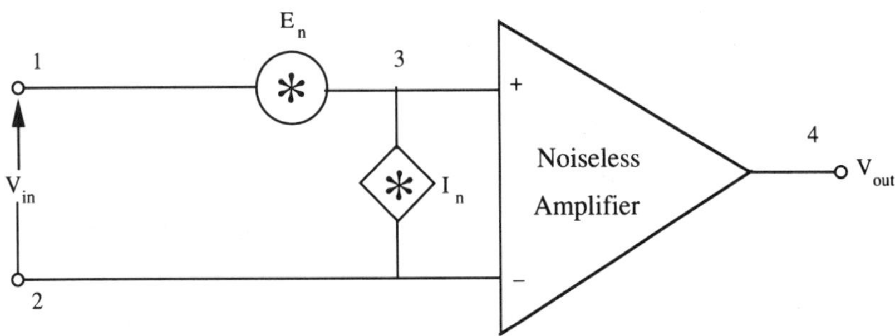

Figure 3.72 Voltage and current noise sources in an amplifier.

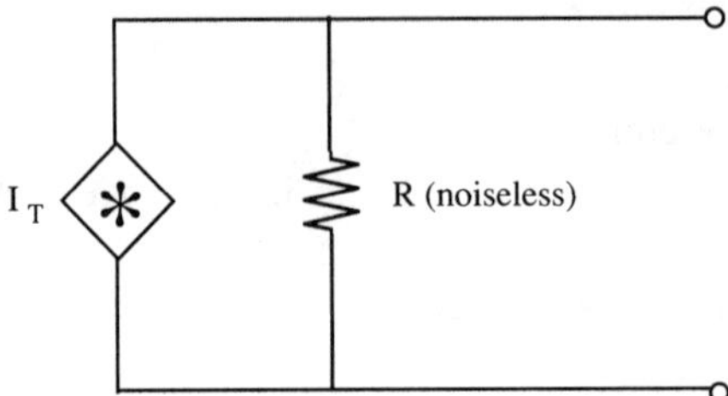

Figure 3.73 SPICE representation of thermal noise in a resistor.

SPICE code to reflect noise sources to the amplifier nodes of Figure 3.72

```
RNE 100 0 16.56K
VESEN 100 0 DC 0
HEN 1 3 VESEN 3K
RNI 200 0 16.56K
VISEN 200 0 DC 0
FIN 2 3 VISEN 10
```

Polarity does not matter for any of the sources because noise is a squared quantity. The multipliers needed for the H and F dependent sources are $(3\text{nV}/\sqrt{\text{Hz}})/(1\text{pA}/\sqrt{\text{Hz}}) = 3\text{K}$ and $(10\text{pA}/\sqrt{\text{Hz}})/(1\text{pA}/\sqrt{\text{Hz}}) = 10$, respectively.

If we wish to include the amplifier's input impedance into our noise model, an extra noise source will be unintentionally introduced through the resistive component of the input impedance. For example, suppose we model the amplifier's differential input impedance as the parallel combination of R_{in} = 100 KΩ and C_{in} = 100 pF. Also let the amplifier have a voltage gain of 750, as shown in Figure 3.75.

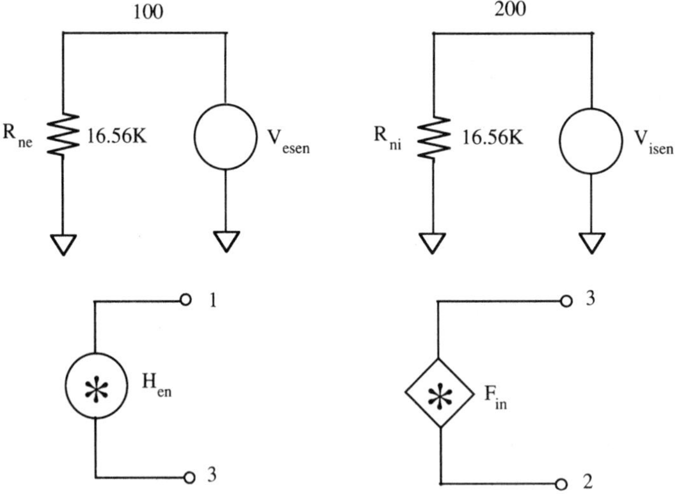

Figure 3.74 Circuit configuration for modeling E_n and I_n noises.

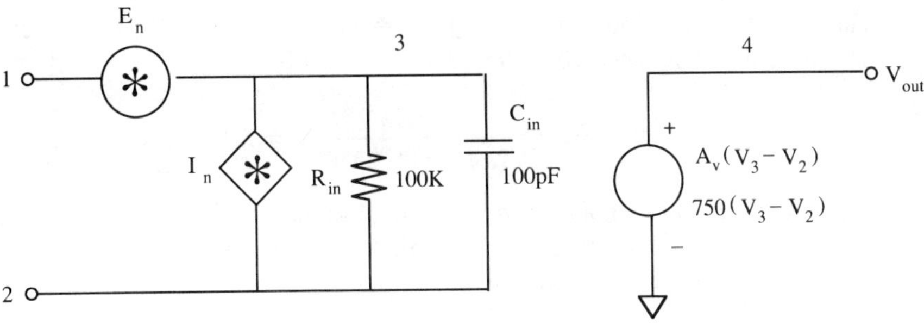

Figure 3.75 Amplifier model with noise sources, input impedance, and voltage gain.

Extra noise will be added through R_{in} unless we replace this resistor with a SPICE element that has the same circuit effect as the resistor but does not produce thermal noise. We can use a voltage controlled current source (VCCS) to model R_{in} as a noiseless resistor. This substitution is shown in Figure 3.76.

SPICE code for Figure 3.76

```
GRID 2 3 2 3 10U
CIN 3 2 100PF
EOL 4 0 3 2 750
```

The 10 U term in GRID is the conductance of R_{in}—that is, the reciprocal of 100 KΩ. When it becomes necessary to include 1/f effects into E_n and I_n, the resistors must be replaced with diodes.[2] SPICE calculates a noise current in a diode according to

$$I_{nd}^2 = \frac{KF(I_{DC})^{AF}}{f} + 2qI_{DC} \tag{3.3}$$

where KF is the flicker noise coefficient in amps, AF is the flicker noise exponent, q is electronic charge (1.602 x 10^{-19} C), f is frequency in Hertz, and I_{DC} is the DC bias current through the diode. As an example, suppose we wish to model E_n noise of $20\text{nV}/\sqrt{\text{Hz}}$ with a noise corner $f_{nce} = 200$ Hz and I_n noise of $0.5\text{pA}/\sqrt{\text{Hz}}$ with a noise corner frequency $f_{nci} = 500$ Hz. We use independent current sources to set a DC diode

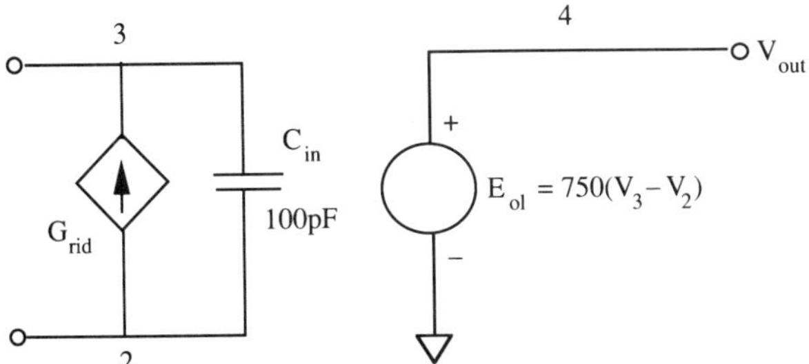

Figure 3.76 Adding a noiseless input resistance to the amplifier model.

current that will produce a standard reference level of $1\text{pA}/\sqrt{\text{Hz}}$ in each diode. This DC current is calculated according to

$$I_{DC} = \frac{10^{-24}}{2 \times 1.602 \times 10^{-19}} = 3.121 \text{ mA} \tag{3.4}$$

Next the flicker noise term is set equal to the $1\text{pA}/\sqrt{\text{Hz}}$ reference noise current at the f_{nc} frequency. For the E_n source with AF = 1, KF is calculated by equating the two right-hand terms of Equation (3.3), or

$$KF = 2qf_{nc} = 6.408\text{x}10^{-17} \text{ A} \tag{3.5}$$

In a like manner, the KF for I_n is determined as 1.602×10^{-16}A. Figure 3.77 shows how the bias current is established in the diode.

The 1 GF capacitor blocks the DC bias current and permits the AC short circuit noise current to be sensed by the V_{SENSE} independent voltage source. Figure 3.78 shows the complete macromodel for the E_n and I_n noise sources, with 1/f variations positioned to match the nodes in the amplifier previously cited in Figure 3.72.

SPICE code for Figure 3.78

```
IEN 0 100 3.121UA
CEN 100 101 1GF
VESEN 101 0 DC 0
DENOISE 100 0 DENOISE
.MODEL DENOISE D (KF=6.408E-17, AF=1)
HEN 1 3 VESEN 20K
IIN 0 200 3.121UA
CIN 200 201 1GF
VISEN 201 0 DC 0
DINOISE 200 0 DINOISE
.MODEL DINOISE D (KF=1.602E-16, AF=1)
FIN 2 3 VISEN 0.5
```

Resistors are known to produce more noise than that attributable solely to temperature.[3] This additional noise, called excess noise, is a function of the DC biasing voltage, the frequency, the wattage rating, and the composition of the material making up the resistor. Excess noise has a 1/f frequency spectrum and is specified according to a noise index (NI) in dB or in microvolt/V_{DC}/decade of frequency. The square of the spectral

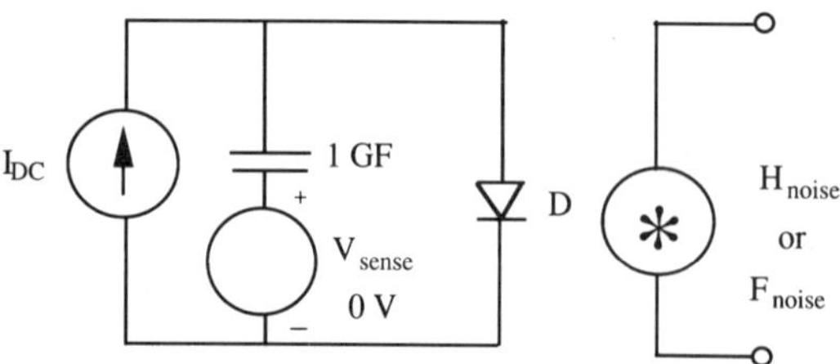

Figure 3.77 Circuit to produce noise with 1/f frequency component.

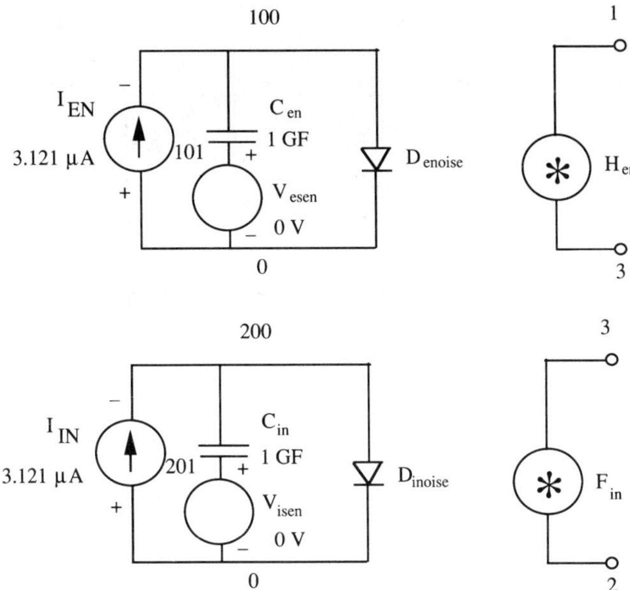

Figure 3.78 E_n and I_n noise sources having 1/f components for inclusion in the amplifier of Figure 3.72.

density of excess noise in V^2/Hz is given by

$$V_{ex}^2 = \left(\frac{10^{NI/10} \times 10^{-12}}{\ln 10}\right)\left(\frac{V_{DC}^2}{f}\right)(\text{units of } V^2/Hz) \tag{3.6}$$

where V_{DC} is the DC voltage across the resistor, f is frequency in Hertz, and NI is the noise index in dB. As a numerical example, a resistor with a NI = −10 dB biased with 5 volts DC produces an excess spectral density noise voltage of $33nV/\sqrt{Hz}$ at a frequency of 1,000 Hertz.

Excess and thermal noises can be modeled in SPICE using the circuit shown in Figure 3.79, where R is the resistor of interest. The dependent current source, F_{NOISE}, produces the excess noise current.

The independent voltage source, V_{ISENSE}, senses the DC current through R. This current is reflected to the diode subcircuit to control the dependent current source, which, in turn, biases one of the diodes, D_1 or D_2, depending on polarity. The gain of this dependent current source is equal to R, so that the diode current has a magnitude equal to V_{DC}, the DC voltage developed across the resistor R. In the diode model, KF is set as follows:

$$KF = \left(\frac{10^{NI/10} \times 10^{-12}}{\ln 10}\right)(\text{units of A}) \tag{3.7}$$

and AF is set equal to 2. Then the diode spectral denisty expression of Equation (3.3) yields an expression whose first term is that of Equation (3.6). For typical NI values of

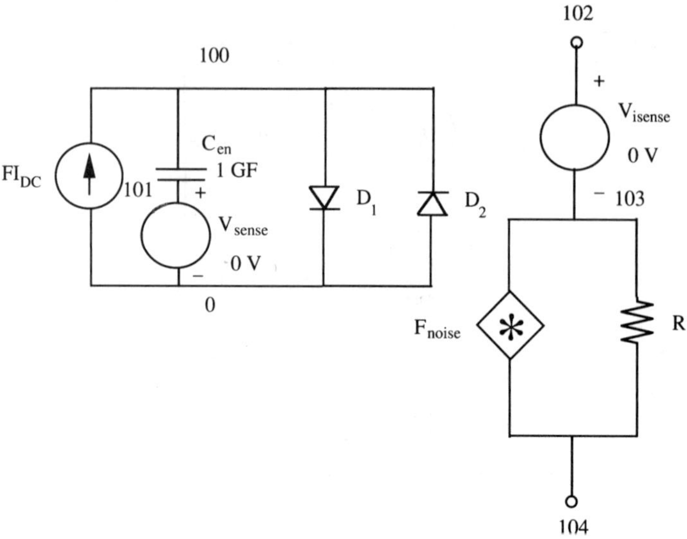

Figure 3.79 Resistor current noise model for SPICE.

-40 dB $<$ NI $< +10$ dB, the shot noise term of the diode spectral density is negligible. The magnitude of the diode flicker noise term is the noise voltage in R caused by resistor current noise. Since the noise element F_{NOISE} is a current source, the gain of this dependent source must be 1/R when the noise is reflected to the main circuit.

As an example consider the circuit shown in Figure 3.80, where the resistor R_{excess} has a noise index of 10 dB.

SPICE control file for Figure 3.80

```
EXCESS NOISE EXAMPLE
VIN 1 0 DC 10V
R1 1 2 1K
R2 2 0 2K
R3 2 3 1K
R4 3 0 2K
R5 3 4 1K
R6 4 0 2K
REXCESS 41 0 2K
VISENSE 4 41 DC 0
FNOISE 41 0 VSENSE 5E-4
FIDC 0 100 VISENSE 2K
CSEN 100 101 1GF
VSENSE 101 0 DC 0
D1 100 0 DEXCESS
D2 0 100 DEXCESS
MODEL DEXCESS D (KF=4.343E-12, AF=2)
.AC DEC 10 1 100
```

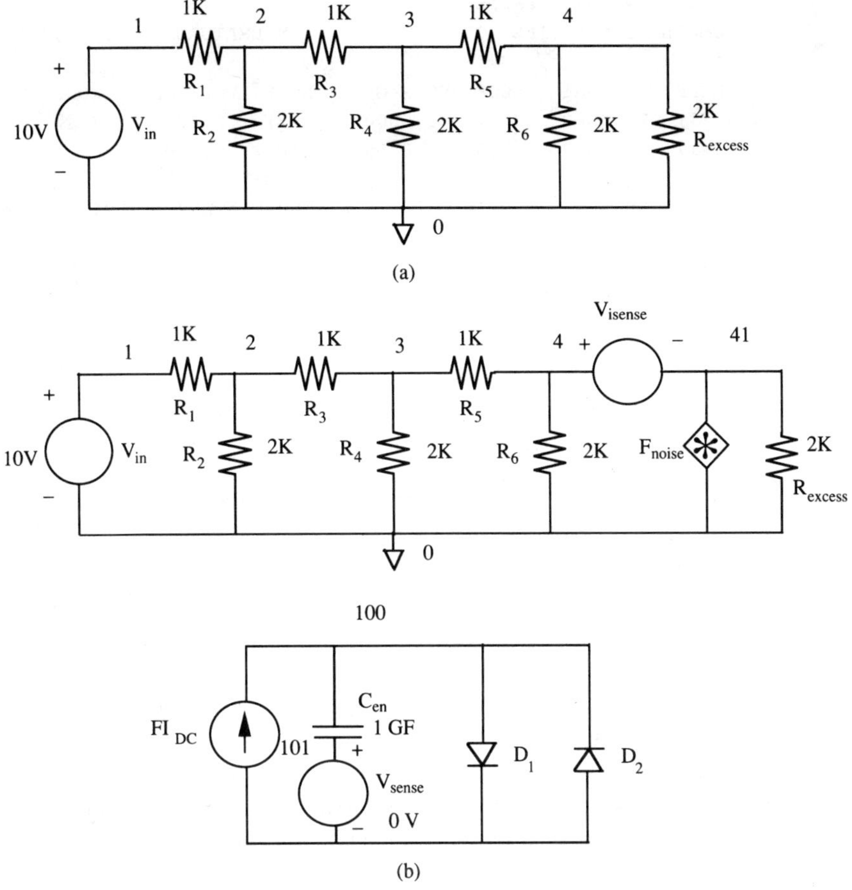

Figure 3.80 Example of modeling excess noise. (a) Example circuit. (b) Circuit with added excess noise source.

```
.NOISE V(41) VIN 10
.PRINT NOISE INOISE ONOISE
.END
```

For simplicity we assume that the excess noises of all other resistors are negligibly small. Next we calculate KF from Equation (3.7) to be 4.343×10^{-12}. Finally we use the excess noise model from Figure 3.79 to produce the complete circuit of Figure 3.80(b). Edited SPICE simulation results, shown in Figure 3.81, match the results obtained by hand calculations using Equation (3.6).

3.2.25 Random Noise Generator

Occasionally we may need to test a circuit or system to see how it would respond to a random input voltage like that produced by Gaussian noise. The .NOISE function in

```
EXCESS NOISE EXAMPLE
SMALL SIGNAL BIAS SOLUTION          TEMPERATURE =   27.000 DEG C

NODE  VOLTAGE    NODE  VOLTAGE     NODE  VOLTAGE     NODE  VOLTAGE
 (1)  10.0000    ( 2)    5.0000    ( 3)    2.5000    ( 4)    1.2500
(41)    1.2500   (100)    .8395    (101)  0.0000

NOISE ANALYSIS                      TEMPERATURE =   27.000 DEG C

FREQUENCY =   1.000E+00 HZ
DIODE SQUARED NOISE VOLTAGES (SQ V/HZ)
                D1          D2
 RS        0.000E+00  0.000E+00
 ID        4.005E-19  0.000E+00
 FN        6.786E-12  0.000E+00
 TOTAL     6.786E-12  0.000E+00

RESISTOR SQUARED NOISE VOLTAGES (SQ V/HZ)
           REXCESS
 TOTAL     3.315E-17

TOTAL OUTPUT NOISE VOLTAGE    =   6.786E-12 SQ V/HZ
                              =     2.605E-06 V/RT HZ
```

Figure 3.81 Edited SPICE output showing 260 nV/Hz excess noise voltage at a frequency of 1 Hertz.

SPICE performs an AC analysis in the frequency domain, but does not have a built-in way to simulate noise in the time domain. The transient analysis option (.TRAN) can be used with a specially configured time domain noise generator (TDNG).

The first step for constructing a TDNG is to find some means of generating a set of random numbers. Many software packages such as MathCAD[4] and Excel[5] have such a function built in as an option. Assume that we can obtain a set of 50 random numbers. It is required that the set of 50 random numbers be in the range from 0 to 1.

Find the arithmetic average, m, and the standard deviation, s, of the random number set. Then perform the following transformation so that the new arithmetic average (mean) and the new standard deviation will be 0 and 1, respectively.

$$\frac{X_i - m}{s} => Y_1, Y_2, Y_3, \cdots Y_n \tag{3.8}$$

Decide on the time interval of interest and split it up into a number of equal segments equal to the number of random numbers in the set. For our example we might be interested in the time response from 0 to 5 μs. We would divide up the segments in 100 ns intervals. Finally, we construct a series combination of three independent PWL voltage sources, each of which will have voltage values at 20 specific instants of time and be zero for all other times. The PWL voltage source model is used as this voltage source. Note that a single PWL source is limited to a maximum of 20 amplitude-time data points.

In Figure 3.82, PWL source V_1 has its random voltage value for the first 20 time steps, from 0 to 1.8 μs, and then is zero at all other times. The second source, V_2, has value for the second set of 20 time steps and is zero elsewhere. This process is repeated for the third PWL source up to the 5 μs limit. If more data points are desired, additional sources, as well as additional random numbers, need to be added in the TDNG model.

SPICE netlist for Figure 3.82

```
Time-Dependent Noise Generator
*
ENGEN 3 0 1 0 [A]          ; [A] is the rms noise level = 20 nV for this example
RL1 1 0 1
RL2 3 0 1T
*
V1 1 100 PWL 0 0 100N -.5059494 200N 1.25103432 300N 1.12552402  400N
+ -1.0173531 500N -.5376212  600N 1.04056649 700N 1.53779272  800N
+ .34721034 900N -.3457178 1000N 1.08154726 1100N .19639468  1200N
+ -.0911127 1300N -.9962207 1400N .4986145 1500N -1.1072326 1600N
-.6822844 + 1700N -.7236932 1800N 0 100000N 0
V2 100 200 PWL 0 0 1700N 0 1800N 1.65212159 1900N .31232854   2000N
+ -1.2090425 2100N -.3849866  2200N -.2828022 2300N -.5988249  2400N
+.28696968 2500N .26043382 2600N -.1844697 2700N  .87739239  2800N
+ .79259537 2900N -.2760077 3000N 1.33406586  3100N .4709716 3200N
```

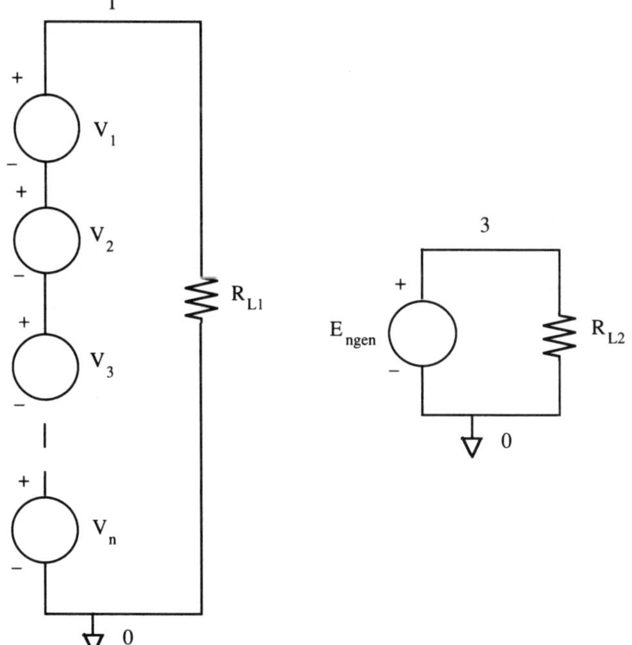

Figure 3.82 Circuit representation for time-dependent noise generator.

```
+ -1.3384583 3300N 1.41324542 3400N 0 100000N 0
V3 200 0 PWL 0 0 3300N 0 3400N .31109804 3500N -.3693647 3600N
+ -.1817947 3700N -.2328334 3800N 1.5594066  3900N -1.3422033 4000N
+ -1.4945704  4100N .96459691 4200N 1.20224259 4300N  -1.476755 4400N
+ -1.7641554  4500N -.3685622 4600N -1.0082046 4700N -.9962207 4800N
+ -.0651654 4900N .84267109 5000N 0 100000N 0
.PROBE
.TRAN 20NS 5US
.END
```

Over the time interval of interest, the output voltage between node 1 and ground in Figure 3.82 will have a zero mean value and a standard deviation voltage (rms noise voltage) of 1 volt. The VCVS reflects and scales this 1 volt noise level to the desired new level. For example, if a noise level of 20 nV is desired as the output level between node 3 and ground, the multiplier is set to 20 N, as shown. Figure 3.83 shows the output from the TDNG model where the time interval was chosen as 0 to 5 μs and the rms noise level was 20 nV.

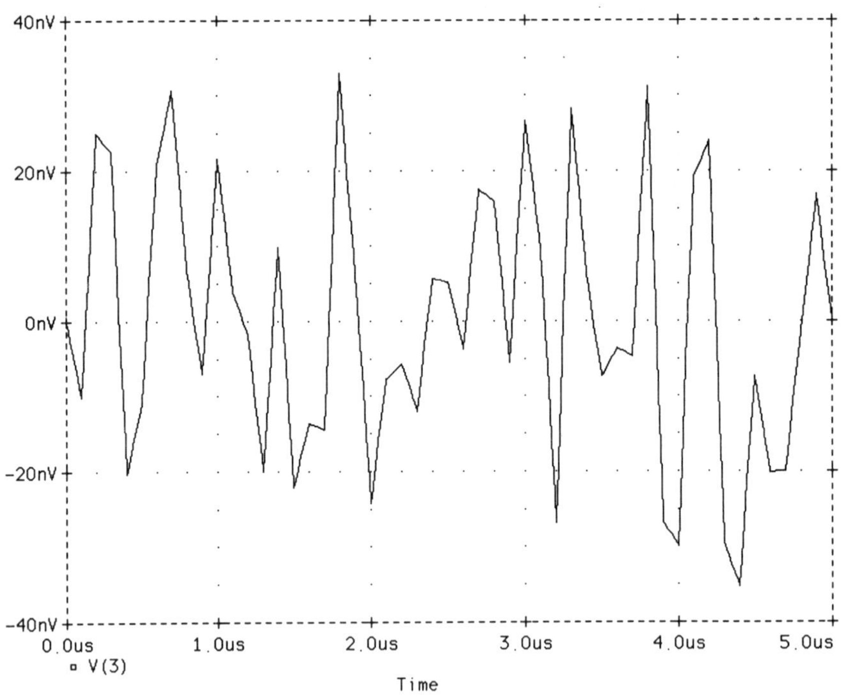

Figure 3.83 Output of time-dependent noise generator.

3.2.26 Voltage Controlled Oscillator (Sine, Square, and Triangle Waveforms)

The voltage controlled oscillator (VCO) is a critical functional block in a PLL system. A VCO often is used as a function and waveform generator. Although widely used in modern electronic circuits, complete VCO circuits often have simulation difficulties, with convergence problems and long simulation times. This macromodel, as shown in Figure 3.84, solves these problems, making it possible to verify many VCO designs and applications through SPICE simulations. For an explanation of the derivation of this macromodel, refer to Section 4.3.

Terminology for the VCO

f_0' = Free running frequency when V_{in} is zero.
K_v = Voltage-to-frequency conversion gain (Units = Radians/second/volt)
R_{time} = Timing resistor
C_{time} = Timing capacitor
$V_{p\text{-}pT}$ = Peak-to-peak voltage of triangle wave
$V_{p\text{-}pS}$ = Peak-to-peak voltage of sine wave
V_{AM} = Amplitude of square wave
V_{PAM} = Peak amplitude of square wave

SPICE code for the VCO

```
RIN 100 101 [A]                         ; [A] = Input resistance
E1 102 0 POLY(1) 100 101 [B] [C]        ; [B] and [C] see notes
R1 102 0 1
E2 103 0 POLY(2) 102 0 107 0
+ 0 0 0 0 -0.1
R2 104 0 1MEG
E3 105 0 104 0 100MEG
RTIME 103 104 [D]                       ; [D] = Timing resistance
CTIME 104 105[E]                        ; [E] = Timing capacitance
E4 106 0 POLY(2)
+ 105 0 107 0 0 -10MEG 1MEG
R3 106 107 1T
```

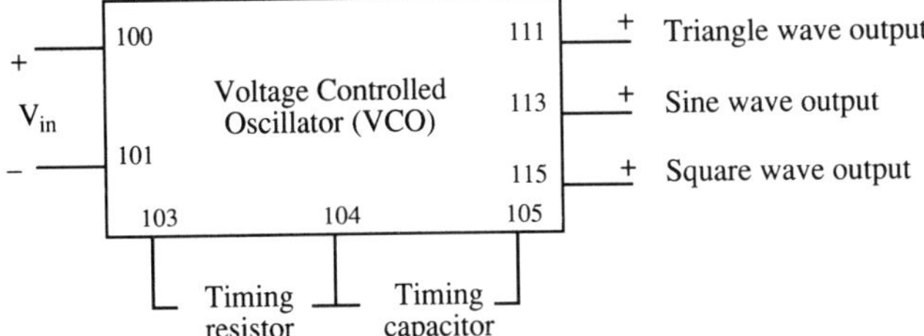

Figure 3.84 Functional block diagram for the voltage controlled oscillator (sine, square, and triangle waveforms).

```
C1 107 0 1E-15
D1 107 108 DIODE
V1 108 0 DC=10
D2 109 107 DIODE
V2 109 0 DC=-10
E5 110 0 POLY(1) 105 0 [F] [G]          ; [F] = DC level of triangle wave
                                        ; [G] = Vp-pT/2
R4 111 0 1G
ROUTT 110 111 [H]                       ; [H] = Output resistance I
E6 112 0 POLY(1) 105 0 [I]              ; [I] = DC level of sine wave
+[J · 1.5708] 0 [-J · 6.4596E-1] 0
+[J · 7.9593E-2] 0 [-J · 4.6818E-3]     ; [J] = Vp-pS/2
R5 113 0 1G
ROUTS 112 113 [K]                       ; [K] = Output resistance II
E7 114 0 POLY(1) 107 0 [L] [M]          ; [L] = VPAM - 0.5 · VAM
R6 115 0 1G                             ; [M] = VAM · 0.05
ROUTSQ 114 115 [N]                      ; [N] = Output resistance III
.MODEL DIODE D(IS=10U N=0.001)
```

Notes

1. $[B]= 4 \cdot R_{time} \cdot C_{time} \cdot f_0'$
   $[C]= 4 \cdot R_{time} \cdot C_{time} \cdot K_v/2\pi$
2. Frequency accuracy depends on the value of C_1. If C_1 is larger than 1E-15F, the accuracy decreases but computational efficiency increases.
3. For initial conditions, use .IC such as .IC V(105) = 0.0 V(107) = −10.0.
4. This macromodel is used for (a) step of frequency source, (b) ramp of frequency source, (c) voltage controlled oscillator, (d) sawtooth oscillator macromodel, and (e) rectangular wave generator.
5. Output resistance I—Output resistance of the triangle wave output
   Output resistance II—Output resistance of the sine wave output
   Output resistance III—Output resistance of the square wave output.

Example

Input resistance= 1 MΩ

Output resistance I, II, and III = 1 KΩ,

$f_0' = 2.5$KHz	$K_v = 13089.97$ rad/sec/V
$R_{time} = 12$KΩ	$C_{time} = 10$nF
$V_{p\text{-}pT} = 4$V (DC level = 4 V)	$V_{p\text{-}pS} = 4$V (DC level = 4 V)
$V_{AM} = 5.7$V	$V_{PAM} = 5$V

Type of input signal = Step of voltage

Node 101 is grounded in this example.

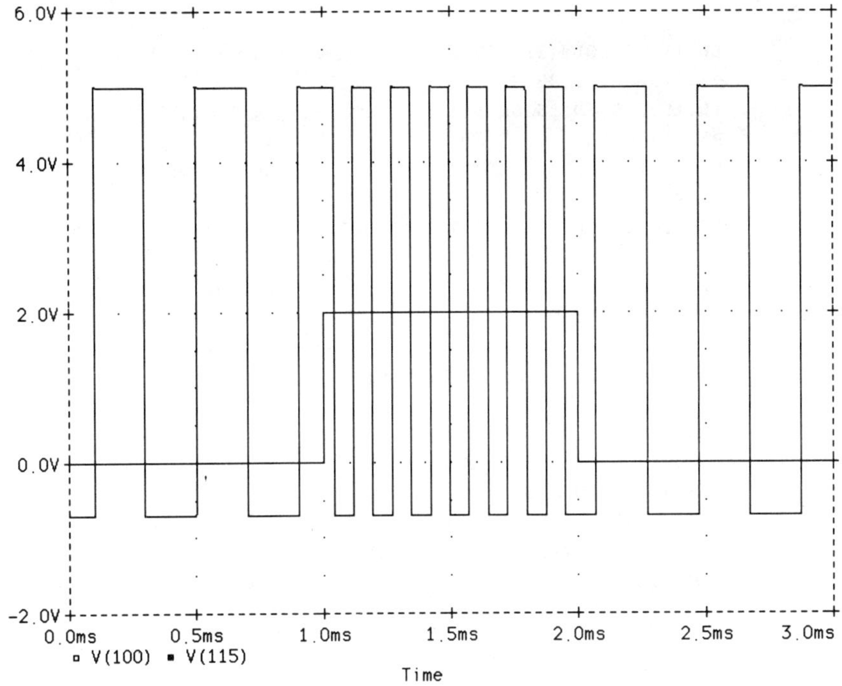

Figure 3.85 VCO square wave output.

```
RIN 100 0 1MEG                        ; Input resistance = 1 MΩ
E1 102 0 POLY(1) 100 0 1.2 1          ; [B] = 1.2, [C] = 1
R1 102 0 1
E2 103 0 POLY(2) 102 0 107 0 0 0 0 0 -0.1
R2 104 0 1MEG
E3 105 0 104 0 100MEG
RTIME 103 104 12K                     ; Timing resistance
*                                        = 12 KΩ
CTIME 104 105 10N                     ; Timing capacitance
*                                        = 10 nF
E4 106 0 POLY(2) 105 0 107 0
+ 0 -10MEG 1MEG
R3 106 107 1T
C1 107 0 1E-15
D1 107 108 DIODE
V1 108 0 DC=10
D2 109 107 DIODE
V2 109 0 DC=-10
E5 110 0 POLY(1) 105 0 4 2            ; DC level = 4 V,
*                                        Vp-pT/2 = 2
R4 111 0 1G
ROUTT 110 111 1K                      ; Output resistance I
```

```
*                                       = 1 KΩ
E6 112 0 POLY(1) 105 0 4            ; DC level = 4 V
+3.1416 0 -1.2919 0
+0.1592 0 -9.3636E-3                ; Vp-pS/2 = 2 V
R5 113 0 1G
ROUTS 112 113 1K                    ; Output resistance II
*                                       = 1 KΩ
E7 114 0 POLY(1) 107 0 2.15 0.285       ; VPAM - 0.5 · VAM
*                                           = 2.15 V
R6 115 0 1G                         ; VAM · 0.05 = 0.285 V
ROUTSQ 114 115 1K                   ; Output resistance III
*                                       = 1 KΩ
.MODEL DIODE D(IS=10U N=0.001)

VIN 100 0
+ PWL(0 0 1M 0 1.001M 2 2M 2 2.001M 0 3M 0)
.TRAN 10U 3M UIC
.IC V(105)=0 V(107)=-10
.PRINT TRAN V(111) V(113) V(115)
.PROBE                              ; PSpice case
.END
```

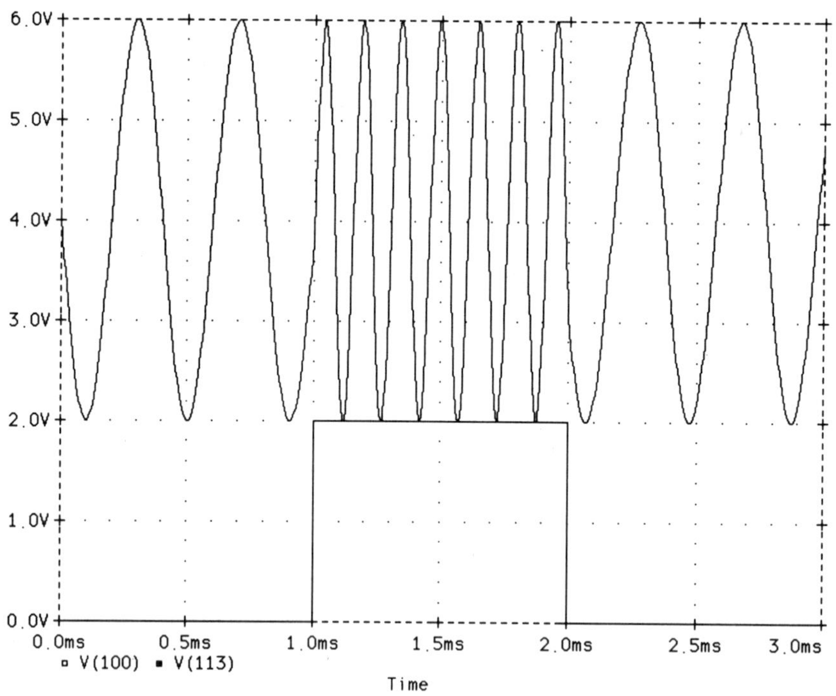

Figure 3.86 VCO sine wave output.

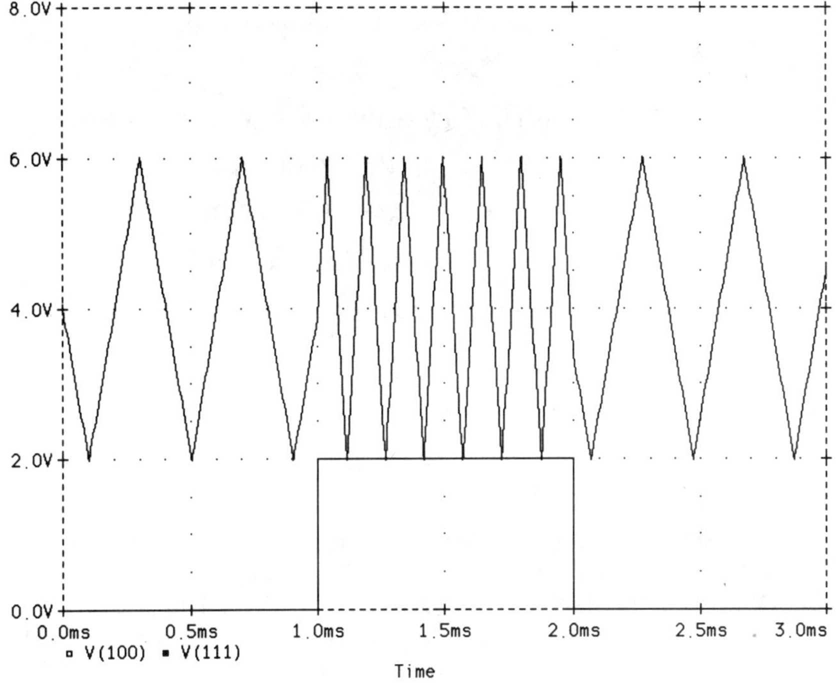

Figure 3.87 VCO triangular wave output.

3.2.27 Frequency Counter: Frequency-to-Voltage Converter

This macromodel, shown in Figure 3.88, functions as a frequency-to-voltage converter, where the output voltage is linearly dependent on the input signal frequency.

$$V_{out} = K(f_{in})$$

The conversion gain K has units of volts/Hz and is constant. This model works only for sinusoidal inputs with a known magnitude. When the input frequency changes, the output voltage becomes valid after a maximum of one period of V_{in}. It should be noted that this macromodel uses switch functions that are available in PSpice (Version 4.02 and later) and other simulators but are not available in SPICE II or III.

The structure of the macromodel, shown in Figure 3.89, is based on the sample-and-hold principle. It converts frequency to voltage and samples it once every period.

Figure 3.88 Functional block for the frequency counter—frequency-to-voltage converter.

$$V_{in} = (V_{max})[\sin(\omega t + \beta)]$$

$$E_1 = V_{in}$$

$$I(C_1) = C_1 d(V_{in})/dt = C_1 \omega (V_{max}) \cos(\omega t + \beta)$$

$$H_1 = K_0[I(C_1)]$$

$$K_o = -1/(2\pi V_{max})$$

$$H_1 = -f_{in}[\cos(\omega t + \beta)]$$

Diode D combined with capacitor C_2 would hold the maximum value of H_1 for $\cos(\omega t + \beta) = -1$.

$$H_{1max} = f_{in}$$

$$E_2 = V(104) = f_{in}$$

$$E_{out} = KV(104) = Kf_{in}$$

where K is a constant.

Note that the voltage stored on capacitor C_2 must be discharged at the end of each period. The figures illustrate the discharging function. It is assumed that $V_{in} = 1V \sin[2\pi(100\ KHz)t]$ and $E_3 = 200\ V_{in} = 200$ V peak to peak.

For $E_3 \geq 80$ V, switch S_1 closes. Then C_2 will be discharged while $V(103) < 0$. During this period S_2 is open, which prevents the transfer of 0 V to C_3. Then S_1 opens while V(105) is reaching its maximum value of 100 KV. When $E_3 \leq 10V$, S_2 closes and capacitor C_3 samples the new value of V(104). About half a period later, when $E_3 \geq 80$ V, S_1 closes, C_2 discharges, and the cycle repeats.

SPICE code for Figure 3.89

```
Rin     100    0      [A]                ; [A] = Input resistance
E1      101    0      100   0  1
Vtest   101    102    0
C1      102    0      1
H1      103    0      Vtest    [Ko]      ; Ko = -1/(2πVmax)
R1      103    0      1
Ds      103    104    Diode
```

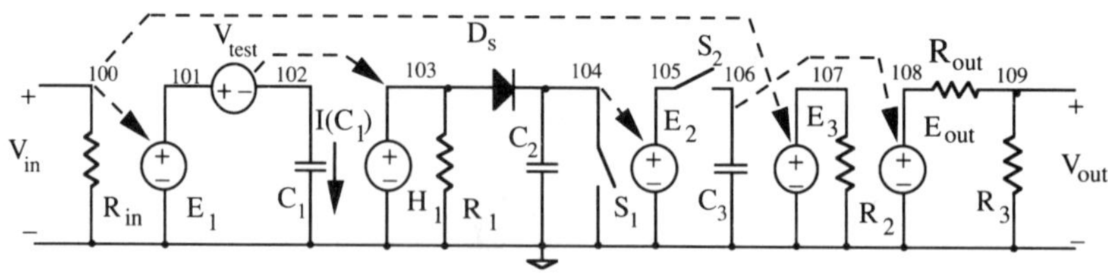

Figure 3.89 Equivalent circuit for the frequency counter—frequency-to-voltage converter.

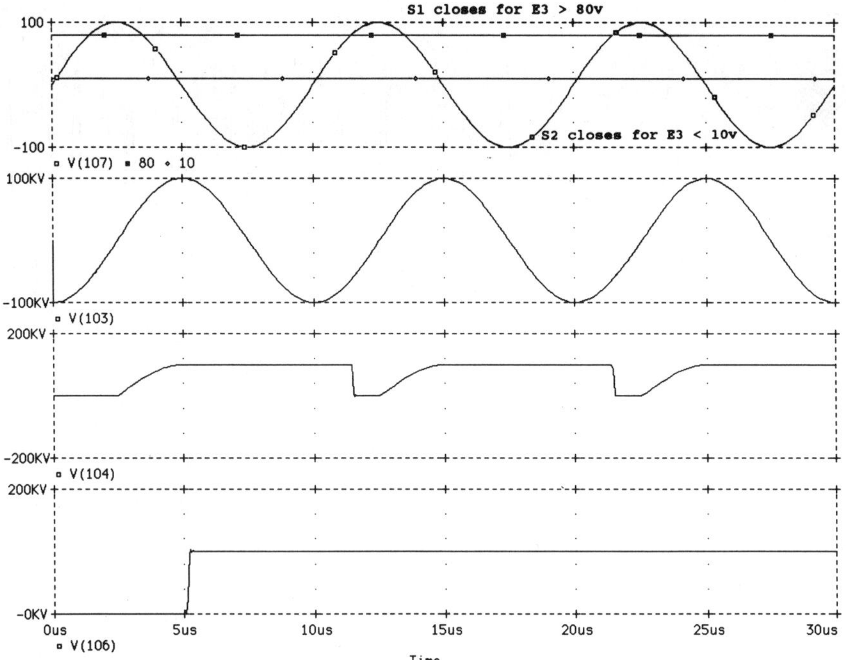

Figure 3.90 Simulation results for the frequency counter with a constant input frequency.

```
C2     104   0     200N
S1     104   0     107  0   Reset1
E2     105   0     104  0   1
S2     105   106   0  107   Reset2
C3     106   0     200N
E3     107   0     100  0   [B]      ; [B] = 100/Vmax
R2     107   0     1
Eout   108   0     106  0   [K]      ; [K] = Conversion gain
Rout   108   109   [C]               ; [C] = Output resistance
R3     109   0     1T
.Model Reset1 Vswitch(Von=80 Voff=75 Ron=0.01)
.Model Reset2 Vswitch(Von=10 Voff=5 Ron=0.01)
.Model Diode D(Is=10U N=.001)
.Options Gmin=1E-18 ITL5=0 ITL4=300
```

Notes

1. The open circuit resistance in the SPICE program is set to $1E + 18\Omega$ by setting $G_{min} = 1E - 18$.
2. E_3 controls the two switches S_1 and S_2. Set it to 200 V peak to peak by adjusting the value of B.

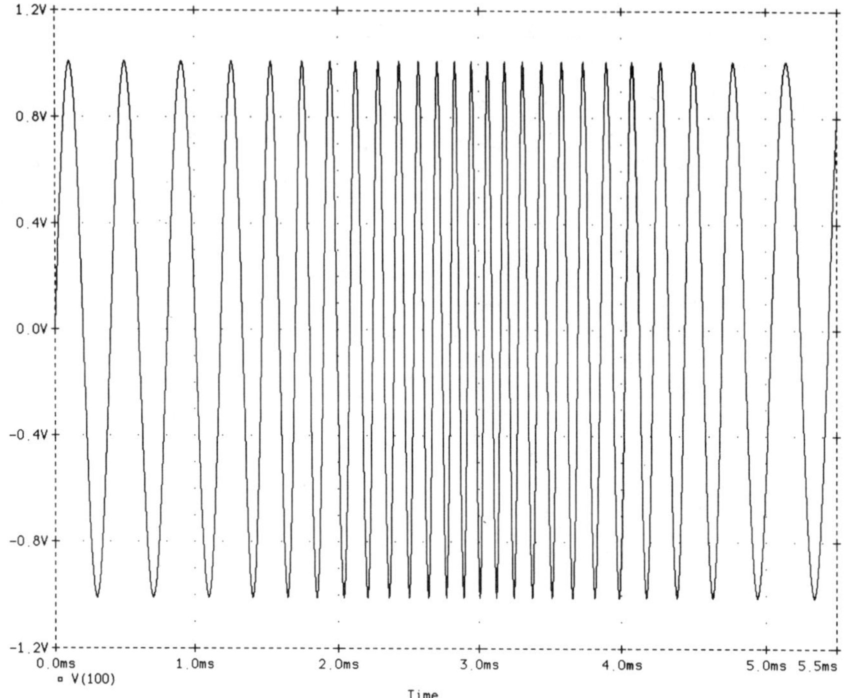

Figure 3.91 VCO input to the frequency counter.

3. Transient analysis will give better results if the maximum step size is set to about 1% or less of the input period.

Example. In this example the VCO macromodel from Section 3.2.26 is used as the input signal source for the frequency-to-voltage converter. The VCO macromodel has the following characteristics:

V_{out} = 1 V sin [(2.5 KHz)2π + V_{in}(13089.97 rad/s/volt)t]
Output resistance = 1 KΩ
Input resistance = 1 MΩ

The specifications of the frequency-to-voltage converter are:

Output resistance = 1 KΩ
Input resistance = 1 MΩ
Gain conversion = K = 1 V/Hz

```
*The VCO Model (Nodes of VCO are from 201 through 214)
Rin1    201     0       1MEG
```

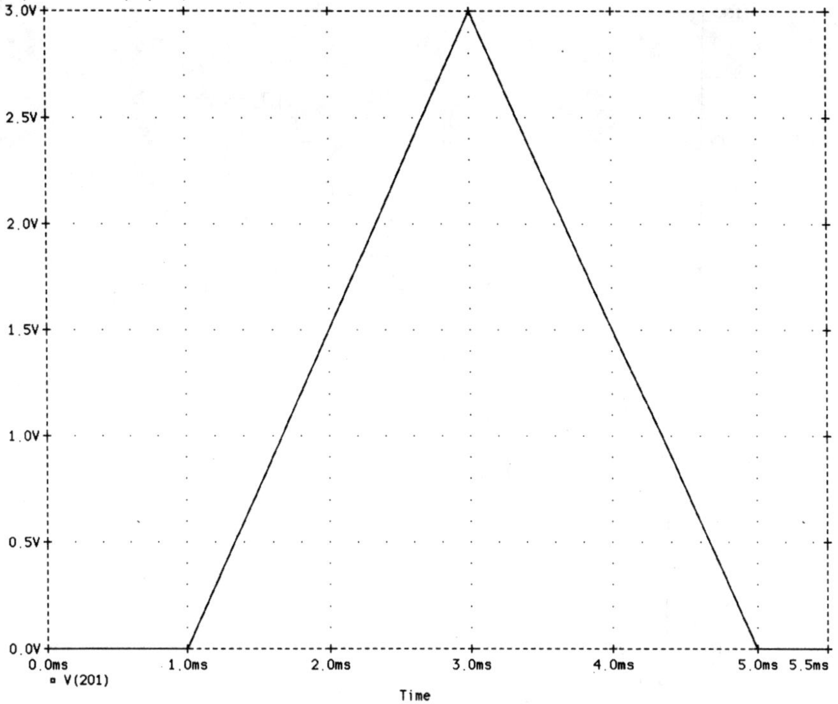

Figure 3.92 VCO control voltage.

```
E11      203     0       POLY(1)  201  0 1.2  1
R11      203     0       1
E21      204     0       POLY(2)  203  0 208 0
+0 0 0 0 -0.1
R21      205     0       1MEG
E31      206     0       105  0  100MEG
Rtime    204     205     12K
Ctime    205     206     10N
E41      207     0       POLY(2)  206  0 208 0
+0 -10MEG 10MEG
R31      207     208     1T
C11      208     0       1E-15
D11      208     209     Diode
V11      209     0       10
D21      210     0       Diode
V21      210     0       -10
E61      213     0       POLY(1)  206  0
+0    1.57080    0       -0.64596
+0    7.9593E-2  0       4.6818E-3
Rout1    213     214     1K
R51      214     0       1T
```

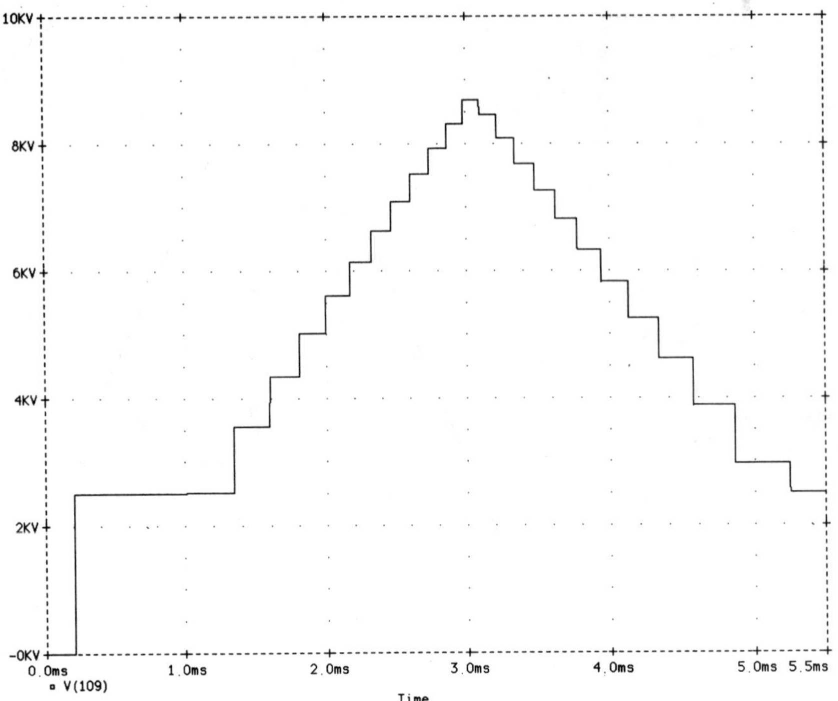

Figure 3.93 Simulation results for the frequency counter with a swept frequency input signal.

```
*The Frequency Counter
Ein     100     0       214  0  1
Rin     100     0       1MEG      ; Input resistance = 1 MΩ
E1      101     0       100  0  1
Vtest   101     102     0
C1      102     0       1
H1      103     0       Vtest    -0.1591549 ; -1/(2πVmax) = -0.1591549
R1      103     0       1
Ds      103     104     Diode
C2      104     0       200N
S1      104     0       107  0  Reset1
E2      105     0       104  0  1
S2      105     106     0  107  Reset2
C3      106     0       200N
E3      107     0       100  0  100       ; 100/Vmax = 100
R2      107     0       1
Eout    108     0       106  0  1         ; Conversion gain = 1
Rout    108     109     1K                ; Output resistance = 1 KΩ
R3      109     0       1T
```

```
.Model Reset1 Vswitch(Von=80 Voff=75 Ron=0.01)
.Model Reset2 Vswitch(Von=10 Voff=5 Ron=0.01)
.Model Diode D(Is=10U N=.001)
.Options Gmin=1E-18 ITL5=0 ITL4=300
Vin      201    0        PWL(0 0 1M 0 3M 3 5M 0)

.TRAN    10U    5.5M    0        1U       UIC
.IC      V(208) = -10      V(203) = 1.2
.PROBE V(201) V(208) V(100) V(103) V(104) V(105) V(106) V(109)
.END
```

3.3 FUNCTIONAL MACROMODELS

3.3.1 Voltage Comparators

A comparator, as its name implies, compares a signal voltage on one input with a known or reference voltage on the other input. In its simplest form, with two analog inputs and a digital output, the output will be either at a logic high voltage level corresponding to the positive saturation voltage, or at a logic low level equal to the negative saturation voltage. The digital output will be high if the voltage on the positive input is larger than that on the negative input. Comparators are used as analog to digital interfaces, Schmitt triggers, discriminators, voltage-level detectors, and oscillators.

A simple macromodel for a comparator includes the basic transfer characteristic, the maximum and minimum output voltages, the input resistance and input offset voltage, and the voltage gain. If a highly accurate comparator macromodel is required, see Getreu et al.[6]

The voltages at each node as defined from the functional block diagram of Figure 3.94 are:

$$V_a = V_{in} + V\text{ (input offset)} - V\text{ (reference)}$$

$$V_b = \text{Voltage gain} \cdot V_a = \text{Voltage gain} \cdot [V_{in} + V\text{ (input offset)} - V\text{ (reference)}]$$

$$V_{out} = K(V) \text{ if and only if } V_b > 0$$

$$= -K(V) \text{ if and only if } V_b < 0$$

where +K (or −K) is the maximum (or minimum) output voltage.

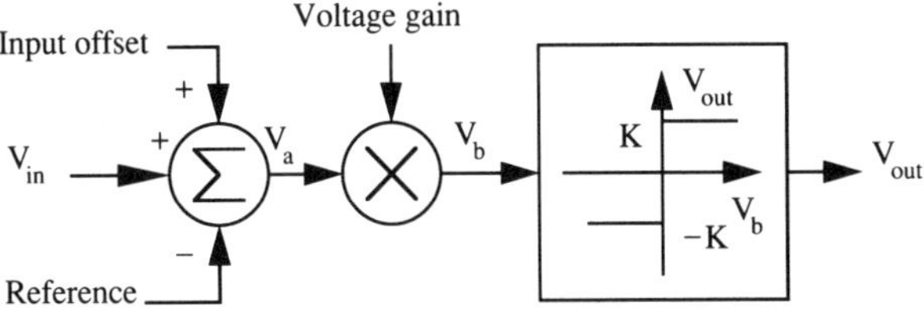

Figure 3.94 Functional block diagram of a comparator macromodel.

The equations for the comparator macromodel of Figure 3.95 are:

$$V(102,0) = \text{Voltage gain} \cdot [V_{in} - V(\text{Reference}) - V(\text{Input offset})] + 10^{-15} \cdot V(104)$$
$$= \text{Voltage gain} \cdot [V(100,0) - V(101,0) - P_0 \text{ in } E_1] + 10^{-15} \cdot V(104)$$
$$V(104) = 10^6 \cdot V(102,0)$$

SPICE code for Figure 3.95

```
RIN 100 101 [A]             ; [A] = Input resistance
E1 102 0 POLY(3) 100 0 101 0 104 0
+ [P0] [P1] [P2] [P3]       ; [P0] = −(Input offset voltage · Voltage gain)
                            ; [P1] = Voltage gain
                            ; [P2] = −(Voltage gain)
                            ; [P3] = 1E-15
R1 102 0 1
E2 103 0 102 0 1MEG
R2 103 104 1T
C1 104 0 1E-15
D1 104 105 DIODE
V1 105 0 [B]                 ; [B] = Maximum output voltage = K1
D2 106 104 DIODE
V2 106 0 DC=[C]             ; [C] = Minimum output voltage = −K2
.MODEL DIODE D(N=0.001)
```

Example

Input resistance = 1 MΩ

Voltage gain = 100

Input offset voltage = 0.5 V

Reference voltage = 1 V

Maximum output voltage = 10 V

Minimum output voltage = −10 V

```
RIN 100 101 1MEG            ; Input resistance = 1 MΩ
E1 102 0 POLY(3) 100 0 101 0 104 0
```

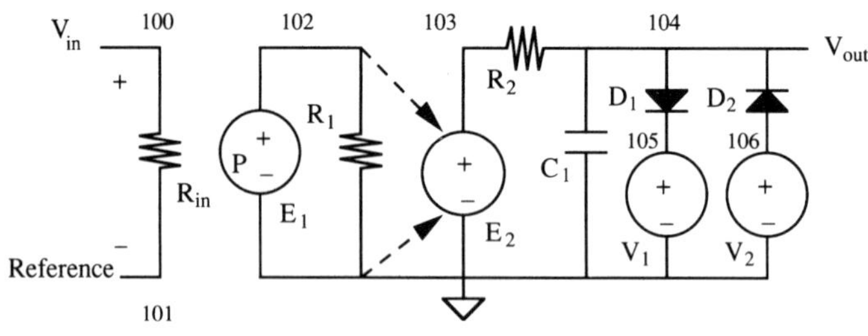

Figure 3.95 Equivalent circuit for a voltage comparator.

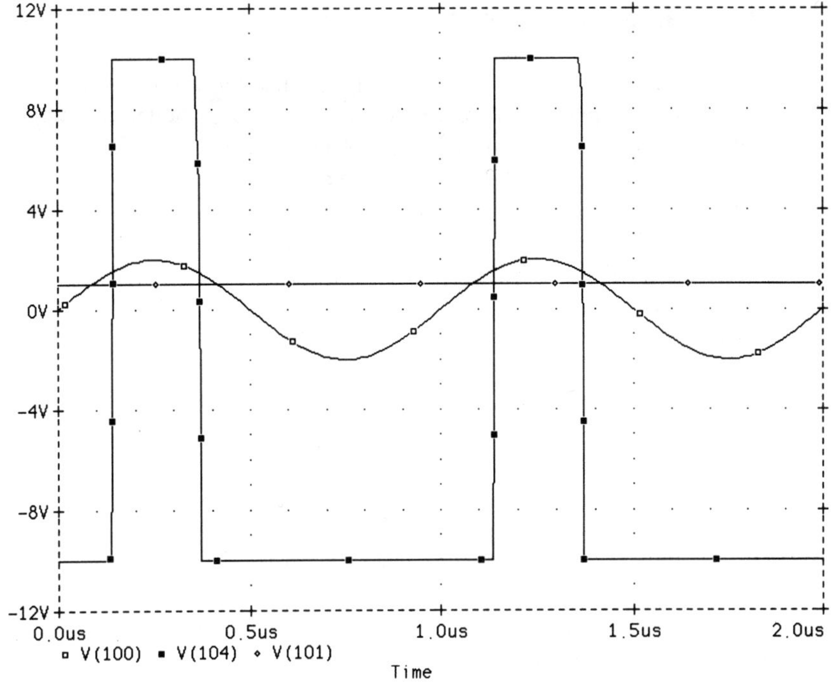

Figure 3.96 Simulation results for the voltage comparator.

```
+ -50 100 -100 1E-15        ; [P0] = -(Input offset voltage × Voltage gain) = -50
                            ; [P1] = Voltage gain = 100
                            ; [P2] = -100
                            ; [P3] = 1E-15
R1 102 0 1
E2 103 0 102 0 1MEG
R2 103 104 1T
C1 104 0 1E-15
D1 104 105 DIODE
V1 105 0 10                 ; Maximum output voltage = 10 V
D2 106 104 DIODE
V2 106 0 DC=-10             ; Minimum output voltage = -10 V
.MODEL DIODE D(N=0.001)

VIN 100 0 SIN(0 2 1MEG)
VREF 101 0 DC=1.0
.IC V(104)=0
.TRAN 10N 3U 0 10N UIC
.PRINT TRAN V(104)
.PROBE                      ; PSpice case
.END
```

3.3.2 Operational Amplifiers

Introduction. A generalized amplifier macromodel is introduced. The simplicity and lack of component interdependence enables this model to emulate virtually any amplifier regardless of its operating frequency range. Therefore, the macromodel is useful for simulating most any type voltage amplifier from operational amps to microwave amplifiers. Performance parameters available from measurements or data sheets can be used to create a specific amplifier model. Because the operational amplifier is such a widely used device, several op amp models will be introduced; however, these models are not limited to op amps. The introductory op amp model described in Chapter 1 is further expanded to include a more general model.

Many excellent references on op amp macromodels can be found.[7] The Boyle macromodel, one of the earliest models, uses a differential pair to model the front end of an op amp.[8] A modified version of this model is used by several manufacturers, such as Texas Instruments, to create macromodels of their product line. These models are available on diskettes.[9] Another excellent reference may be found in E. J. Kennedy's book *Operational Amplifier Circuits*.[10]

Simplicity, efficiency, and flexiblity are the main objectives of this amplifier macromodel. The emphasis is on separating each important data sheet parameter into a subcircuit independent of the other parameters. The macromodel is broken down into five functional stages: input, frequency response, gain, slew rate, and an output stage, as shown in Figure 3.97. Each stage contains data sheet parameters to be modeled. In the following discussion, each stage is broken down to show the dependence of each parameter on the components. Several examples of implementing this macromodel to emulate commercially available op amps are given. A detailed derivation of the μA741 model is given, and macromodels for the LF351, TLC2201, and OP-10 op amp are introduced. A comparison is also made between the simulated and the actual characteristics of these models. A program for generating the macromodel circuit control files automatically is available in Chapter 6.

The SPICE primitive elements for the input stage are shown in Figure 3.98. This stage models the differential and common mode input resistance, the input bias and offset currents, the offset voltage, and the input capacitance.

The differential and common mode input resistances are modeled by R_1, R_2, and R_3 according to

$$R_{dm} = R_2 \tag{3.9}$$

$$R_{cm} = R_1 = R_3 \tag{3.10}$$

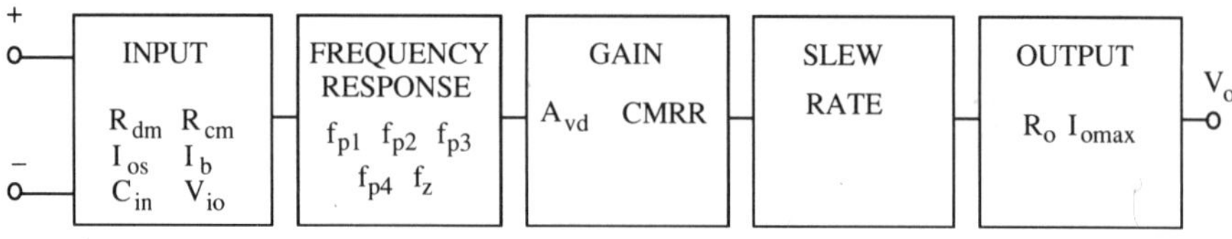

Figure 3.97 Amplifier functional block.

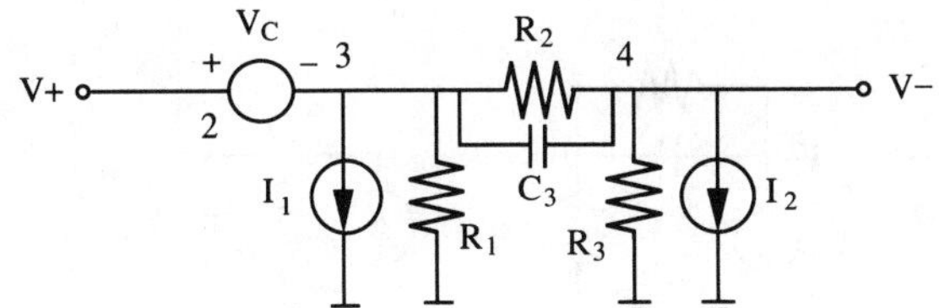

Figure 3.98 Input stage.

The input bias and offset currents are modeled by the two current sources I_1 and I_2.

$$I_b = \frac{1}{2}(I_1 + I_2) \tag{3.11}$$

$$I_{os} = I_1 - I_2 \tag{3.12}$$

The input offset voltage is modeled by V_c and the input capacitance by C_3.

$$V_{os} = V_c \tag{3.13}$$

$$C_{in} = C_3 \tag{3.14}$$

If a more accurate input stage using bipolar junction transistors is desired, diodes D_1 and D_2 can be added to the model as shown in Figure 3.99. For an npn input, the diodes have the orientation shown. For a pnp input, the diodes and the current sources, I_1 and I_2, should be reversed. The diodes are modeled as ideal (N = .001). When this front-end stage is used, a bandwidth restriction is placed on the circuit as given by Equation (3.15).

$$\text{B.W.} = \frac{500q}{\pi k T C_3 \left[\frac{1}{I_b - \frac{I_{os}}{2}} + \frac{1}{I_b + \frac{I_{os}}{2}}\right]} \tag{3.15}$$

where q is electronic charge, k is Boltzmann's constant, and T is temperature in Kelvin.

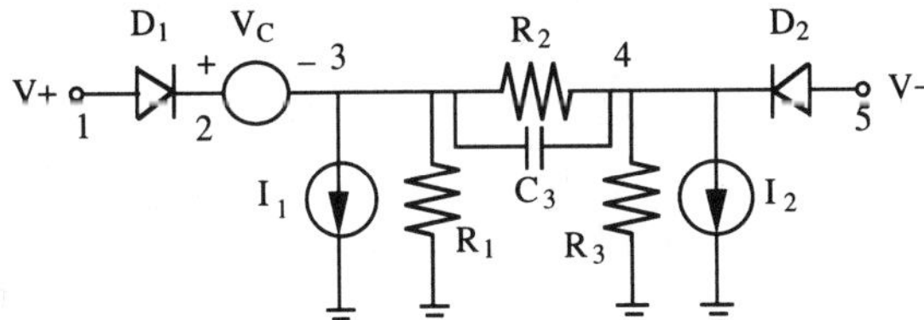

Figure 3.99 Modified input stage for NPN input transistors.

Frequency response. The differential mode frequency response stage is shown in Figure 3.100. This stage models the three poles, f_{p2}, f_{p3}, and f_{p4}. A zero, f_z, is also modeled. The dominant pole, f_{p1}, is modeled in the gain stage. The transfer function is:

$$A_{vd}(s) = \frac{A_{vd}(0)(s/\omega_z + 1)}{(s/\omega_{p1} + 1)(s/\omega_{p2} + 1)(s/\omega_{p3} + 1)(s/\omega_{p4} + 1)} \tag{3.16}$$

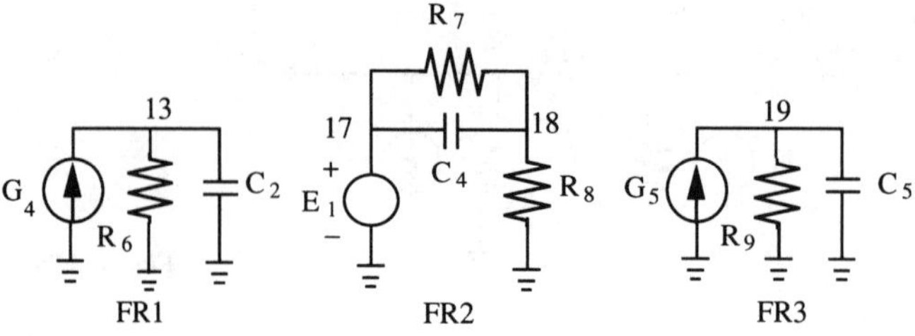

Figure 3.100 Frequency response stage.

where $A_{vd}(0)$ is the DC differential mode gain. The common mode frequency response is modeled by:

$$A_{vcm}(s) = \frac{A_{vcm}(0)}{(s/\omega_{p1} + 1)} \tag{3.17}$$

where $A_{vcm}(0)$ is the DC common mode gain.

Substage FR1 models the second pole, f_{p2}. The gain of each frequency response stage is unity. This is done by making

$$G_4 = g_4 \cdot V(3,4) \tag{3.18}$$

with $g_4 = 1/R_6$. The pole is located at

$$f_{p2} = \frac{1}{2\pi(C_2)(R_6)} \tag{3.19}$$

Substage FR2 models the third pole, f_{p3}, and the zero, f_z. The frequency of the zero must be less than the frequency of the pole. If only a three-pole response is desired without a zero, the third pole is used to cancel the zero. The equations for this middle section are

$$E_1 = e_1 \cdot V(13,0) \tag{3.20}$$

$$e_1 = \frac{R_8 + R_7}{R_8} \tag{3.21}$$

$$f_z = \frac{1}{2\pi(R_7)(C_4)} \tag{3.22}$$

$$f_{p3} = \frac{R_7 + R_8}{2\pi(R_7)(R_8)(C_4)} \tag{3.23}$$

$$f_{p3} = f_z(\text{ If } R_8 >> R_7) \tag{3.24}$$

Substage FR3 models the fourth pole according to

$$G_5 = g_5 \cdot V(18,0) \tag{3.25}$$

$$g_5 = \frac{1}{R_9} \tag{3.26}$$

$$f_{p4} = \frac{1}{2\pi(C_5)(R_9)} \tag{3.27}$$

Gain stage. The gain stage is shown in Figure 3.101. This stage models the DC differential mode gain, $A_{vd}(0)$, and the CMRR.

The dominant pole, f_{p1}, is also modeled in this stage because it is common to both the common mode and the differential mode frequency response. The difference and common modes $A_{vd}(s)$ and $A_{vcm}(s)$ are combined in this stage. The defining equations are

$$G_1 = g_1 \cdot V(19) \tag{3.28}$$

$$A_{vd}(0) = (g_1)(R_4) \tag{3.29}$$

$$G_2 = g_2 \cdot (V(3) + V(4)) \tag{3.30}$$

$$CMRR = \frac{g_1}{2g_2} \tag{3.31}$$

$$f_{p1} = \frac{1}{2\pi(R_4)(C_1)} \tag{3.32}$$

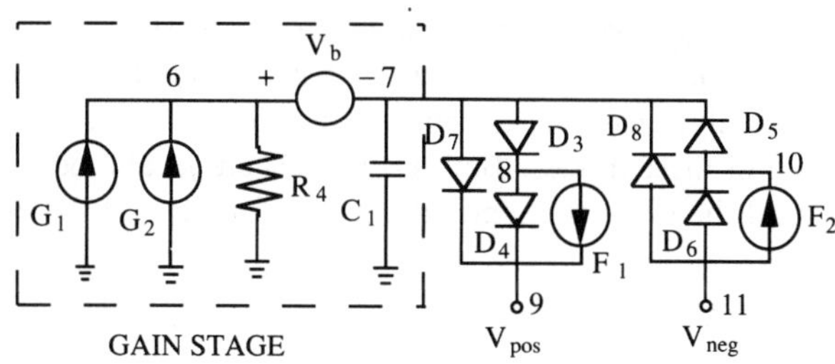

Figure 3.101 Gain stage.

Slew rate stage. The slew rate stage is shown in Figure 3.102. This stage limits the current in C_1 to produce the desired slew rate.

This stage also limits the dynamic range of the output voltage to that of the positive and negative power supplies. Both F_1 and F_2 are polynomial sources. V_b is a 0 V DC source used to sense the current in the capacitor. The governing equations are

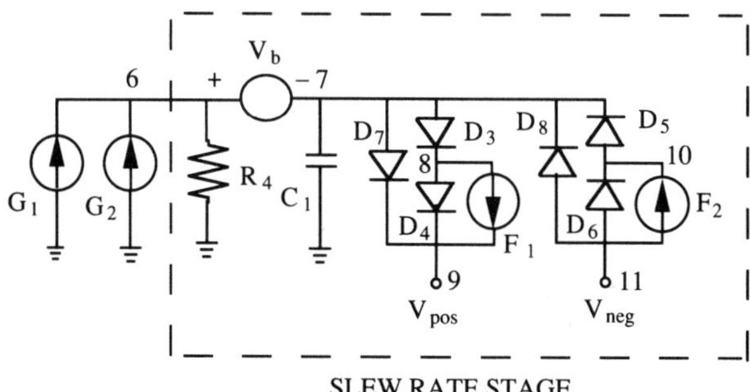

Figure 3.102 Slew rate stage.

$$F_1 = I(V_b) - I_{max+} \quad (3.33)$$

$$F_2 = -I(V_b) - I_{max-} \quad (3.34)$$

$$SR+ = \frac{I_{max+}}{C_1} \quad (3.35)$$

$$SR- = \frac{I_{max-}}{C_1} \quad (3.36)$$

The positive slew rate is modeled by F_1, D_3, and D_4. If the current in C_1 is less than the maximum current allowed in C_1, then the current in F_1 is negative with respect to its positive direction, and the current circulates in D_4; therefore, no current limiting occurs. Additionally, D_3 is turned off. If the current in C_1 is momentarily greater than the maximum current allowed in C_1, then the current in F_1 is positive with respect to its orientation, and the excess current is removed through D_3 by F_1. Furthermore, D_4 is turned off. The negative slew rate is modeled in a similar fashion, and F_2 controls the maximum negative current through V_b and C_1. D_1 and D_2 are both ideal diodes (N = 0.001).

The supply voltages V_{pos} and V_{neg} plus diodes D_7 and D_8 keep the slew rate circuit functional. Clamping the voltage at node 7 to V_{pos} ensures that D_3 and D_4 will not be turned on except by F_1. Additionally, clamping the voltage at node 7 to V_{neg} ensures that only F_2 will control the state of these diodes. Both D_7 and D_8 are ideal diodes (N = 0.001). Again, V_{pos} and V_{neg} limit the maximum and minimum excursions of the output voltage according to

$$V_o(max) = V_{pos} \quad (3.37)$$

$$V_o(min) = V_{neg} \quad (3.38)$$

Output stage. The output stage is shown in Figure 3.103.

This stage models the output resistance and the output current limiting characteristics of the amplifier. The functional elements and equations are

$$R_{out} = R_5 \quad (3.39)$$

$$F_3 = I(V_a) - I_{max}(\text{source}) \quad (3.40)$$

$$F_4 = -I(V_a) - I_{max}(\text{sink}) \quad (3.41)$$

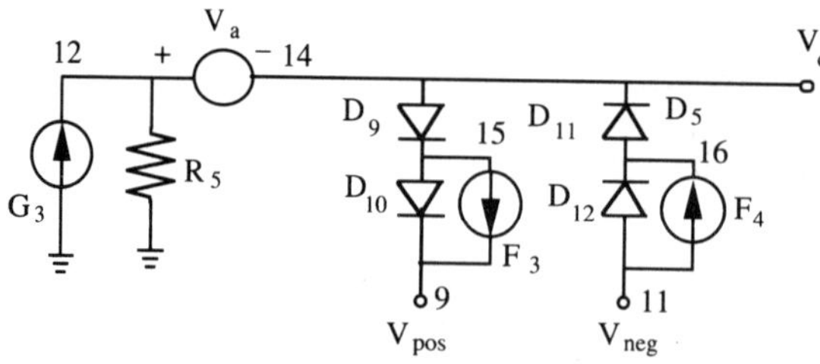

Figure 3.103 Output stage.

SPICE code for Operational Amplifier

```
R1 3 0 [A]          ; [A] = Common mode input resistance
R2 3 4 [B]          ; [B] = Differential mode input resistance
R3 4 0 [A]
R4 6 0 1E3
R5 12 0 [C]         ; [C] = Output resistance
R6 13 0 1E3
R7  17 18 10E3
R8 18 0 [D]         ; [D] = See note below
R9 19 0 1E3         ;
I1 3 0 [E]          ; [E] = Ib + (Ios/2)
I2 4 0 [F]          ; [F] = Ib - (Ios/2)
C1 7 0 [G]          ; [G] = 1/(2,000πfp1)
C2 13 0 [H]         ; [H] = 1/(2,000πfp2)
C3 3 4 [I]          ; [I] = Cin
C4 17 18 [J]        ; [J] = 1/(20,000πfz)
C5 19 0 [K]         ; [K] = 1/(2,000πfp4)
G1 0 6 18 0 [L]                          ; [L] = Avd(0)/(1,000)
G2 0 6 POLY(2) (3,0) (4,0) 0 [M] [M]     ; [M] = Avd(0)/(2,000CMRR)
G3 0 12  7 0 [N]                         ; [N] = 1/Rout
G4 0 13 3 4 .001                         ;
G5 0 19 18 0 .001                        ;
VB 6 7 DC 0                              ;
VA 12 14 DC 0                            ;
F1 8 9 POLY(1) VB [O] 1                  ; [O] = -(SR+)[G]
F2 11 10 POLY(1) VB [P] -1               ; [P] = -(SR-)[G]
F3 15 9 POLY(1) VA [Q] 1                 ; [Q] = - |Maximum output source current|
F4 11 16 POLY(1) VA [R] -1               ; [R] = - |Maximum output sink current|
E1 17 0 13 0 [S]                         ; [S] = ([D] + 10,000)/[D]
VC 2 3 [T]                               ; [T] = Input offset voltage.
D3 7 8 DX
D4 8 9 DX
D5 10 7 DX
D6 11 10 DX
D7 7 9 DX
D8 11 7 DX
D9 14 15 DX
D10 15 9 DX
D11 16 14 DX
D12 11 16 DX
.MODEL DX D(N=.001)
```

Notes

[A] = Common mode input resistance

[B] = Differential mode input resistance

[C] = Output resistance

$[D] = \frac{10{,}000}{\frac{f_{p3}}{f_z} - 1}$

$[E] = I_b + \frac{I_{os}}{2}$

$[F] = I_b - \frac{I_{os}}{2}$

$[G] = \frac{1}{2{,}000\,\pi(f_{p1})}$

$[H] = \frac{1}{2{,}000\,\pi(f_{p2})}$

$[I] = C_{in}$

$[J] = \frac{1}{20{,}000\,\pi(f_z)}$

$[K] = \frac{1}{2{,}000\,\pi(f_{p4})}$

$[L] = \frac{A_{vd}(0)}{1{,}000}$

$[M] = \frac{A_{vd}(0)}{2{,}000\,\text{CMRR}}$

$[N] = \frac{1}{R_{out}}$

[O] = −(SR+)[G]

[P] = −(SR−)[G]

[Q] = Maximum output source current

[R] = Maximum output sink current

$[S] = \frac{[D]+10^4}{[D]}$

[T] = Input offset voltage

Example. The following are typical data sheet values for a μA741 op amp:

$A_{vd}(0)$ = 106 dB	R_{cm} = 2 GΩ	R_{dm} = 2 MΩ
CMRR = 90 dB	I_{os} = 20 nA	I_b = 80 nA
R_{out} = 75 Ω	SR(+) = 0.5 V/μs	SR(−) = 0.5 V/μs
I_{max}(source) = 25 mA	I_{max}(sink) = 25 mA	V_{io} = 1 mV
V_{outmax} = 13 V	V_{outmin} = −12V	C_{in} = 1.4 pF
f_{p1} = 5 Hz	f_{p2} = 2 MHz	f_{p3} = 20 MHz
f_{p4} = 100 MHz	f_z = 5 MHz	

$[A] = R_{cm} = 2\ \text{G}\Omega$

$[B] = R_{dm} = 2\ \text{M}\Omega$

$[C] = R_{out} = 75\ \Omega$

$[D] = \frac{10{,}000}{\frac{20}{5} - 1} = 3.333\ \text{K}\Omega$

$[E] = 80E - 9 + \frac{20E-9}{2} = 90\ \text{nA}$

$[F] = 80E - 9 - \frac{20E-9}{2} = 70\ \text{nA}$

$[G] = \frac{1}{2{,}000\,\pi(5)} = 31.83\ \mu F$

$[H] = \frac{1}{2{,}000\,\pi(2E6)} = 79.58\ pF$

$[I] = C_{in} = 1.4\ pF$

$[J] = \frac{1}{20{,}000\,\pi(5E6)} = 3.183\ pF$

$[K] = \frac{1}{2{,}000\,\pi(100E6)} = 1.592\ pF$

$[L] = \frac{2E5}{1{,}000} = 200S$

$[M] = \frac{2E5}{(2{,}000)(31{,}623)} = 3.162\ mS$

$[N] = \frac{1}{75} = 13.33\ mS$

$[O] = -(0.5E6)(31.83E-6) = -15.92\ A$

$[P] = -(0.5E6)(31.83E-6) = -15.92\ A$

$[Q] = I_{max}(source) = 25\ mA$

$[R] = I_{max}(sink) = 25\ mA$

$[S] = \frac{3{,}333+10^4}{3{,}333} = 4.000$

$[T] = V_{io} = 1\ mV$

μA741 test results

Parameter	Data Sheet Value	Measured Value
R_{cm}	2 GΩ	2 GΩ
R_{dm}	2 MΩ	2 MΩ
$A_{vd}(0)$	106 dB	105.9 dB
CMRR	90 dB	90.1 dB
I_{os}	20 nA	20.09 nA
I_b	80 nA	80.09 nA
R_{out}	75 Ω	74.44 Ω
f_{p1}	5 Hz	5 Hz
f_{p2}	2 MHz	2 MHz
f_{p3}	20 MHz	20 MHz
f_{p4}	100 MHz	100 MHz
f_z	5 MHz	5 MHz
SR+	0.5 V/μs	0.498 V/μs
SR-	0.5 V/μs	0.498 V/μs
I_{maxso}	25 mA	25 mA
I_{maxsi}	25 mA	25 mA
V_{os}	1 mV	1 mV

Op amp macromodels. The following is the SPICE code generated by MACRO.EXE for a μA741, a LF351, an OP-10, and a TLC2201:

```
* μA741 Macromodel
* 1   = V+ Input Node,         5 = V- Input Node
* 9   = V+ Power Supply   11 = V- Power Supply
* 14 =  Output Node
*
* Nodes 2-19 used
*
R1 3 0 2E9
R2 3 4 2E6
R3 4 0 2E9
R4 6 0 1E3
R5 12 0 75
R6 13 0 1E3
R7  17 18 10E3
R8 18 0 1E15
R9 19 0 1E3
I1 3 0 90E-9
I2 4 0 70E-9
C1 7 0 .318E-4
C2 13 0 15.92E-12
C3 3 4 3E-12
C4 17 18  .159E-12
C5 19 0 .0159E-12
G1 0 6 18 0  200
G2 0 6 POLY(2) (3,0) (4,0) 0   3.16E-3 3.16E-3
G3 0 12  7 0 0.013333
G4 0 13 3 4 .001
G5 0 19 18 0 .001
VB 6 7 DC 0
VA 12 14 DC 0
F1 8 9 POLY(1) VB -.159 1
F2 11 10 POLY(1) VB -.159 -1
F3 15 9 POLY(1) VA -.025 1
F4 11 16 POLY(1) VA -.025 -1
E1 17 0 13 0 1
VC 2 3 1E-3
D3 7 8 DX
D4 8 9 DX
D5 10 7 DX
D6 11 10 DX
D7 7 9 DX
D8 11 7 DX
D9 14 15 DX
D10 15 9 DX
D11 16 14 DX
D12 11 16 DX
.MODEL DX D(N=.001)
```

```
* LF 351 Macromodel
* 1   = V+ Input Node,         5 = V- Input Node
* 9   = V+ Power Supply   11 = V- Power Supply
* 14 =  Output Node
*
* Nodes 2-19 used
*
R1 3 0 1E14
R2 3 4 1E13
R3 4 0 1E14
R4 6 0 1E3
R5 12 0 75
R6 13 0 1E3
R7  17 18 10E3
R8 18 0 10E3
R9 19 0 1E3
I1 3 0 6.25E-11
I2 4 0 3.75E-11
C1 7 0 1.061033E-5
C2 13 0 2.652582E-11
C3 3 4 1.4E-12
C4 17 18  1.591549E-12
C5 19 0 1.591549E-12
G1 0 6 18 0  199.5262
G2 0 6 POLY(2) (3,0) (4,0) 0  9.976312E-4 9.976312E-4
G3 0 12  7 0 0.013333
G4 0 13 3 4 .001
G5 0 19 18 0 .001
VB 6 7 DC 0
VA 12 14 DC 0
F1 8 9 POLY(1) VB -137.9343 1
F2 11 10 POLY(1) VB -137.9343 -1
F3 15 9 POLY(1) VA -.02 1
F4 11 16 POLY(1) VA -.02 -1
E1 17 0 13 0 2
VC 2 3 .005
D3 7 8 DX
D4 8 9 DX
D5 10 7 DX
D6 11 10 DX
D7 7 9 DX
D8 11 7 DX
D9 14 15 DX
D10 15 9 DX
D11 16 14 DX
D12 11 16 DX
.MODEL DX D(N=.001)
```

```
* TLC2201  Macromodel
* 1   = V+ Input Node,        5 = V- Input Node
* 9   = V+ Power Supply   11 = V- Power Supply
* 14 =  Output Node
*
* Nodes 2-19 used
*
R1 3 0 1E14
R2 3 4 1E13
R3 4 0 1E14
R4 6 0 1E3
R5 12 0 75
R6 13 0 1E3
R7  17 18 10E3
R8 18 0 10E3
R9 19 0 1E3
I1 3 01.25E-12
I2 4 0 7.5E-13
C1 7 0 1.768388E-5
C2 13 0 2.652582E-11
C3 3 4 1.4E-12
C4 17 18  1.591549E-12
C5 19 0 1.591549E-12
G1 0 6 18 0  199.5262
G2 0 6 POLY(2) (3,0) (4,0) 0  3.154786E-4 3.154787E-4
G3 0 12  7 0 0.013333
G4 0 13 3 4 .001
G5 0 19 18 0 .001
VB 6 7 DC 0
VA 12 14 DC 0
F1 8 9 POLY(1) VB -47.74648 1
F2 11 10 POLY(1) VB -47.74648 -1
F3 15 9 POLY(1) VA -.02 1
F4 11 16 POLY(1) VA -.02 -1
E1 17 0 13 0 2
VC 2 3 80E-6
D3 7 8 DX
D4 8 9 DX
D5 10 7 DX
D6 11 10 DX
D7 7 9 DX
D8 11 7 DX
D9 14 15 DX
D10 15 9 DX
D11 16 14 DX
D12 11 16 DX
.MODEL DX D(N=.001)
```

```
* OP-10 Macromodel
* 1   = V+ Input Node,        5 = V- Input Node
* 9   = V+ Power Supply   11 = V- Power Supply
* 14 =  Output Node
*
* Nodes 2-19 used
*
R1 3 0 2E11
R2 3 4 6E7
R3 4 0 2E11
R4 6 0 1E3
R5 12 0 60
R6 13 0 1E3
R7  17 18 10E3
R8 18 0 3333.333
R9 19 0 1E3
I1 3 0 4.4E-9
I2 4 0 1.6E-9
C1 7 0 3.183099E-5
C2 13 0 7.957747E-11
C3 3 4 1.4E-12
C4 17 18  3.183099E-12
C5 19 0 1.591549E-12
G1 0 6 18 0  199.5262
G2 0 6 POLY(2) (3,0) (4,0) 0  5E-5 5E-5
G3 0 12  7 0 1.666667E-2
G4 0 13 3 4 .001
G5 0 19 18 0 .001
VB 6 7 DC 0
VA 12 14 DC 0
F1 8 9 POLY(1) VB -5.411268 1
F2 11 10 POLY(1) VB -5.411268 -1
F3 15 9 POLY(1) VA -.015 1
F4 11 16 POLY(1) VA -.015 -1
E1 17 0 13 0 1
VC 2 3 .0002
D3 7 8 DX
D4 8 9 DX
D5 10 7 DX
D6 11 10 DX
D7 7 9 DX
D8 11 7 DX
D9 14 15 DX
D10 15 9 DX
D11 16 14 DX
D12 11 16 DX
.MODEL DX D(N=.001)
```

3.4 SUMMARY

In this chapter, various drop-in macromodels including behavioral and functional have been introduced, equations of operation presented, and example simulation results shown for each model. These macromodels can be used individually or as parts of a large system to verify operation without requiring large simulation times or developmental efforts. Many of the models presented in this chapter are available for automatic generation following the procedures given in Chapter 6.

REFERENCES

1. Tom Terlizzi, "Here Are Simpler Circuit Models," *Electronic Design*, September 14, 1972, pp. 144–147; Graeme R. Boyle, Barry M. Cohn, Donald O. Pederson, and James E. Solomon, "Macromodeling of Integrated Circuit Operational Amplifiers," *IEEE Journal of Solid-State Circuits*, Vol. SC-9, No. 6, December 1974, pp. 353–363; Barry M. Cohn and Donald O. Pederson, "Macromodeling of Operational Amplifiers," *Proceedings of International Solid-State Circuit Conference 1974*, pp. 42–43; M. Glesner and C. Weisang, "Computer Aided Macromodeling of Integrated Circuit Operational Amplifiers," *IEEE Proceedings of International Symposium on Circuits and Systems/76*, 1976, pp. 255–258; Ibrahim Mandour and Rashid El Awady, "Macromodeling of Operational Amplifiers and Its Application to Active Filter," *Proceedings of MELECON '83 (Mediterranean Electrotechnical Conference) IEEE*, Vol. 1, May 1983, pp. B1.0711–B1.0712; Claudio Turchetti and Guido Masetti, "A Macromodel for Integrated All-MOS Operational Amplifiers," *IEEE Journal of Solid-State Circuits*, Vol. SC-18, No. 4, August 1983, pp. 389–394; Maciej J. Ogorzalek and Andrzej R. Scibich, "Improved Macromodel of Operational Transconductance Amplifier (OTA)," *Proceedings of 1984 IEEE International Symposium on Circuits and Systems*, Montreal, Canada, 1984, Vol. 3, pp. 1474–1476; Sherif Michael and Patrick Gariano, "Micro and Macromodeling of Integrated Operational Amplifiers and Their Use in Developing and Evaluating New Composite Opamps," *Proceedings of 28th Midwest Symposium on Circuits and Systems*, Louisville, Kentucky, 1985, pp. 308–311; Gerald Cotreau and Brian Mathews, "Simplify Circuit Analysis with Easy OpAmp Models," *Electronic Design*, September 22, 1988, pp. 91–96; Ian E. Getreu, Andreas D. Hadiwidjaja, and Johan M. Brinch, "An Integrated-Circuit Comparator Macromodel," *IEEE Journal of Solid-State Circuits*, Vol. SC-11, No. 6, December 1976, pp 826–833.
2. G. J. Scott and T. M. Chen, "Addition of Excess Noise in SPICE Circuit Simulations," *IEEE SOUTHEASTCON Conference Proceedings*, Tampa, Florida, April 5–8, 1987, Vol. 1, pp. 186–190.
3. Curt D. Motchenbacher and F. C. Fitchen, *Low Noise Electronic Design* (New York: Wiley Interscience, 1973), pp. 171–179.
4. Richard B. Anderson, *The Student Edition of MathCAD* (Reading, MA: Addison-Wesley/Benjamin Cummings, 1988).
5. Microsoft Corporation, *Getting Started with Microsoft Excel*, Version 2.2, 1989.
6. Getreu et al., "An Integrated-Circuit Comparator Macromodel."
7. J. E. Solomon, "The Monolithic Op Amp: A Tutorial Study," *IEEE Journal of Solid-State Circuits*, Vol. SC-9, No. 6, December 1974; Walter G. Jung, "An LT1013 Op Amp Macromodel,"

Linear Technology Design Note No. 13, July 1988; Jung, "A SPICE Op Amp Macromodel for the LT1012," Linear Technology Design Note No. 28, November 1989; G. Krajewska and F. E. Holmes, "Macromodeling of FET/Bipolar Operational Amplifiers," *IEEE Journal of Solid-State Circuits*, Vol. SC-14, No. 6, December 1979; Jung, "Using Op Amp Macromodels," *Electronic Products*, June 1990, pp. 25–312; Mark Alexander and Derek F. Bowers, "New SPICE Compatible Op Amp Model Boosts ac Simulation Accuracy," *EDN*, February 15, 1990, pp. 143–154; Alexander and Bowers, "Op Amp Macromodel Proves Superior in High-Frequency Regions," *EDN*, March 1, 1990, pp. 155–164.

8. Boyle et al., "Macromodeling of Integrated Circuit Operational Amplifiers" (see reference 1).
9. Texas Instruments, *Operational Amplifier Macromodels Linear Circuits Data Manual*, Dallas, 1990.
10. E. J. Kennedy, *Operational Amplifier Circuits Theory and Applications* (New York: Holt, Rinehart and Winston, 1988).

FURTHER READINGS

1. Paolo Antognetti and Giuseppe Massobrio, *Semiconductor Device Modeling with SPICE* (New York: McGraw-Hill, 1988).
2. Joe Buxton, "OP-42 Advanced SPICE Macro-Model—Application Note AN-117," Precision Monolithics Inc., 1989.
3. Joe Buxton, "OP-64 Advanced SPICE Macro-Model—Application Note AN-110," Precision Monolithics Inc., 1989.
4. Bert Epler, "SPICE2 Application Notes for Dependent Sources," *IEEE Circuits and Devices Magazine*, September 1987, pp. 36–44.
5. Brian Kerridge, "Analog SPICE Simulation Models—SPICE Models Enjoy Multiple Sources," *EDN*, September 3, 1990, pp. 79–84.
6. John Till, "Flexible Op-Amp Model Improves SPICE," *Electronic Design*, June 22, 1989, pp. 115–118.

4

The Phase Locked Loop Macromodel

4.1 INTRODUCTION

The phase locked loop (PLL) is a feedback system composed of a phase comparator, error amplifier, low-pass filter, and voltage controlled oscillator (VCO).[1] The basic PLL concept has been known since the early 1930s,[2] and has been widely used in many diverse applications. Before IC technology matured, PLL usage was limited by the cost and complexity of discrete component systems.

The development of the analog IC in the late 1960s changed the situation rapidly. The development of monolithic PLL ICs opened up a wide range of applications for PLL techniques, particularly in the area of consumer, industrial, and military communication equipment. Today, low-cost monolithic PLL circuits have become basic building blocks of AM and FM detectors, modems, FSK demodulators, stereo demodulators, tone decoders, frequency synthesizers, clock regenerators, waveform generators, and television display systems.[3]

Although the monolithic PLL is regarded as a basic building block in various applications, and its basic theory and principles of operation are rather straightforward, the mathematical analysis of the loop characteristics to determine key performance indexes is quite difficult, especially when the nonlinearities of the PLL are considered.[4] More difficulty arises if the effect of noise at the input of PLL is considered. In such cases, conventional analytical methods are almost impossible to use for computing the transient responses. Furthermore, the equations normally encountered in the analysis of a PLL are no longer valid for the Hybrid Phase-Locked Loop (HPLL) recently proposed by D. Asta and D. N. Green.[5] This type of PLL comprises analog functional blocks for the phase detector and the VCO but uses switched capacitor techniques for the loop filter.

Computer models represent a viable solution to overcome the analytical difficulties in the analysis of PLL. The many different models proposed to verify PLL performance can be broadly divided into two different categories: (1) mathematical structure[6] and (2) equivalent circuit structure.[7] Each type has limitations and difficulties, for example in accuracy, operating region, and simulation language when used to verify the PLL itself or in various PLL applications.

A generalized equivalent circuit structure macromodel for Analog Phase Locked Loops (APLL) is introduced. This APLL macromodel is constructed by combining the sub-macromodels of a phase detector, a VCO, and a low pass filter. This type of equivalent circuit model is very flexible and useful because it models APLL performance in all four regions: unlocked state, signal acquisition, locked mode, and the loss of lock region.

In this chapter, the electrical performance parameters to be modeled are discussed and defined. On the basis of the transfer function and the terminal characteristics, functional block diagrams for the phase detector and the VCO are developed. The block diagrams are reduced to the equivalent circuit structure models whose elements are SPICE primitives. The equivalent circuit models are tested and verified with PSpice under various operating conditions.

4.2 ELECTRICAL AND PERFORMANCE PARAMETERS TO BE MODELED

System performance parameters. Before developing a macromodel, careful partitioning of the functions performed by the various parts of the APLL is required. In a typical monolithic PLL system, the two basic functional blocks of the phase detector and VCO are on the chip, while the loop filter is usually external. Normally, key system parameters such as the loop conversion gain, VCO free-running frequency, PLL tracking and capture ranges, the loop damping factor, and the closed-loop bandwidth are set by external components such as timing resistors, timing capacitors, and specific loop filter configurations. Based on the characteristics of the monolithic APLL system, the requirements of the equivalent circuit structure APLL macromodel are broken down as follows:

1. The APLL macromodel must have two basic sub-macromodels of the phase detector and the VCO.
2. The APLL macromodel should have two or more open terminals for inserting almost any arbitrary loop filter. Thus, arbitrary order and arbitrary type PLLs can be simulated without extensive mathematical calculations.
3. The macromodel will have "drop-in" capability so that various APLL applications can be verified and simulated.
4. Using the macromodel, the transition behavior from the unlocked to locked state, or vice versa, will be predicted and simulated. This means that the APLL macromodel is valid in both the unlocked and locked regions.
5. The key performance parameters such as capture range, lock range, lock-up time, overshoot, loop damping factor, and critical input specifications will be predicted by the macromodel.

6. The APLL macromodel will be kept simple in order to guarantee reasonable computer run times with sufficient accuracy. In general, a target accuracy of ±5% will be sought for general macromodel evaluations. Specifications will be tightened on key APLL applications to guarantee ±1% target accuracies in order to determine computer run time requirements.

Phase detector. In APLLs, the phase detector is almost always realized as a balanced modulator because this circuit block has a differential configuration readily suited to monolithic fabrication. The transfer properties of balanced modulators can be either linear or nonlinear. A simplified circuit configuration of the switched-type phase detector using the balanced-modulator structure is shown in Figure 4.1(a). This type essentially functions as a synchronous switch, chopping the input signal in synchronization with the VCO signal. It is also referred to as a multiplier-type phase detector because its basic mathematical operation is the multiplication of two periodic input signals. The output frequency is made up of the sum and difference components of the input frequencies.

When the two input signals have the same frequency but some phase difference, the phase detector provides a DC output voltage that is proportional to the phase difference of two periodic input signals. Thus, the transfer characteristic between the input phase difference and the output DC voltage is referred to as the conversion gain, K_d, whose units are V/radian. When both input signals are assumed to be large enough to switch all transistors in Figure 4.1(a) fully on and off, the phase-to-voltage transfer characteristics can be calculated from the waveform shown in Figure 4.1(b). The average value of the output is given by the difference between areas A_1 and A_2, or, mathematically,

$$\begin{aligned}(V_d)_{av} &= -\frac{(\text{Area } A_1 - \text{Area } A_2)}{\pi} \\ &= I_{EE} \cdot R_L \cdot \left(\frac{2 \cdot \theta_d}{\pi} - 1\right) \\ &= \frac{2 \cdot I_{EE} \cdot R_L}{\pi} \cdot \theta_d - I_{EE} \cdot R_L \\ &= K_d \cdot \theta_d - I_{EE} \cdot R_L\end{aligned} \tag{4.1}$$

As Equation (4.1) shows, the average output voltage is both linearly proportional to θ_d and insensitive to the amplitude of the input as long as the input signal is sufficiently large to switch the transistors fully on and off. The waveforms in Figure 4.1(b) show that the fully switched balanced-modulator circuit has two output levels: low when both inputs have the same polarity and high when both input signals have different polarities. The performance characteristics to be modeled for the phase detector macromodel include:

1. Insensitivity to the amplitude of the input signal
2. Multiplication of two periodic signals
3. Linear conversion gain
4. An optional sinusoidal characteristic of the conversion gain

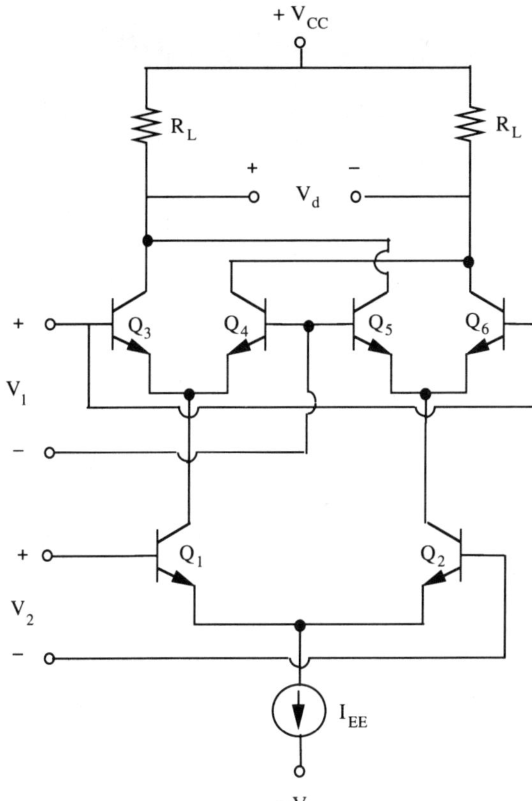

Figure 4.1 A simplified balanced modulator. (a) Simplified circuit schematic diagram. (b) Timing diagram of input and output waveforms.

Voltage controlled oscillator. The voltage controlled oscillator is the most critical block within the monolithic APLL system. The PLL center frequency and the frequency discrimination properties are set by the VCO performance parameters. The VCO produces an output signal whose frequency is linearly proportional to a controlling DC input voltage. The proportionality coefficient is called the conversion gain, K_v, having units of radian/s/V. However, a limited dynamic range exists in practical VCOs because the output frequency becomes saturated to maximum and minimum values at certain input voltage levels. The performance characteristics to be modeled for the VCO macromodel are:

1. Adjustable free-running frequency by external timing resistors and capacitors
2. Triangle, rectangular, and sine wave outputs
3. Frequency saturation property (maximum and minimum frequencies)
4. Voltage-to-frequency conversion gain
5. Amplitude control of each output waveshape

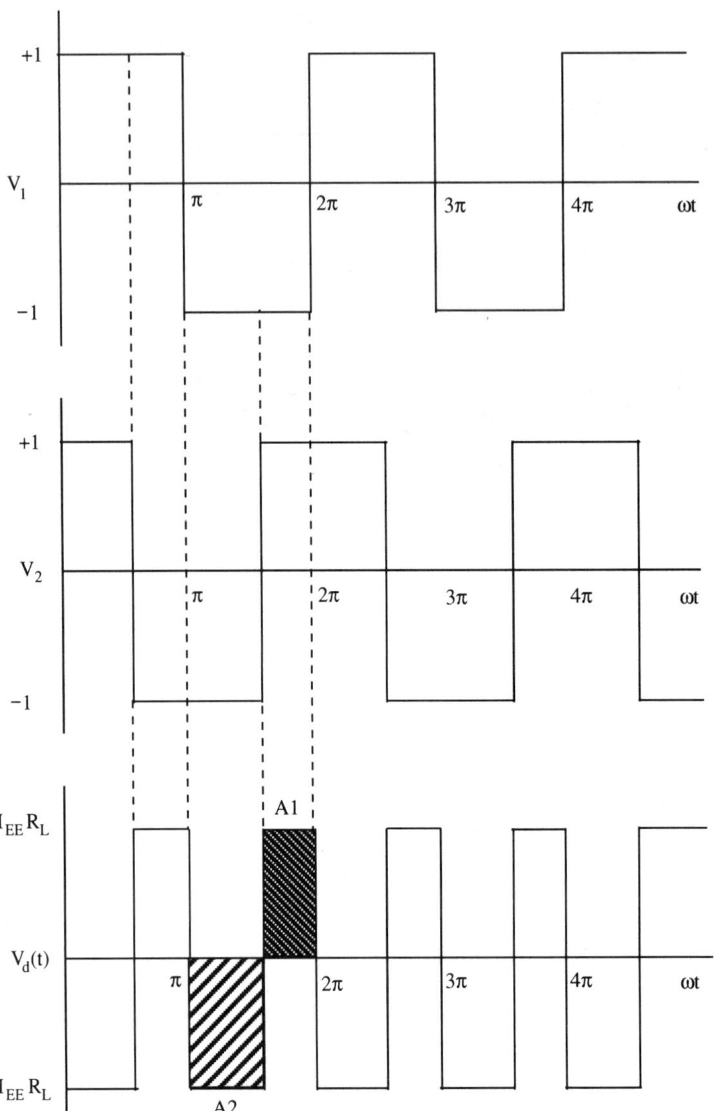

Figure 4.1(b)

4.3 CHOOSING THE DROP-IN MODELS

Phase detector macromodel. Figure 4.2 illustrates a functional block diagram for the general balanced-modulator type phase detector macromodel developed from the performance characteristics. The functional block diagram contains a sine-to-square-wave converter (SSqWC), two analog multipliers (M-I and M-II), and a triangle-to-sine-wave converter (TSWC).

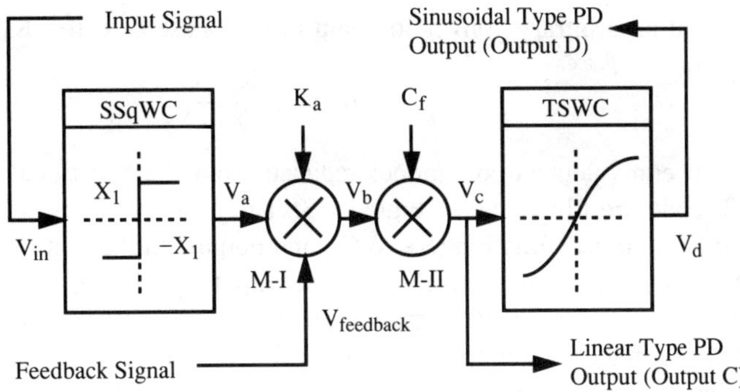

Figure 4.2 A functional block diagram for a balanced-modulator type phase detector macromodel.

The SSqWC block produces a rectangular output wave with a fixed amplitude ($\pm X_1$ V) regardless of the amplitude of the input signal. The analog multiplication function is realized by the multiplier block M-I having two inputs: the output signal of the SSqWC and the feedback signal from the VCO. The multiplication factor K_a sets an arbitrary conversion gain of the phase detector. The other constant C_f is a correction factor and is used to reduce the error between an actual phase detector and its macromodel. This C_f factor is set to unity for the initial simulation, and then its value is determined after comparing measured data and macromodel simulation data. Output C is for the linear type phase detector. When the conversion gain of the phase detector has a sinusoidal transfer characteristic, the TSWC block is added. This TSWC block converts the linear type phase detector model to a nonlinear type macromodel. Therefore, the voltage at the output D is proportional to $\cos\theta_d$ (nonlinear type), while the voltage at output C is proportional to θ_d (linear type).

Now consider the transfer characteristics of the functional block diagram. For initial conditions, assume that the input signal is a sine wave with arbitrary amplitude but larger than the threshold voltage set by the SSqWC. The feedback signal is a square wave with a $\pm X_2$ V amplitude. Under these initial conditions, the waveform at the output of the SSqWC, V_a, is a rectangular wave with a 50% duty cycle and an amplitude fixed to X_1 V. The multiplication blocks M-I and M-II produce rectangular waves whose pulse width is modulated by the phase difference. Thus the average voltage at the branch point C is given by

$$\begin{aligned}(V_c)_{av} &= K_a \cdot X_1 \cdot X_2 \cdot C_f \cdot \left(\frac{2\cdot\theta_d}{\pi} - 1\right) \\ &= \frac{2\cdot K_a \cdot X_1 \cdot X_2}{\pi} \cdot \theta_d - K_a \cdot X_1 \cdot X_2 \qquad (4.2) \\ &= K_d \cdot \theta_d - K_a \cdot X_1 \cdot X_2\end{aligned}$$

where K_a is an arbitrary constant and C_f is unity. Thus the average voltage at the output of the M-II is linearly proportional to the phase difference between the input signals. To implement an arbitrary conversion gain of the phase detector, K_a is defined by

$$K_a = \frac{K_d \cdot \pi}{2 \cdot X_1 \cdot X_2} \tag{4.3}$$

An equivalent circuit model reduced from the functional block diagram and its SPICE code are illustrated in Figure 4.3. The primitive phase detector macromodel has three stages: input, function (SSqWC), and output. In the input stage the resistances R_{in1} and R_{in2} model the input resistances of the input signal port and the feedback signal port, respectively. The equivalent circuit for the SSqWC block is constructed by using the generalized macromodel approach for PWL functions described in Chapter 2. It consists of a two-dimensional polynomial dependent VCVS, E_1 (whose controlling voltages are the voltage across R_{in1} and the voltage at node 5, the output of the SSqWC), resistor R_1, capacitor C_1, two diodes D_1 and D_2, and two independent voltage sources V_1 and V_2.

SPICE code for Figure 4.3

```
RIN1 1 3 [A]                    ; [A] = Input resistance I (Input signal)
RIN2 2 3 [B]                    ; [B] = Input resistance II (Feedback signal)
E1 4 0 POLY(2) 1 3 5 0
+ 0 10MEG -1E-11
R1 4 5 1T
C1 5 0 1E-15
D1 5 6 SWITCH
V1 6 0 DC=[X1]                  ; [X1] = Amplitude of square wave
D2 7 5 SWITCH
V2 7 0 DC=[-X1]
EOUT 8 10 POLY(2) 5 0 2 3
+ 0 0 0 0 [C]                   ; [C] = Ka * Cf
R2 9 10 1G
ROUT 8 9 [D]                    ; [D] = Output resistance
.MODEL SWITCH    (N = .001)
```

The functional behavior of the SSqWC model will be described later, because the SSqWC model is derived from the Schmitt trigger model, with some slight modifications.

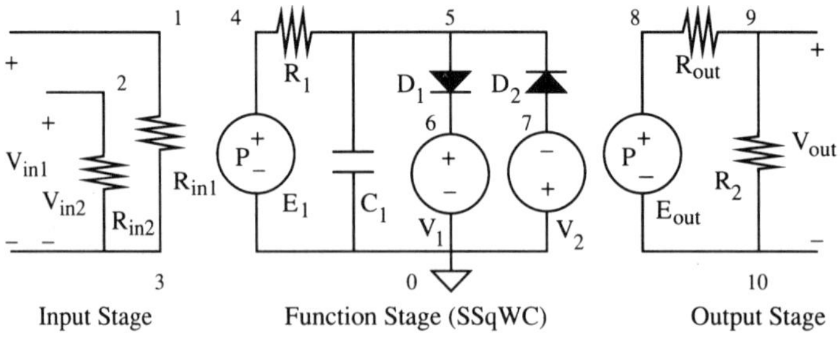

Figure 4.3 An equivalent circuit structure macromodel for balanced-modulator type phase detector.

The amplitude of the SSqWC output is determined by the voltage levels of V_1 (= X_1) and V_2 (= $-X_1$). For example, if the voltage at node 4 is larger than X_1, the diode D_1 is "ON" while D_2 is "OFF," fixing the voltage at node 5 to X_1 V. If the voltage at node 4 is less than $-X_1$, the diode D_1 is "OFF" and D_2 is "ON." Thus, the voltage at node 5 becomes $-X_1$ V. When the node 4 voltage is between $-X_1$ and X_1 V, both diodes D_1 and D_2 are "OFF." Then the output of the SSqWC is linearly proportional to the input voltage, but the waveform is not square. In order to avoid this situation, the gain of E_1 is made very large so that the voltage at node 4 switches sharply from X_1 to $-X_1$ and vice versa. Therefore, the SSqWC output V(5) is equivalent to V_a in Figure 4.2.

The multiplier blocks M-I and M-II are realized through a two-dimensional polynomial VCVS, E_{out}, in the output stage. Its controlling voltages are the voltage at node 5 (the output of the SSqWC) and the voltage across R_{in2}, the feedback signal from the VCO. The voltage at node 8 in Figure 4.3(a) is equivalent to V_c in Figure 4.2 and is given by

$$\begin{aligned} V(8,10) &= P_0 + P_1 \cdot V(5) + P_2 \cdot V(2,3) + P_3 \cdot V(5)^2 + P_4 \cdot V(5) \cdot V(2,3) + \cdots \\ V(8,10) &= K_a \cdot C_f \cdot V_a \cdot V_{feedback} \\ V(8,10) &= V_c \end{aligned} \tag{4.4}$$

Therefore, the coefficients of the polynomial dependent VCVS, E_{out}, are defined by

$$\begin{aligned} &P_0 = 0,\ P_1 = 0,\ P_2 = 0,\ P_3 = 0,\ P_4 = K_a \cdot C_f \\ &P_n = 0\ (n \geq 5) \end{aligned} \tag{4.5}$$

Consider an example phase detector system which has the following electrical parameters:

- Linear type phase detector
- Conversion gain = 0.66 V/rad
- Input resistance I = 1 MΩ
- Input resistance II = 1 MΩ
- Output resistance = 1 KΩ
- Input signal = $\sin(2\pi \cdot 2{,}500 \cdot t + 90°)$ [Figure 4.4(a)]
- Feedback signal = 2.5 KHz square wave with 10 V amplitude [Figure 4.4(b)]

The element values of this macromodel as shown in Figure 4.3 are determined by

- R_{in1} = [A] = 1 MΩ
- R_{in2} = [B] = 1 MΩ
- V_1 = [X_1] = 10V (recommended value)
- Coefficient P_4 in $E_{out} = K_a \cdot C_f = \frac{K_d \cdot \pi}{2 \cdot X_1 \cdot X_2} = \frac{0.66 \cdot \pi}{2 \cdot 10 \cdot 10} \approx 1.0367E-2$
- R_{out} = [D] = 1 KΩ

The phase difference between these two signals is variable. Regardless of θ_d, the duty-cycle of the SSqWC output wave is almost 50%, and its amplitude is fixed to 10 V peak as shown in Figure 4.4(c). Figure 4.4(d) illustrates the output of the linear type phase

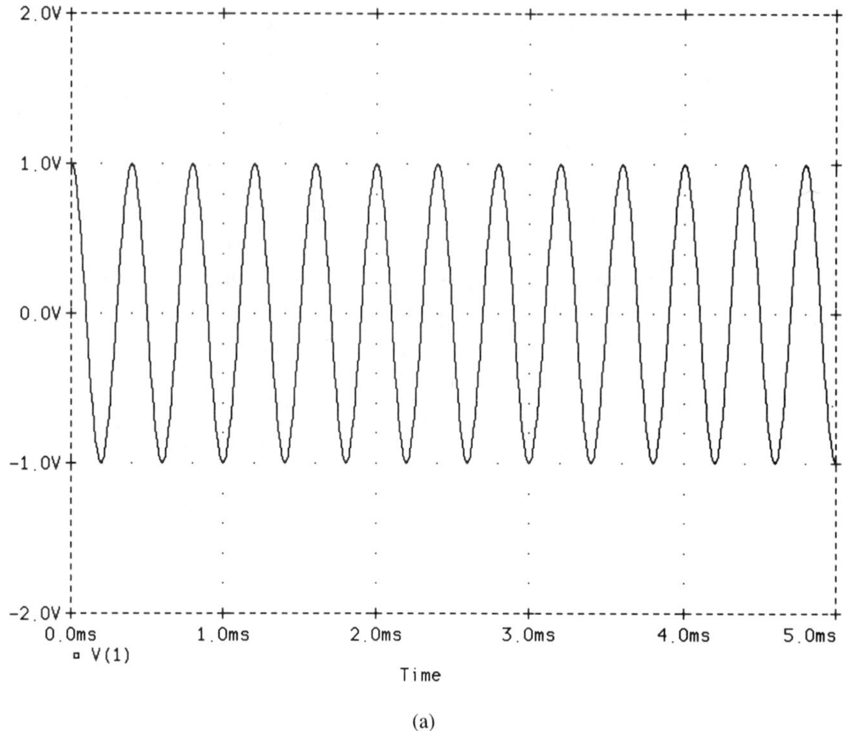

(a)

Figure 4.4 PSpice simulation results of the phase detector macromodel. (a) Input signal (at node 1). (b) Feedback signal (at node 2). (c) Output waveform of SSqWC (at node 5). (d) Output of linear phase detector macromodel (at node 9).

detector macromodel when the input phase difference is 90 degrees. The average value in the steady state has some output offset voltage because of an imperfect square wave at the SSqWC output and the feedback signal. This error offset voltage can be controlled by adjusting C_f. Figure 4.5 shows the DC voltage between nodes 9 and 10 when the phase differences are 0, 45, 90, 135, and 180 degrees. The DC voltage decreases linearly for increasing phase difference. Thus, the DC voltage is largest when $\theta_d = 0$, zero when $\theta_d = \pi/2$, and smallest when $\theta_d = \pi$.

Voltage controlled oscillator macromodel. Figure 4.6 shows a functional block diagram of a VCO macromodel whose functional behavior is similar to a voltage controlled bistable multivibrator. It is composed of a voltage limiter, integrator, Schmitt trigger, summer (S-I), and multipliers (M-I, M-II, and M-III). The transfer function of the voltage limiter block includes two linear segments whose slopes are different and are centered about a threshold voltage, V_x. If the input controlling voltage, V_{in}, is less than V_x, the output voltage V_a is identical to the input voltage since the slope of the first linear segment is defined as unity. When the input voltage is larger than V_x, the output voltage is reduced because the slope of the second linear segment is less than

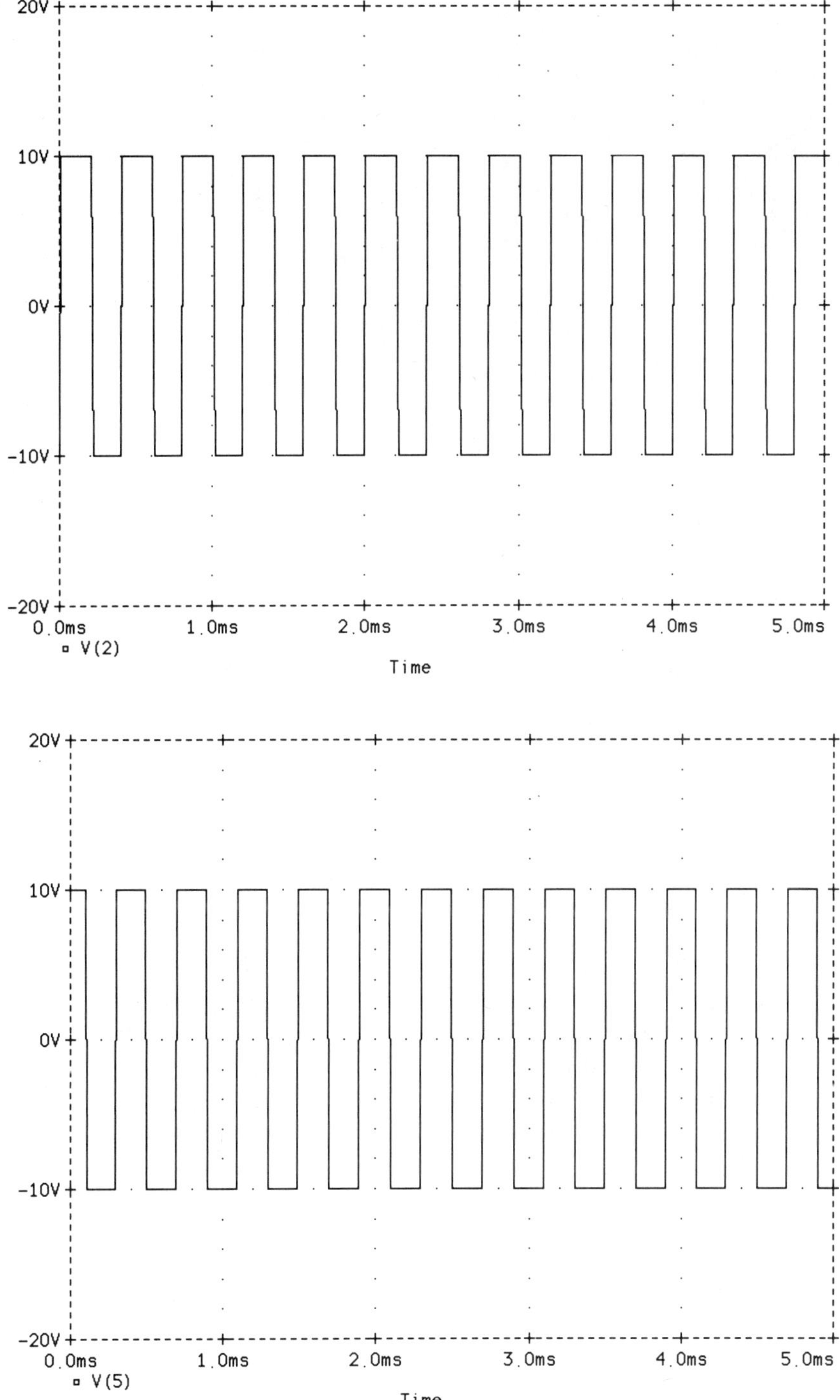

Figure 4.4(b) and (c)

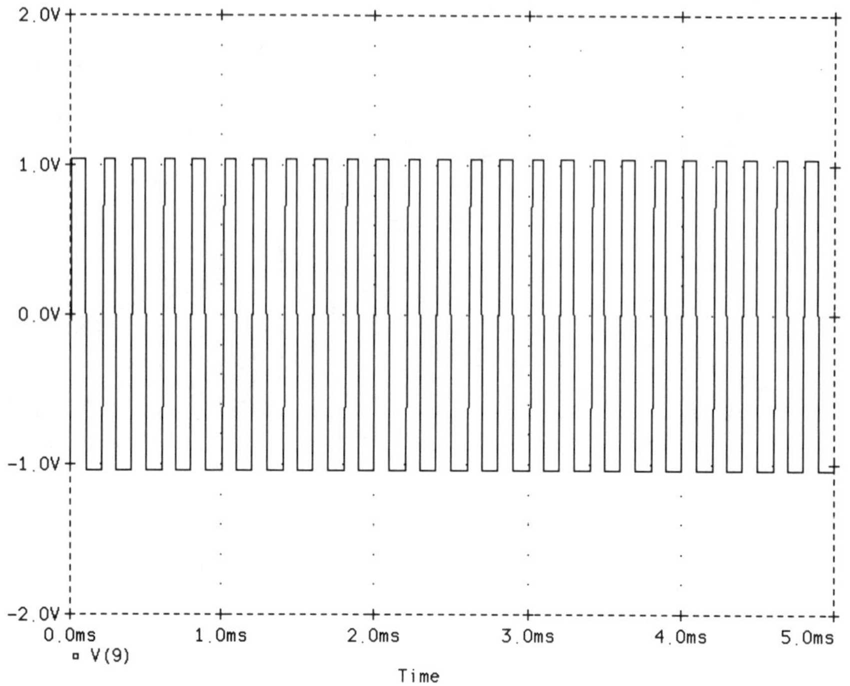

Figure 4.4(d)

unity. Thus, the output of the voltage limiter V_a has two operational regions—normal and saturated. Eventually, these two regions are made to cover the dynamic range of the VCO. The inverting integrator functional block with a time constant τ creates a triangle wave by integrating the DC input controlling voltage, V_e. The voltage V_e is produced from the limiter output, V_a, constants K_b and K_c, and the Schmitt trigger output V_f, and by adding V_{free}. The Schmitt trigger block controls not only the peak-to-peak voltage of the output of the integrator, V_{out1}, but also produces a rectangular wave V_{out2}, with an amplitude X_2 V. Whenever the voltage level of the triangle wave reaches the upper trigger point (V_p) and lower trigger point (V_n) values, the output voltage of the Schmitt trigger toggles from the high (X_2) to the low level ($-X_2$), or vice versa. Thus it changes the polarity of the input signal, which drives the integrator.

Now consider the mathematical characteristics of each functional block in the VCO macromodel. First assume that the VCO macromodel works in the normal operation region; that is, the input voltage is less than the threshold voltage V_x. The amplitude of the Schmitt trigger output is defined as X_2 V, and the lowest peak voltage of the triangle wave is V_n for the initial conditions. As mentioned earlier, the output voltage of the voltage limiter model is identical to the input voltage in the normal operation region.

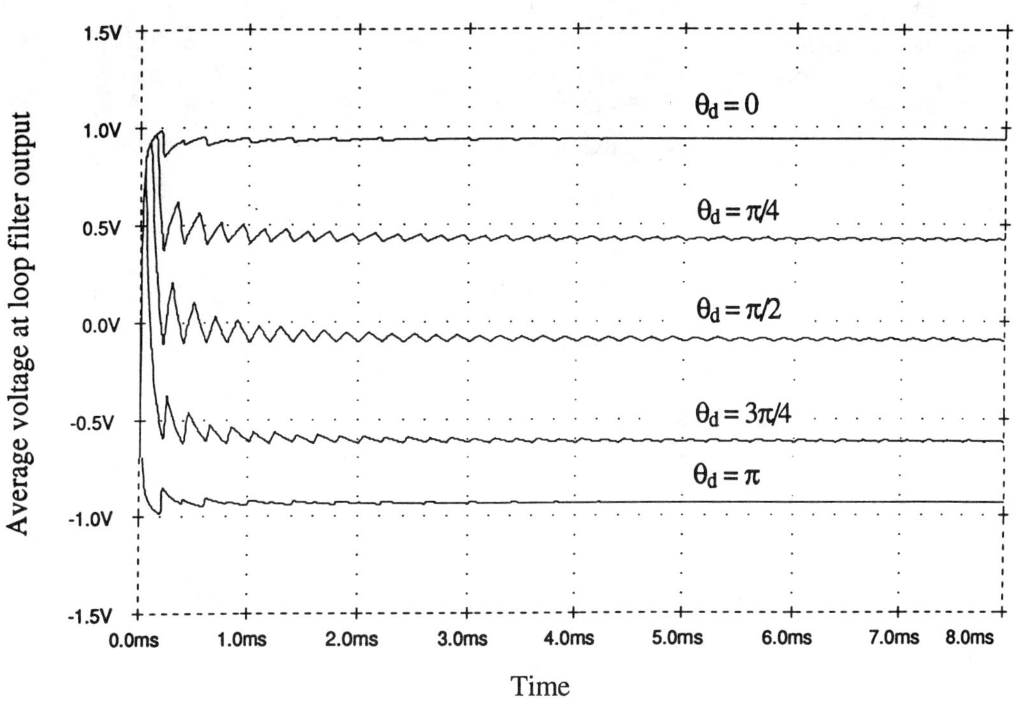

Figure 4.5 DC output voltages of the phase detector macromodel for different phase errors.

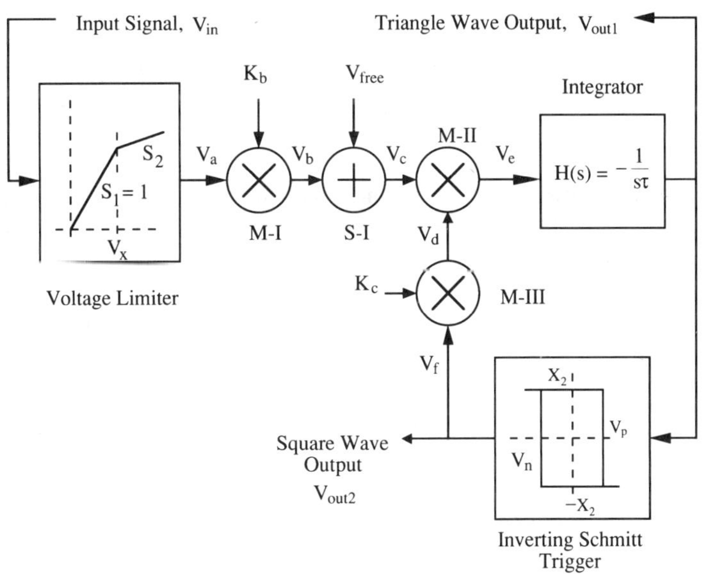

Figure 4.6 A functional block diagram for the voltage controlled oscillator macromodel.

Thus,

$$\begin{aligned}
V_a &= S_1 \cdot V_{in} = V_{in} \\
V_b &= K_b \cdot V_a = K_b \cdot V_{in} \\
V_c &= V_b + V_{free} = K_b \cdot V_{in} + V_{free} \\
V_d &= K_c \cdot V_f = K_c \cdot X_2 \\
V_e &= V_c \cdot V_d = (K_b \cdot V_{in} + V_{free}) \cdot (K_c \cdot X_2) \text{ if } K_c = -1/X_2 \\
&= -(K_b \cdot V_{in} + V_{free})
\end{aligned} \tag{4.6}$$

The output voltage of the integrator block in the continuous time domain is given by

$$\begin{aligned}
V_{out}(t) &= -\frac{1}{\tau}\int V_e dt \\
V_{out}(t) &= \frac{1}{\tau}\int (K_b \cdot V_{in} + V_{free}) dt \\
V_{out}(t) &= \frac{K_b \cdot V_{in} + V_{free}}{\tau} \cdot t + C
\end{aligned} \tag{4.7}$$

Since the initial conditions are already defined as

$$\text{Output of the integrator } V_{out}(0) = V_n V$$

$$\text{Output of the Schmitt trigger } V_f(0) = X_2 V$$

the output voltage level of the integrator increases for increasing time. At t = T/2, the output of the integrator reaches the peak voltage V_p, where T is the period of triangle wave. Thus

$$V_{out}\left(\frac{T}{2}\right) - V_{out}(0) = V_p - V_n$$

$$V_{out}\left(\frac{T}{2}\right) - V_{out}(0) = \frac{K_b \cdot V_{in} + V_{free}}{\tau} \cdot t + C\Big|_{t=0}^{t=T/2}$$

$$V_{out}\left(\frac{T}{2}\right) - V_{out}(0) = \frac{K_b \cdot V_{in} + V_{free}}{\tau} \cdot \frac{T}{2}$$

$$T = \frac{2 \cdot (V_p - V_n) \cdot \tau}{K_b \cdot V_{in} + V_{free}} \tag{4.8}$$

Since frequency is reciprocal of the period of the signal, the output frequency of the integrator model can be expressed by

$$f_{out} = \frac{1}{T} = \frac{K_b \cdot V_{in} + V_{free}}{2 \cdot (V_p - V_n) \cdot \tau} \text{ (Hz)}$$

$$f_{out} = \frac{1}{T} = \frac{K_b}{2 \cdot (V_p - V_n) \cdot \tau} \cdot V_{in} + \frac{V_{free}}{2 \cdot (V_p - V_n) \cdot \tau} \text{ (Hz)} \tag{4.9}$$

$$f_{out} = \frac{1}{T} = K_{vnor.} \cdot V_{in} + f_0' \text{ (Hz)}$$

Equation (4.9) shows that the output frequency of the VCO macromodel is linearly proportional to the input controlling voltage with coefficient, K_{vnor}. Since the free-running frequency, f_0', is defined as the output frequency for zero input controlling voltage, the constant term in Equation (4.9) is the free-running frequency of the VCO macromodel. Thus

$$\text{Free-running frequency } f_0' = \frac{V_{free}}{2 \cdot (V_p - V_n) \cdot \tau} \text{ (Hz)} \tag{4.10}$$

The conversion gain of the VCO is the deviation of the output frequency with respect to the input controlling voltage (units of rad/s/V). Thus the slope of Equation (4.9) is the conversion gain and is given by

$$\text{Conversion gain } K_{vnor.} = \frac{2 \cdot \pi \cdot K_b}{2 \cdot (V_p - V_n) \cdot \tau} \text{ (rad/s/V)} \tag{4.11}$$

where K_b is an arbitrary constant to adjust the value of the conversion gain.

Now consider how the frequency saturation characteristic of an actual VCO is realized in the macromodel. Figure 4.7 shows a simplified nonlinear characteristic of a common VCO device. The mathematical expression of the output frequency is

$$f_{out} = \begin{cases} K_{vnor.} \cdot V_{in} + f_0' & \text{when } V_{in} \leq V_x \\ K_{vsat.} \cdot (V_{in} - V_x) + f_0 & \text{when } V_{in} > V_x \end{cases} \tag{4.12}$$

where

$$f_0 = K_{vnor.} \cdot V_x + f_0'$$

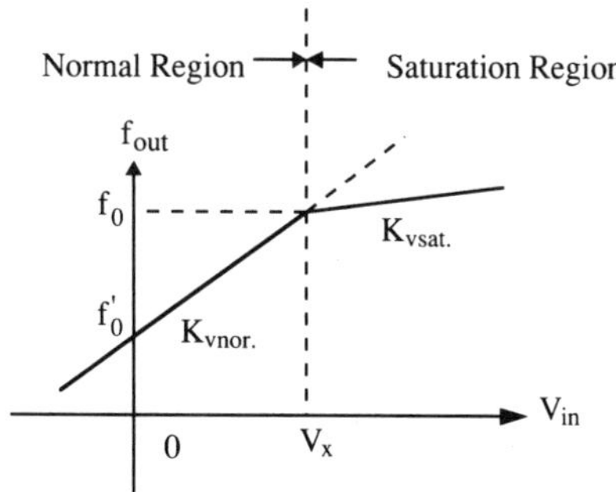

Figure 4.7 Simplified nonlinear characteristics of a practical VCO.

The output frequency in the saturation operation region is alternately described by

$$f_{out} = K_{vsat} \cdot (V_{in} - V_x) + K_{vnor} \cdot V_x + f_0'$$

$$f_{out} = K_{vsat} \cdot V_{in} + K_{vnor} \cdot \left(1 - \frac{K_{vsat.}}{K_{vnor}}\right) \cdot V_x + f_0' \tag{4.13}$$

In the normal operation region the output frequency of the VCO macromodel is set by Equation (4.9). Thus, V_a in Figure 4.6 is identical to the input voltage V_{in}, since the slope of the first linear segment in the voltage limiter block is unity. However, when V_{in} is larger than V_x, the output voltage of the voltage limiter block is given by

$$V_a = S_2 \cdot (V_{in} - V_x) + V_x \tag{4.14}$$

Under this situation, the output frequency of the VCO macromodel is

$$f_{out} = \left.\frac{K_b \cdot V_a + V_{free}}{2 \cdot (V_p - V_n) \cdot \tau}\right|_{V_a = S_2 \cdot (V_{in} - V_x) + V_x} \quad \text{(Hz)}$$

$$f_{out} = \frac{K_b \cdot S_2}{2 \cdot (V_p - V_n) \cdot \tau} \cdot V_{in} + \frac{K_b \cdot (1 - S_2)}{2 \cdot (V_p - V_n) \cdot \tau} \cdot V_x + \frac{V_{free}}{2 \cdot (V_p - V_n) \cdot \tau} \quad \text{(Hz)} \tag{4.15}$$

$$f_{out} = S_2 \cdot K_{vnor} \cdot V_{in} + (1 - S_2) \cdot K_{vnor} \cdot V_x + f_0' \text{ (Hz)}$$

Comparing Equation (4.13) and Equation (4.15) gives

$$1 - S_2 = 1 - \frac{K_{vsat}}{K_{vnor}}$$

$$\therefore\ S_2 = \frac{K_{vsat}}{K_{vnor}} \tag{4.16}$$

Therefore, the arbitrary nonlinear VCO conversion gain can be modeled by defining the slope of the second linear segment in the voltage limiter block.

The functional block diagram is transformed to a SPICE equivalent circuit structure macromodel in Figure 4.8. The model consists of three subfunctional blocks, the input voltage limiter, the integrator, and the Schmitt trigger. In the input and the voltage limiter part, the resistor R_{in} represents the input resistance of the VCO. The voltage limiter model is created by using the piecewise linear approximation approach described in Chapter 2. It consists of linear dependent sources E_1 and H_1; a dummy voltage source V_{TEST}; resistors R_1, R_2, and R_3; diode D_1; and independent voltage sources V_1 and V_2. The controlling node voltage of E_1 is the voltage across R_{in}. If E_1 has unity gain, the voltage at node 2 is identical to the input voltage. The dummy voltage source V_{TEST}, defined with DC = 0 and AC = 0, is used to sense the output current of E_1. The output current of E_1 depends on the status of the diode D_1 whose V-I characteristic is nearly ideal. For example, when the voltage at node 3 is less than V_2, diode D_1 is "ON." Thus, the current through the V_{TEST} becomes

$$I(V_{TEST}) = \left.\frac{V(3) - V_1}{R_1} + \frac{V(3) - V_2}{R_2}\right|_{V(3) = V_{in}}$$

when gain of E_1 is unity and $V_{TEST} = 0$.

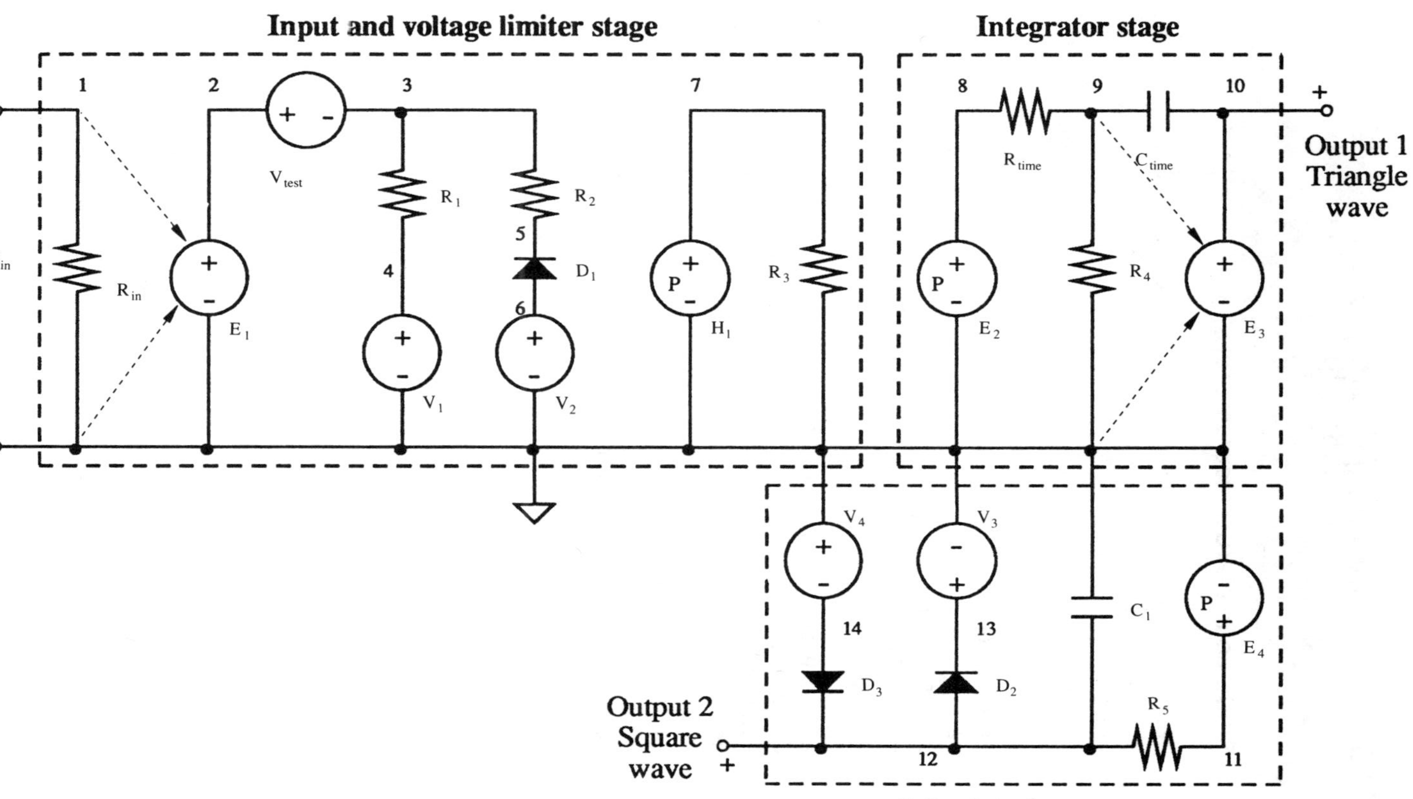

Figure 4.8 An equivalent circuit structure macromodel of a VCO.

$$I(V_{TEST}) = \left(\frac{1}{R_1} + \frac{1}{R_2}\right) \cdot V_{in} - \left(\frac{V_1}{R_1} + \frac{V_2}{R_2}\right) \tag{4.17}$$

SPICE code of VCO macromodel

```
RIN 1 0 [A]                    ; [A] = Input resistance
E1 2 0 1 0 1
VTEST 2 3 DC=0 AC=0
R1 3 4 [B]                     ; [B] = 1/[S1 - (1/R2)]
V1 4 0 DC=[C]                  ; [C] = -(R1/R2) * V2
R2 5 3 [D]                     ; [D] = 1/(S1 - S2)
D1 6 5 DIODE
V2 6 0 DC=[E]                  ; [E] = Threshold voltage Vx
H1 7 0 POLY(1) VTEST
+ [F] [G]                      ; [F] = Vfree, [G] = Kb
R3 7 0 1

E2 8 0 POLY(2) 7 0 12 0
+ 0 0 0 0 [H]                  ; [H] = Kc
RTIME 8 9 [I]                  ; [I] = Timing resistance
R4 9 0 1MEG
CTIME 9 10 [J]                 ; [J] = Timing capacitance
E3 10 0 9 0 100MEG

E4 11 0 POLY(2) 10 0 12 0
+ 0 -10MEG [K]                 ; [K] = Switching voltage
                               ;     = (Vp-p triangle wave/2)

R5 11 12 1T
C1 12 0 1E-15
D2 12 13 DIODE
V3 13 0 DC=[L]                 ; [L] = Amplitude of square
                               ;       wave

D3 14 12 DIODE
V4 14 0 DC=[-L]
.MODEL DIODE D(IS=10U N=0.001)
```

When the voltage at node 3 is larger than V_2, diode D_1 is "OFF" and the current $I(V_{TEST})$ is

$$I(V_{TEST}) = \frac{V_{in} - V_1}{R_1} = \frac{1}{R_1} \cdot V_{in} - \frac{V_1}{R_1} \tag{4.18}$$

Equations (4.17) and (4.18) show that the current through V_{TEST} is mainly affected by V_2, since V_2 acts as the breakpoint. Therefore, the independent voltage source V_2 adjusts the threshold voltage V_x. The operational region of the VCO is divided into the normal and saturation regions at V_2. For instance, assume that the conversion gain in the saturation region is half that in the normal operation region. The value of S_2 is

$$S_2 = \frac{K_{vsat}}{K_{vnor}} = \frac{1}{2} \tag{4.19}$$

If the threshold voltage is defined as 10 V, the resistor R_2 and the independent voltage source V_2 become

$$R_2 = \frac{1}{S_1 - S_2} = 2(\Omega) \tag{4.20}$$

$$V_2 = V_x = \text{ Threshold voltage } = 10(V)$$

To determine R_1 and V_1 values, compare the voltage limiter block in Figure 4.6 with Equation (4.17). Since S_1 is always unity, the slope of the linear equation (4.17) should be

$$S_1 = \frac{1}{R_1} + \frac{1}{R_2}$$

$$\therefore\ R_1 = \frac{R_2}{R_2 - 1} = 2(\Omega) \tag{4.21}$$

When the input voltage is zero, $I(V_{TEST})$ must be zero. Thus

$$0 = \frac{V_1}{R_1} + \frac{V_2}{R_2}$$

$$\therefore\ V_1 = -\frac{R_1}{R_2} \cdot V_x = -10(V) \tag{4.22}$$

The current through V_{TEST} is converted to a voltage by the one-dimensional polynomial dependent CCVS, H_1. Thus, the transfer function of the voltage limiter stage remains as a voltage-to-voltage type converter. The second important function of H_1 is the implementation of the multiplier M-I and the summer S-I in Figure 4.6. If the voltage at node 7 is equivalent to V_c in the functional block diagram, the polynomial equation of H_1 is given by

$$V(7) = P_0 + P_1 \cdot I(V_{TEST}) + P_2 \cdot I(V_{TEST})^2 + \cdots$$

$$V(7) = V_{free} + K_b \cdot V_a$$

$$V(7) = V_c \tag{4.23}$$

The constant K_b in the functional block diagram becomes the transimpedance gain, P_1, in the equivalent circuit model, and V_a is equivalent to $I(V_{TEST})$. Thus, the coefficients of the polynomial dependent CCVS, H_1, are determined as follows:

$$P_0 = V_{free}$$

$$P_1 = K_b$$

$$P_n = 0 \qquad (n \geq 2) \tag{4.24}$$

The two-dimensional polynomial dependent VCVS, E_2, in the integrator block implements the M-II and M-III multipliers. It is controlled by voltages at node 7 and at node 12 and its polynomial equation is given by

$$V(8) = P_0 + P_1 \cdot V(7) + P_2 \cdot V(12) + P_3 \cdot V(7)^2 + P_4 \cdot V(7) \cdot V(12) + \cdots \quad (4.25)$$

where the voltage at node 7 is equivalent to V_c and V(12) is equivalent to V_f in Figure 4.6. Since the voltage at node 8 is identical to V_e,

$$\begin{aligned} V(8) &= P_0 + P_1 \cdot V(7) + P_2 \cdot V(12) + P_3 \cdot V(7)^2 + P_4 \cdot V(7) \cdot V(12) + \cdots \\ V(8) &= V_c \cdot (K_c \cdot V_f) \\ V(8) &= K_c \cdot (V_c \cdot V_f) \\ V(8) &= V_e \end{aligned} \quad (4.26)$$

Thus, the coefficients of the nonlinear dependent VCVS, E_2, are determined by

$$\begin{aligned} P_4 &= K_c \\ P_n &= 0 \ (n = \text{integer}, n \neq 4) \end{aligned} \quad (4.27)$$

The next important functional module is the integrator with the time constant τ. An equivalent circuit model for this block can be derived from the pure analog integrator model as shown in Figure 2.5. The transfer function of the integrator model in the s-domain is given by

$$H(s) = \frac{V_{out}(s)}{V_{in}(s)} = \frac{-K}{s \cdot R_1 \cdot C_1} = -\frac{1}{s \cdot R_{time} \cdot C_{time}} = -\frac{1}{s \cdot \tau} = \frac{V_{10}(s)}{V_8(s)} \quad (4.28)$$

$$\begin{aligned} K &= 1 \\ R_1 &= R_{time} \\ C_1 &= C_{time} \\ \tau &= R_{time} \cdot C_{time} \end{aligned}$$

If resistor R_1 and capacitor C_1 in the integrator model are replaced by the timing element R_{time} and C_{time} respectively, Equation (4.28) is identical to the transfer function of the integrator block. The resistor R_4 (R_4 = 1 MEG) and linear dependent VCVS, E_3 (gain = 100 MEG), actually represent the "ideal" op amp model. Thus, nodes 9 and 10 are equivalent to the inverting input and the output terminals of the "ideal" op amp.

Now consider the Schmitt trigger section. This section is a critical functional block in the VCO macromodel since very sharp voltage transitions occur at every half period of the triangle wave (the output of the integrator model). This situation frequently produces convergence problems because of a small internal time step in the transient analysis. The main function of the Schmitt trigger is that its output changes the polarity of the input of the integrator at every half period of the output frequency and its triggering voltage determines the peak-to-peak voltage of the triangle wave output. It also produces the rectangular wave whose amplitude is controlled by the V_3 and V_4 voltages.

As shown in the Schmitt trigger section in Figure 4.8(a), the equivalent circuit model consists of a two-dimensional nonlinear dependent VCVS, E_4, resistor R_5, capacitor C_1, diodes D_2 and D_3, and independent voltage sources V_3 and V_4. The dependent voltage source, E_4, is controlled by the output voltage of the integrator, V(10), and the output voltage of the Schmitt trigger, V(12). The main function of E_4 is to implement the triggering voltages V_n and V_p. If the voltage gain of E_4 is defined as A_v, the voltage at node 11 is given by

$$V(11) = A_v \cdot (P_0 + P_1 \cdot V(10) + P_2 \cdot V(12) + \cdots) \qquad P_0 = 0,\ P_1 = -1,$$

$$P_n = 0\ (n \geq 3)$$

$$V(11) = A_v \cdot [-V(10) + P_2 \cdot V(12)] \tag{4.29}$$

If the diode D_2 is ideal and initially "ON," V(12) is X_2 V. When V(10) linearly increases from V_n to V_p, V(11) is given by

$$V(11) = A_v \cdot [-V(10) + P_2 \cdot X_2] \tag{4.30}$$

when $V_n < V(10) < V_p$.

In order that the diode D_2 remains "ON," V(11) must be kept as a positive voltage. Thus,

$$-V(10) + P_2 \cdot X_2 > 0 \tag{4.31}$$

when $V_n < V(10) < V_p$.

When V(10) reaches the upper triggering voltage V_p, V(11) must be zero so that the diode D_2 turns "OFF." Thus, we have

$$-V_p + P_2 X_2 = 0 \tag{4.32}$$

The condition to satisfy both Equations (4.31) and (4.32) is

$$P_2 = \frac{V_p}{X_2} \tag{4.33}$$

When V(10) decreases linearly from V_p to V_n, V(11) must stay a negative voltage so that the diode D_2 remains "OFF" and diode D_3 remains "ON." Thus, V(12) remains at X_3 V. When V(11) reaches V_n, the diode D_3 must be "OFF." To satisfy these two conditions, P_2 must be

$$P_2 = \frac{V_n}{X_3} \tag{4.34}$$

As shown in Equations (4.33) and (4.34), the coefficient P_2 is defined as two different values. However, if

$$|V_p| = |V_n| \quad \text{and} \quad |X_2| = |X_3|$$

then

$$P_2 = \frac{|V_p|}{|X_2|} \tag{4.35}$$

The gain of E_4 is another important factor because it makes one of the two diodes always "ON" and the other always "OFF." When V(10) increases linearly from V_n to V_p, the diode D_2 is "ON" while the diode D_3 is "OFF" under the previous initial conditions. Thus the current through R_5 is given by

$$I(R_5) = I(D_2) + I(C_1) = \frac{V(11) - V(12)}{R_5}$$

$$I(R_5) \approx I_S \cdot \exp\left[\frac{V(12) - V(13)}{N \cdot V_T}\right] + s \cdot C_1 \cdot X_2 = \frac{V(11) - X_2}{R_5} \tag{4.36}$$

If C_1 is very small such as a 0.001 pF, the current $I(C_1)$ is negligible. Furthermore, since the V-I characteristic of the diode D_2 is nearly ideal, the exponential term of $I(D_2)$ becomes almost unity because the voltage drop at the diode is very small $[V(12)-V(13) \approx 0]$. Thus the reverse saturation current, I_S, of the diode model becomes

$$I_S = \frac{V(11) - X_2}{R_5} \qquad V(11) = A_v \cdot \{-V(10) + [V_p \cdot V(12)]/X_2\}$$

$$I_S = \frac{A_v}{R_5} \cdot \left[-V(10) + \frac{V_p}{X_2} \cdot V(12)\right] - \frac{X_2}{R_5} \qquad V(12) = X_2 \text{ and } R_5 \text{ is very large}$$

$$I_S \approx \frac{A_v}{R_5} \cdot [V_p - V(10)]$$

$$\therefore\ A_v \approx \frac{I_S \cdot R_5}{V_p - V(10)} \tag{4.37}$$

The RC circuit composed of R_5 and C_1 plus the internal diode resistances has a certain time constant. Thus the rectangular wave output V(12) will have rise and fall times. The time constant of the Schmitt trigger is given by

$$\tau_{\text{Schmitt trigger}} = C_1 \cdot R_5 || R_{D2} || R_{D3} \tag{4.38}$$

where R_{D2} and R_{D3} are the internal resistances of diodes D_2 and D_3, respectively. As shown in Equation (4.38), the time constant mainly depends on the capacitor C_1 if the parallel resistance term is constant. When the C_1 increases, the time constant also increases. This means that the Schmitt trigger model converges more easily at the voltage transition point because the increment of the time constant gives a shallower slope. However, the accuracy of the output frequency is decreased. When the resistor R_5 is extremely large (R_5 = 1 TΩ), the capacitor C_1 is extremely small (C_1 = 0.001 pF), and the diodes D_1 and D_2 are ideal (one of two diodes is always "ON"), the time constant of the Schmitt trigger is

$$\tau_{\text{Schmitt trigger}} \approx 0 \tag{4.39}$$

In this case, the rise and fall times of the square wave output are negligible and the Schmitt trigger macromodel does not have a convergence problem. To simulate the ideal diode model, the emission coefficient N should be small (a value of 0.001 is recommended), resulting in the V-I characteristic of the diode D_2 and D_3 being almost ideal. Thus, the internal resistance of the diodes and the voltage drop across the diodes are virtually zero.

As a practical example, consider a VCO macromodel simulated by PSpice under the following assumptions:

- Free running frequency (f_0') = 2.5 KHz
- Conversion gain ($K_{vnor.}$) = 6,392 rad/s/V≈1017 Hz/V
- Conversion gain ($K_{vsat.}$) = 3,196 rad/s/V≈508.7 Hz/V
- Input threshold voltage = 10 V
- Peak voltage of triangle wave = 1.2 V_{peak}
- Amplitude of square wave = 10 V
- Input resistance = 1 MΩ
- Timing resistor = 12 KΩ
- Timing capacitor = 10 nF

From these initial conditions, the element values of the VCO macromodel shown in Figure 4.8 are calculated by using the equations derived previously. The results are:

- $R_{in} = [A] = 1(M\Omega)$
- $R_2 = [D] = \frac{1}{S_1 - S_2} S_1 = 1,\ S_2 = (K_{vsat.}/K_{vnor.}) = 0.5$
- $R_2 = 2(\Omega)$
- $V_2 = [E] = \text{ Input threshold voltage } = 10\ (V)$
- $R_2 = 2(\Omega)$
- $V_1 = [C] = -\frac{R_1}{R_2} \cdot V_2 = -10(V)$
- $R_{time} = [I] = \text{ Timing resistance } = 12\ (K\Omega)$
- $C_{time} = [J] = \text{ Timing capacitor } = 10\ (nF)$
- Coefficient P_0 in $H_1 = [F] = V_{free} = 2 \cdot V_{p\text{-}p} \cdot R_{time} \cdot C_{time} \cdot f_0' = 1.44$
- Coefficient P_1 in $H_1 = [G] = K_b = \frac{K_{vnor.} \cdot 2V_{p\text{-}p} \cdot R_{time} \cdot C_{time}}{2\pi} = 0.586$
- Coefficient P_4 in $E_2 = [H] = K_c = \frac{1}{\text{Amplitude of square wave}} = -0.1$
- $R_5 = \text{ Very large resistance } = 1\ (T\Omega) \cdots$ recommended value
- $C_1 = \text{ Very small capacitance } = 1\ (fF) \cdots$ recommended value
- $I_S = \text{ Saturation current of diodes } D_2 \text{ and } D_3 = 10\ (\mu A) \cdots$ recommended value
- Voltage gain of $E_4 = I_S {}^* R_5 = 1E7$
- Coefficient P_1 in $E_4 = -A_v = -1E7$
- Coefficient P_2 in $E_4 = [K] = \frac{V_{p\text{-}p}/2}{\text{Amplitude of square wave}} \cdot A_v = 0.12E7$
- $V_3 = [L] = \text{ Amplitude of square wave } = 10(V)$
- $V_4 = [-L] = -V_3 = -10(V)$

Figure 4.9 shows simulation results when the VCO macromodel is in its free-running state. The output frequency of both the triangle and the square wave is exactly

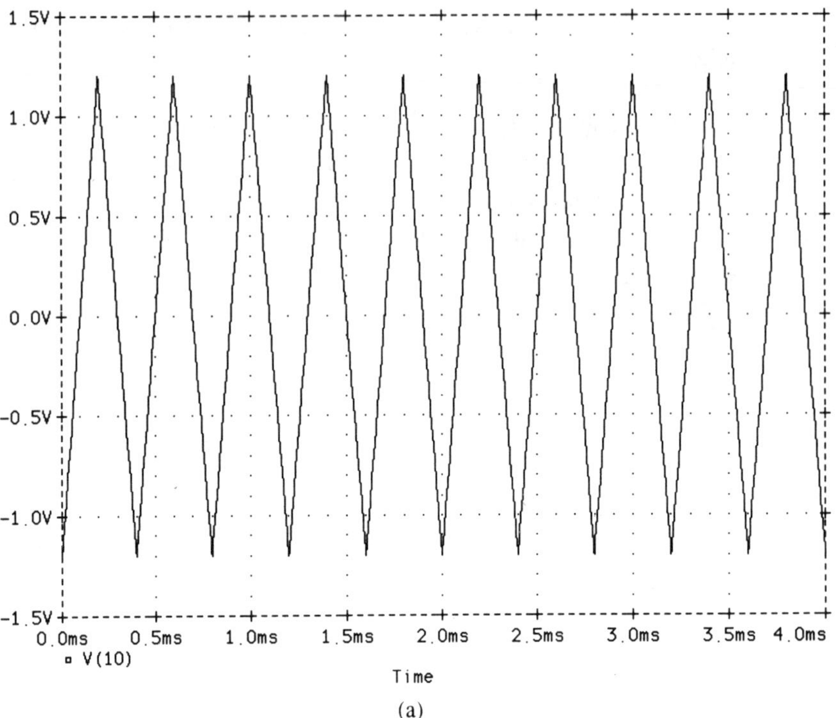

(a)

Figure 4.9 PSpice simulation results of the VCO macromodel. (a) Triangle wave output (free-running state). (b) Square wave output (free-running state). (c) Transfer characteristic of the Schmitt trigger model.

2.5 KHz, the switching voltages of the Schmitt trigger are ± 1.2 V, and the amplitude of the square wave output is 10 V. When the controlling input changes from zero to 5 V at t = 2 msec (step of voltage), the output frequency responds suddenly, changing from the free-running 2.5 KHz to a new frequency, as shown in Figures 4.10(a) and (b). The expected new frequency after 2 ms is

$$f_{out} = K_{vnor} \cdot V_{in} + f_0' \text{ (Hz)}$$

$$f_{out} \approx 1017 \cdot 5 + 2500 = 7585 \text{ (Hz)} \quad (4.40)$$

From the Fourier analysis of Figure 4.10(a), the output frequency of the macromodel changes from 2,500 Hz to 7,500 Hz. Thus, the percentage error is

$$\% \text{ Error of } f_{out} = \frac{7{,}500 - 7{,}585}{7{,}585} \cdot 100(\%) \approx -1.3\% \quad (4.41)$$

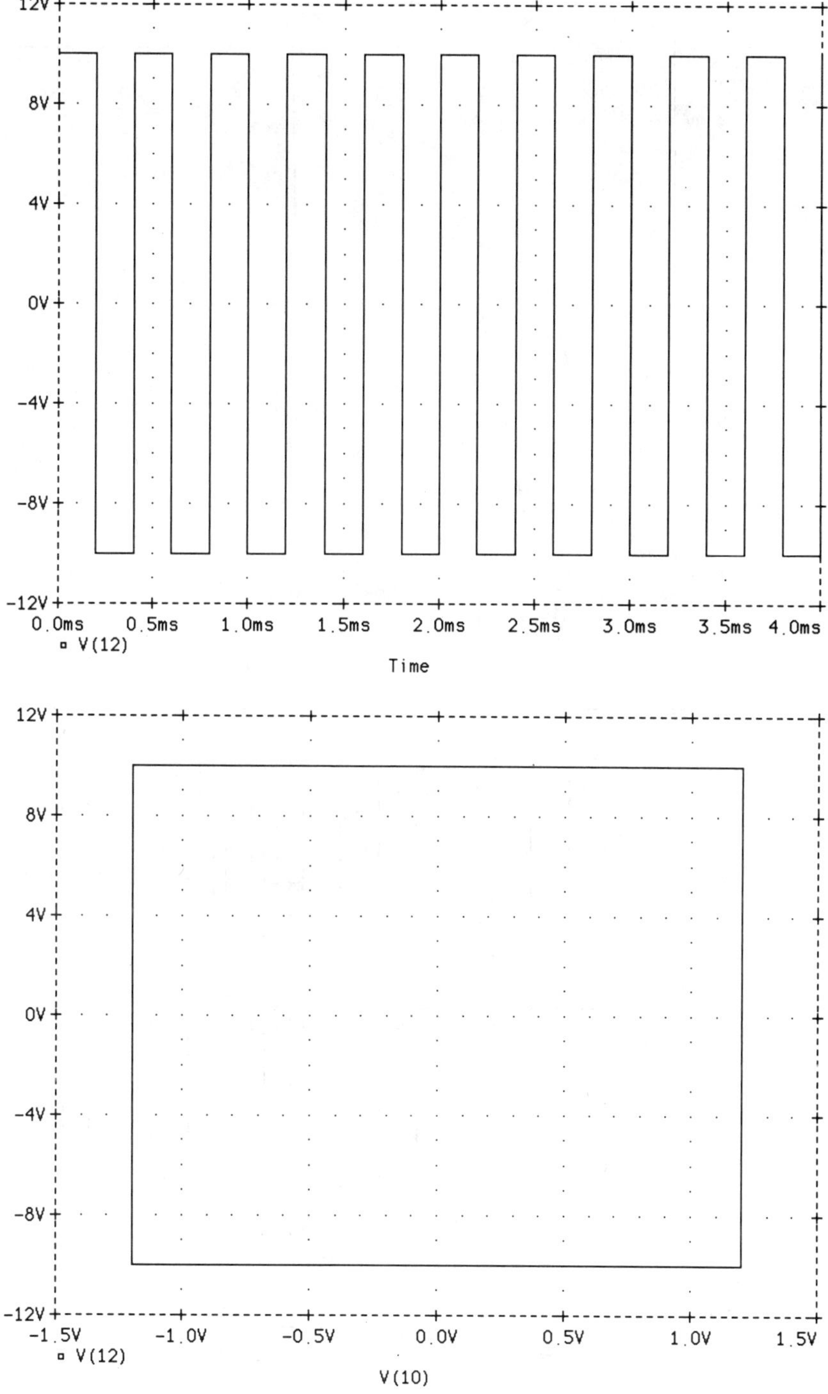

Figure 4.9(b) and (c)

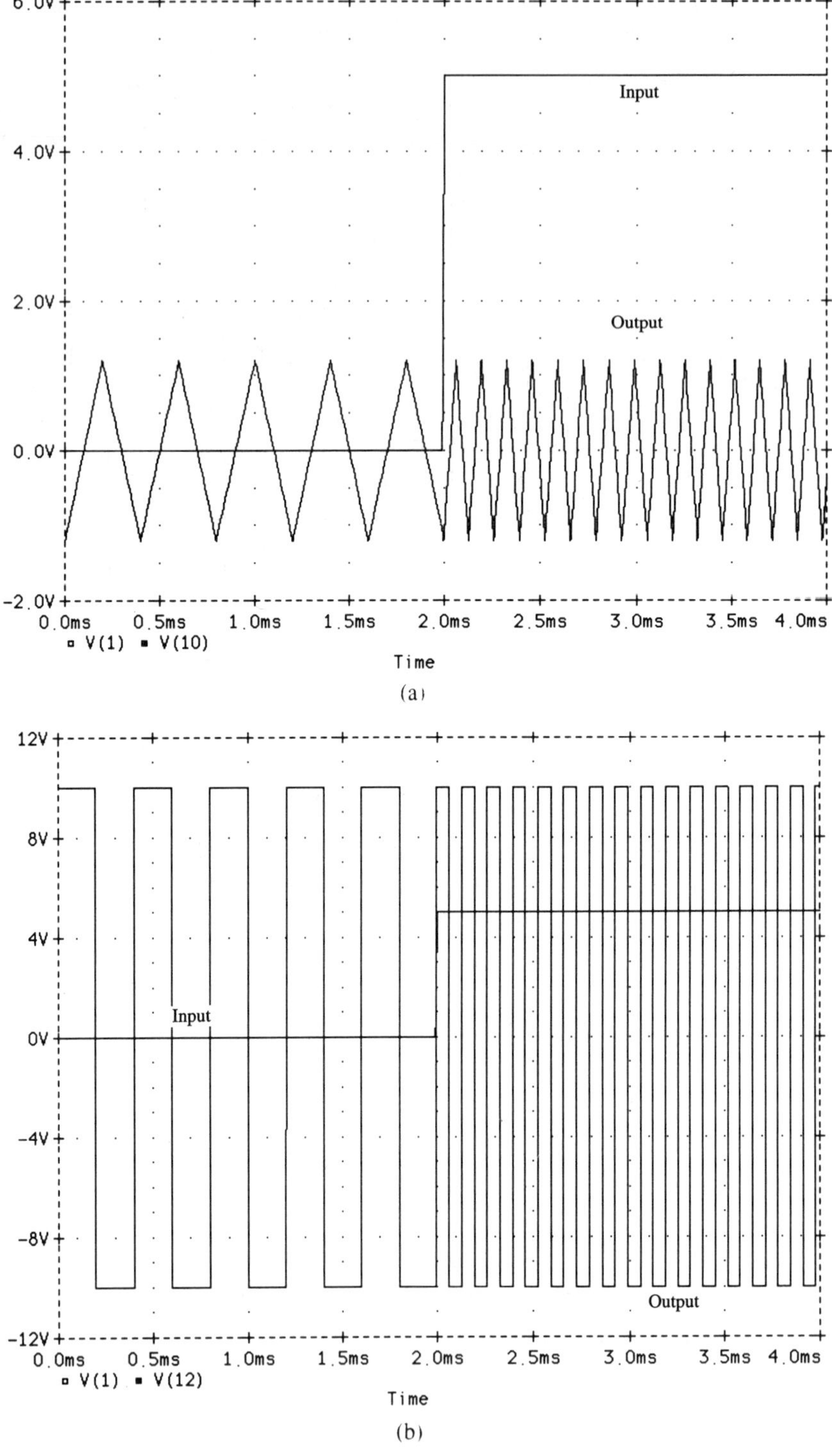

Figure 4.10 Transient responses of the VCO macromodel for a step of voltage input. (a) Triangle wave output. (b) Square wave output.

When the input is a ramp of voltage realized by the PWL source, the output frequency of the VCO macromodel also increases linearly, as shown in Figure 4.11. Thus, the VCO macromodel can be used as a function generator macromodel since the output frequency responds linearly to the input controlling voltage. For simulating the dynamic performance of an APLL by means of the macromodel, a step of frequency and a ramp of frequency inputs are required. However, the SPICE program does not include these types of independent sources. Thus, the VCO macromodel operating as a function generator is used to produce step and ramp frequencies.

Analog phase locked loop macromodel. The phase detector, the low-pass filter circuit, and the VCO models are interconnected to form the complete APLL macromodel, as shown in Figure 4.12. Some simplifications are possible by removing several circuit elements because some components are doubly defined. For example, the output stage of the phase detector macromodel is not necessary for the complete APLL macromodel because it acts as a buffer. Thus, the total number of elements of the APLL macromodel decreases without loss of functionality or accuracy, thereby increasing the computational efficiency.

The new APLL macromodel has eight "external pins" to provide access for the input signal, the triangle and rectangular wave outputs, the timing resistor and capacitor, and any type of loop filter configuration. The feedback path from the VCO to the phase detector is established internally. Each element value can be determined using the equations presented in Table 4.1. Table 4.2 illustrates the setting capability of electrical parameters into the APLL macromodel.

TABLE 4.1 SIMPLE EQUATIONS FOR ELEMENT VALUES OF THE APLL MACROMODEL

[A] = Input resistance of input signal port

$$[B] = \frac{K_d \cdot \pi}{20 \cdot V_{am}} = \frac{K_d \cdot \pi}{20 \cdot [L]}$$

[C] = Resistance in simple RC low-pass filter

[D] = Capacitance in simple RC low-pass filter

$$[E] = 2 \cdot V_{p\text{-}p\ \text{of triangle wave}} \cdot R_{time} \cdot C_{time} \cdot f_0' = 2 \cdot V_{p\text{-}p} \cdot [I] \cdot [J] \cdot f_0'$$

$$[F] = \frac{K_{vnor} \cdot 2 \cdot V_{p\text{-}p} \cdot R_{time} \cdot C_{time}}{2\pi} = \frac{K_{vnor} \cdot 2 \cdot V_{p\text{-}p} \cdot [I] \cdot [J]}{2\pi}$$

$$[G] = \frac{1}{\text{Threshold voltage in VCO input}} = \frac{1}{V_T}$$

$$[H] = -\frac{1}{V_{am}} = -\frac{1}{[L]}$$

[I] = R_{time} = Timing resistance

[J] = C_{time} = Timing capacitance

$$[K] = \frac{V_{p\text{-}p}}{2 \cdot V_{am}} \cdot 10^7 = \frac{V_{p\text{-}p}}{2 \cdot [L]} \cdot 10^7$$

[L] = V_{am} = Amplitude of square wave output

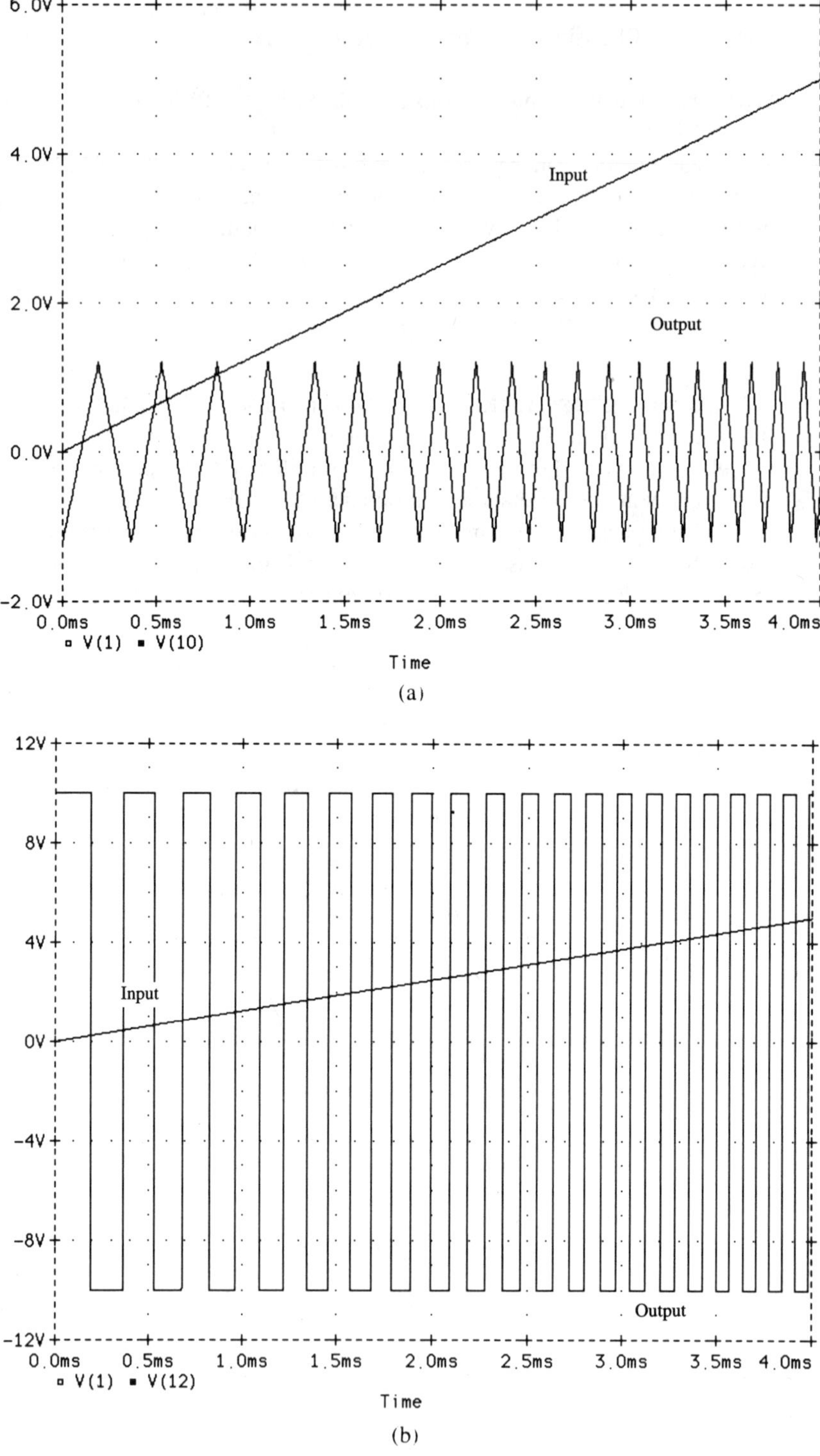

Figure 4.11 Transient response of the VCO macromodel for a ramp of voltage input. (a) Triangle wave output. (b) Square wave output.

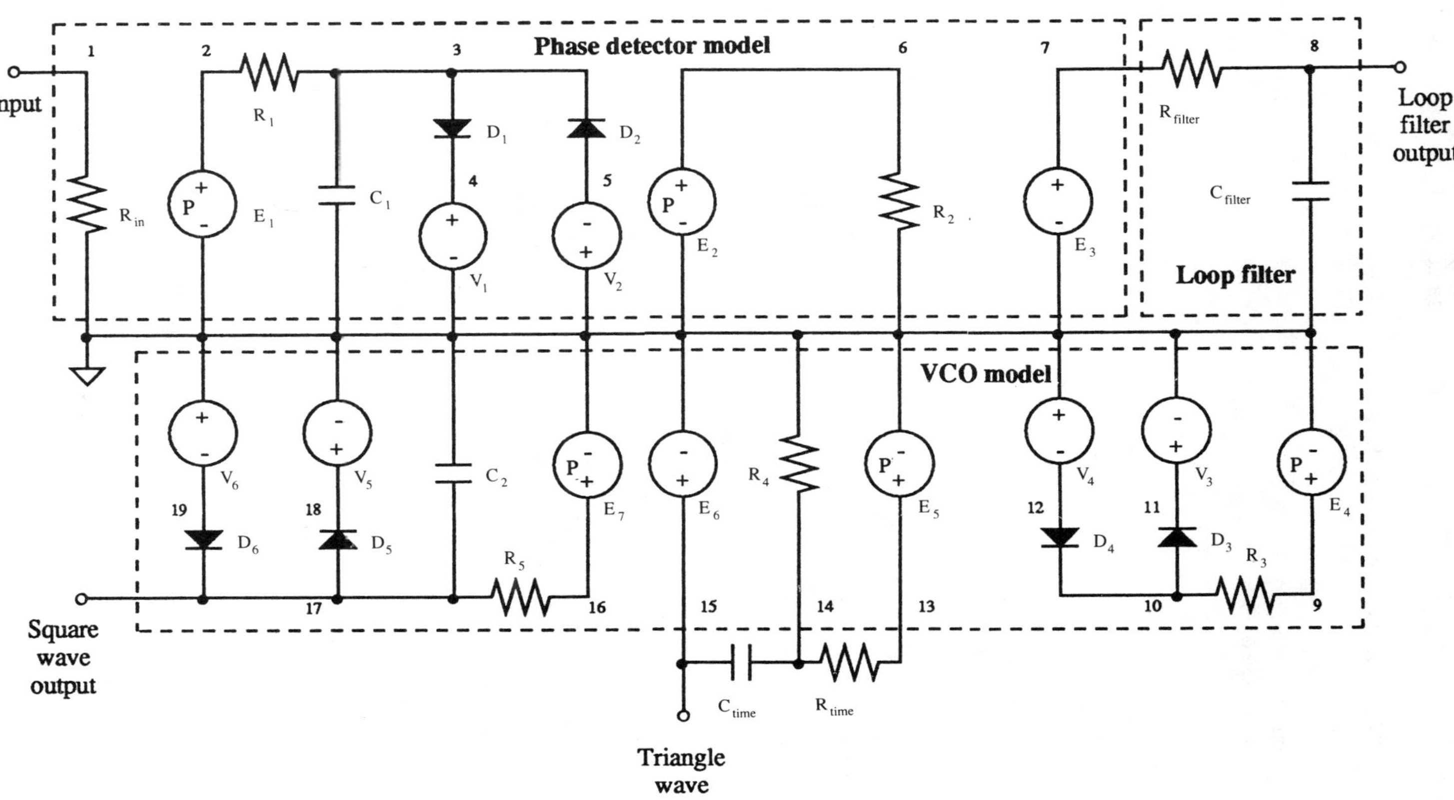

Figure 4.12 An equivalent circuit structure macromodel for analog phase locked loop.

SPICE code for Figure 4.12

```
RIN 1 0 [A]                                  ; [A]
E1 2 0 POLY(2) 1 0 3 0
+ 0 10MEG -1E-11
R1 2 3 1T
C1 3 0 1E-15
D1 3 4 DIODE
V1 4 0 DC=10
D2 5 3 DIODE
V2 5 0 DC=-10
E2 6 0 POLY(2) 3 0 17 0 0 0 0 0 [B]          ; [B]
R2 6 0 1
E3 7 0 6 0 1
RFILTER 7 8 [C]                              ; [C]
CFILTER 8 0 [D]                              ; [D]
E4 9 0 POLY(1) 8 0 [E] [F]                   ; [E], [F]
R3 9 10 1T
D3 10 11 DIODE
V3 11 0 DC=[G]                               ; [G]
D4 12 10 DIODE
V4 12 0 DC=[-G]
E5 13 0 POLY(2) 10 0 17 0
+ 0 0 0 0 [H]                                ; [H]
RTIME 13 14 [I]                              ; [I]
CTIME 14 15 [J]                              ; [J]
R4 14 0 1MEG
E6 15 0 14 0 100MEG
E7 16 0 POLY(2) 15 0 17 0
+ 0 -10MEG [K]                               ; [K]
R5 16 17 1T
C2 17 0 1E-15
D5 17 18 DIODE
V5 18 0 DC=[L]                               ; [L]
D6 19 17 DIODE
V6 19 0 DC=[-L]
.MODEL DIODE D(IS=10U N=0.001)
```

Another equivalent circuit structure macromodel of a PLL is for a linearized PLL system. If the PLL system is locked and the phase difference between the phase detector input signals is small, its performance characteristics can be discussed in the complex frequency domain by applying the Laplace transform. For the linear analysis, it is assumed that the phase detector output is a voltage that is a linear function of the difference in phase between its inputs, that is,

$$V_e = K_d \cdot \theta_d \tag{4.42}$$

TABLE 4.2 SETTING CAPABILITY OF APLL MACROMODEL FOR ACTUAL ELECTRICAL PARAMETERS

Electrical Parameter of Analog Phase Locked Loop	Setting Capability of the Developed APLL Macromodel
Input characteristics	
Input impedance	Yes
VCO characteristics	
VCO conversion gain	Yes
Free-running frequency	Yes
Maximum operating frequency	Yes
Minimum operating frequency	Yes
Drift with temperature	No
Drift with supply voltage	No
Output impedance	Yes
Triangle wave	
Output voltage level	Yes
Peak-to-peak voltage	Yes
Linearity	No
Rectangular wave	
Logic "1" output voltage	Yes
Logic "0" output voltage	Yes
Duty cycle	Fixed (50%)
Rise time	Yes
Fall time	Same as rise time
Output current sink	No
Phase detector	
Conversion gain	Yes
Demodulated output	
Output voltage level	Yes
Maximum voltage swing	Yes
Output offset voltage	Yes

A linear functional block diagram of the PLL in the complex frequency domain is shown in Figure 4.13(a). Its equivalent circuit model can be derived from the generalized macromodeling methodology of the s-domain transfer function. However, the s-domain transfer function must be rederived whenever the loop filter configuration is changed. The linearized PLL macromodel shown in Figure 4.13(b) removes the disadvantage of the s-domain macromodel. Since the loop filter is actually an open block, the user can insert any configuration for the low-pass filter. Therefore, an arbitrary order and type of linearized PLL system can be modeled, subject to SPICE and computer limitations.

In Figure 4.13(b), the resistor, R_{in}, and the two-dimensional nonlinear dependent VCVS, E_{PD}, models the phase detector block. One controlling voltage of E_{PD} is the voltage across R_{in}, and the other is the voltage at node 4 (output of the VCO model). The voltage gain of E_{PD} is identical to the conversion gain, K_d, of the phase detector. The loop filter consisting of R_{LOW} and C_{LOW} is actually open so that any low-pass

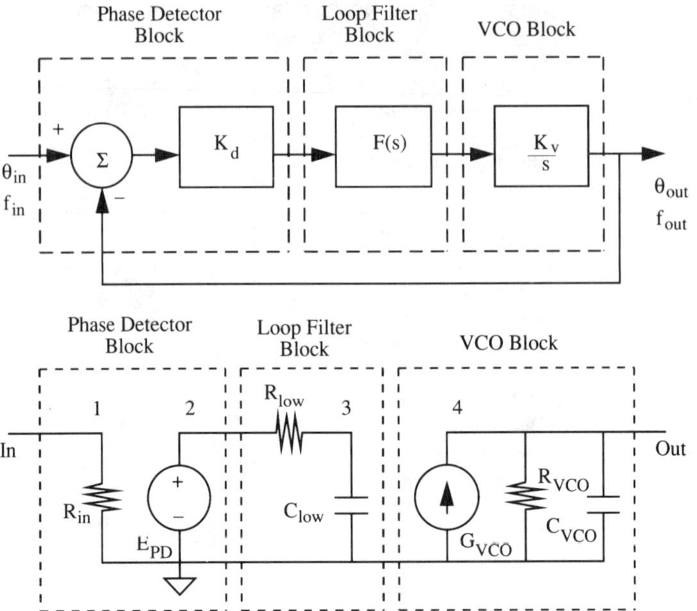

Figure 4.13 An equivalent circuit structure macromodel for linearized PLL systems. (a) Functional block diagram. (b) Circuit structure.

filter structure can be connected. The linear dependent VCCS, G_{VCO}, resistor R_{VCO}, and capacitor C_{VCO} form the integrator model. Its transimpedance function is

$$H(s) = \frac{V_4(s)}{I_{GVCO}(s)} \approx \frac{K}{s} \tag{4.43}$$

where K is the tranconductance of G_{VCO}. If K is equivalent to the conversion gain of the VCO, K_v, the VCO block in Figure 4.13(a) is modeled. The SPICE file is also shown. This linearized PLL macromodel is very useful for checking overshoot and loop stability.

SPICE code for Figure 4.13

```
RIN 1 0 [A]                    ; [A] = Input resistance
EPD 2 0 POLY(2) 1 0 4 0        ; [B] = Conversion gain
+ 0 [B] [-B]                           of phase detector
RLOW 2 3 [C]                   ; [C] = Resistance in LPF
CLOW 3 0 [D]                   ; [D] = Capacitance in LPF
GVCO 0 4 3 0 [E]               ; [E] = Conversion gain
*                                      of VCO
RVCO 4 0 1T
CVCO 4 0 1
```

4.4 SUMMARY

A generalized equivalent circuit structure macromodel for analog phase locked loops has been developed. The macromodel consists of subfunctional blocks of a phase detector, loop filter, and voltage controlled oscillator. Each functional block is modeled to duplicate the main transfer function and the terminal characteristics of the device. Nonlinear effects were included to improve the accuracy of the primitive model . The elements in the model are utilized by the standard SPICE primitives, which makes the model applicable not only to SPICE itself but also to SPICE extensions. Moreover, this equivalent structure macromodel is very flexible and generalized, which permits the user to modify or replace each functional block model as required.

REFERENCES

1. Roland E. Best, *Phase-Locked Loops Theory, Design, and Applications* (New York: McGraw-Hill, 1984).
2. H. De Bellescize, "La Reception Synchrone," *Onde (Electronics)*, Vol. 11, June 1932, pp. 230–240.
3. S. C. Gupta, "Phase-Locked Loops," *Proceedings of the IEEE*, Vol. 63, No. 2, February 1975, pp. 291–306; J. Alvin Connelly, *Phase Locked Loops Design and Application Manual*, Signetics Corporation, 1979, pp. 265–342; Floyd M. Gardner, *Phaselock Techniques*, 2nd ed. (New York: Wiley, 1979); V. Manassewitsch, *Frequency Synthesizers Theory and Design*, 2nd ed. (New York: Wiley, 1980); R. M. Marston, "CMOS PLLs," *Radio-Electronics*, December 1989, pp. 55–58; D. Wulich, "Fast Frequency Synthesis by PLL Using a Continuous Phase Divider," *Proceedings of the IEEE*, Vol. 76, No. 1, January 1988, pp. 85–88.
4. D. Kesner, "Take the Guesswork out of Phase Locked Loop Design," *EDN*, January 5, 1973, pp. 54–60; H. Cohen, "How Phase-Locked Loops Work," *Popular Electronics*, February 1975, pp. 32–34; A. B. Przedpelski, "PLL Primer Part I," *EDN*, March–April 1983, pp. 18-28; "PLL Primer Part II," *EDN*, May–June 1983, pp. 12–28; "PLL Primer Part III," *EDN*, July–August 1983, pp. 48-96; P. Hasan, "Nonlinear PLL Analysis," *Electronic Engineering*, Vol. 56, No. 686, February 1984, pp. 30–32.
5. D. Asta and D. N. Green, "Analysis and Design of a Hybrid Analog/Switched-Capacitor Phase-Locked Loop," *Proceedings of 1986 IEEE International Symposium on Circuits and Systems*, Vol. 1, May 1986; pp. 365-368.
6. E. E. L. Mitchell, "Phase Locked Loop Technique: Interactive Simulation with the Advanced Continuous Simulation Language (ACSL)," *Proceedings of the 1978 Summer Simulation Conference*, July 1978, pp. 444–451; B. Priestley and B. Langley, "Nonlinear Analysis of Phase Locked Loop," *Electronic Engineering*, April 1983, p. 27; B. Labitt, "Phase Locked Loop Simulation," *Proceedings of Twenty-fifth Midwest Symposium on Circuits and Systems*, 1982, pp.200–206; S. Yousef, I. E. Dokknay, and S. A. Monaem, "A Computer Program DESPIL for Design of Phase Locked Loop PLL," *Proceedings of IEEE 1983 National Aerospace and Electronics Conference*, Vol. 1, pp. 362–365; H. Mortier, "BASIC-Program Simuliert PLL-Schaltung," *Elecktronix* (in German), Vol. 33, No. 10, May 18, 1984, pp. 110–112; S. Lubs, "Simple Program Finds Parts Values for

Phase-Locked Loop," *Electronic Design*, Vol. 32, No. 22, October 31, 1984, pp. 319–320; S. Zubiel, "Predict PLL Transient Response with Computer Simulation," *RF Design*, Vol. 8, No. 5, May 1985, pp. 35–38; R. Rippy, "BASIC Program Eases Analysis of Phase Locked Loops," *EDN*, Vol. 31, No. 13, June 26, 1986, pp. 157–162; J. Kovats, "A Simulation Program for High-Order Nonlinear PLLs," *Simulation of Control Systems*. Selected Papers from IFAC symposium, edited by I. T. Kopacek and F. Breitenecker, 1987, pp. 423–426; P. Champagne, "BASIC Program Simulates Digital Phase Locked Loops," *Microwave and RF*, Vol. 27, No. 1, January 1988, pp. 117–120.

7. S. J. Garrett and V. K. Jain, "Macromodel of a Phase Lock Loop Integrated Circuit," *IEEE Proceedings of Southeastcon 1979*, April 1979, pp. 438–441; Ping Yang, editor, and Mark Sitkowski, "Simulation and Modeling—The Macromodel of Phase Locked Loops for the Spice Simulator," *IEEE Circuits and Devices*, Vol. 7, No. 2, March 1991, pp. 11–15.

5

Using SPICE to Design, Analyze, and Test a PLL System

5.1 INTRODUCTION

In the previous chapter, the performance of the phase detector and VCO macromodels were verified for various input signals. The simulation results demonstrated that the operation of the macromodels closely matched the theoretical expectations. In this chapter, the APLL macromodel is examined more closely to determine its accuracy and practicality using SPICE simulations. The key performance parameters of the APLL macromodel are compared to those of the NE 565F analog phase locked loop system produced by Signetics and other IC manufacturers.[1]

Consider the performance, accuracy, and functionality of the APLL macromodel. To verify the performance of the macromodel, three critical input signals of a step of phase, a step of frequency, and a ramp of frequency are applied. A special signal source not included in the standard SPICE primitives is also required. For instance, a ramp of frequency input source is needed to predict the capture and lock ranges. However, SPICE does not have an independent source that provides a ramp of frequency. Here a VCO macromodel serving as a voltage-to-frequency converter becomes the required function generator.

To determine modeling accuracy, the simulated performance parameters of the NE 565F macromodel obtained via simulations are compared with experimental and specification sheet data. The accuracy is expressed as a percentage error according to

$$\%\text{Error of macromodel} = \frac{\text{Macromodel—Experimental}}{\text{Experimental}} \cdot 100\% \qquad (5.1)$$

The capture and lock ranges are further examined with mathematical expressions for the NE 565F developed by the manufacturer since it is very difficult to measure these two parameters accurately. The linear mathematical PLL model also is used to define the accuracy of the loop natural frequency and the loop damping factor. A FSK demodulation system is presented as a typical example application of the APLL macromodel. The NE 565F phase locked loop is often used to demodulate FSK tones and to provide logic "0" and "1" levels to initiate mark and space frequencies.

5.2 SETTING THE MODEL PARAMETERS

An equivalent circuit structure macromodel of NE565 APLL is developed by determining the element values of Figure 4.12 using Table 4.1. The results are described as follows.

Phase detector. The electrical parameters of the phase detector part in NE 565 are:

- Linear type phase detector
- Conversion gain = 0.66 V/rad
- Input resistance = 1 MΩ

The element values of its macromodel shown in Figure 4.12 are determined by

$$[A] = R_{in1} = 1MEG$$

$$[B] = \text{Coefficient } P_4 \text{ in } E_2$$

$$= \frac{K_d \cdot \pi}{20 \cdot \text{ Amplitude of feedback signal}} = \frac{0.66 \cdot \pi}{20 \cdot 10} \approx 1.0367E-2$$

Voltage controlled oscillator. The electrical parameters of the VCO part of the NE 565 APLL are:

- Free running frequency (f_0') = 2.5 KHz
- Conversion gain (K_{vnor}) = 6392 rad/s/V ≈ 1017 Hz/V
- Dynamic range of the VCO input = 10 V
- Peak voltage of triangle wave = 1.2 V_{peak}
- Amplitude of square wave = 10 V
- Timing resistor = 12 KΩ
- Timing capacitor = 10 nF

From these initial conditions, the element values of the VCO macromodel part are calculated by using the equations derived previously. The results are:

$$[E] = \text{Coefficient } P_0 \text{ in } E_4 = 2 \cdot V_{p\text{-}p} \cdot R_{time} \cdot C_{time} \cdot f_0' = 1.44$$

$$[F] = \text{Coefficient } P_1 \text{ in } E_4 = \frac{K_{vnor.} \cdot 2V_{p\text{-}p} \cdot R_{time} \cdot C_{time}}{2\pi} = 0.586$$

$$[G] = \text{Dynamic range of the VCO input voltage} = 10 \text{ V}$$

[H] = Coefficient P_4 in E_5 = $\frac{1}{\text{Amplitude of square wave}} = -0.1$

[I] = R_{time} = Timing resistance = 12 (KΩ)

[J] = C_{time} = Timing capacitor = 10 (nF)

[K] = Coefficient P_2 in E_7 = $\frac{V_{p\text{-}p}/2}{\text{Amplitude of square wave}} \cdot 10^7 = 0.12E7$

[L] = Amplitude of square wave = 10 (V)

Low-pass filter. The APLL dynamic performance depends on the electrical parameters of each functional block, especially the low-pass filter. The low-pass filter serves dual functions in the operation of the loop. First, it attenuates the high-frequency harmonic error components at the output of the phase comparator. Second, it provides a short-term memory for the PLL and ensures a rapid recapture of the signal if the system is thrown out of lock by a noise transient. Decreasing the low-pass filter bandwidth has the following effects on system performance:

1. The capture process becomes slower, and the lock-up time increases.
2. The capture range decreases.
3. Interference-rejection properties of the PLL improve because the error voltage caused by an interfering frequency is attenuated further by the low-pass filter.
4. The transient response of the loop (the response of the PLL to a sudden change of the input frequency or phase within the capture range) becomes underdamped.

Two, low-pass filters (second-order Type I structure), as shown in Figure 5.1, are used to verify the dynamic performance. The loop filter is a simple first-order RC circuit with different cutoff frequencies.

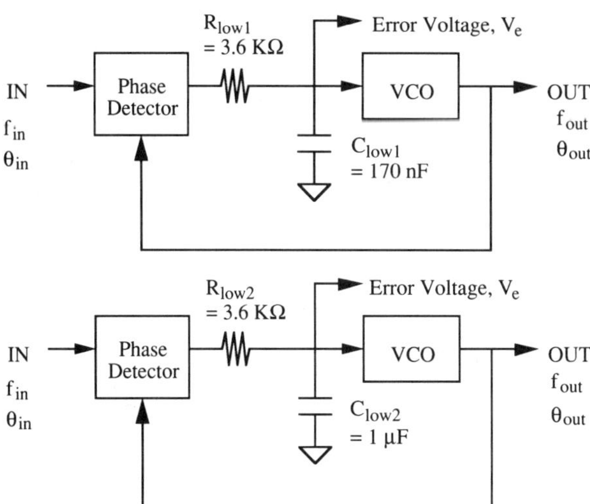

Figure 5.1 Structure of the two APLL test systems. (a) System I. (b) System II.

[C] = Internal resistance of the low pass filter = 3.6 KΩ

[D] = External capacitance in the low pass filter = 170 nF for f_{c1}
= 1 uF for f_{c2}

The cutoff frequencies of the low-pass filters used in this verification are given by

$$f_{c1} = \frac{1}{2\pi \cdot R_{low} \cdot C_{low}} = 260.1\,(\text{Hz})$$

$$f_{c2} = \frac{1}{2\pi \cdot R_{low} \cdot C_{low}} = 44.2\,(\text{Hz}) \tag{5.2}$$

5.3 TESTING THE ANALOG PLL SYSTEM MACROMODEL

Step of phase input. A step of phase input is used to verify how the APLL macromodel is affected by the initial phase error between the input and feedback signals. This initial phase error is a dominant factor influencing the lock-up time t_L. If the initial phases are not synchronized, the phase error voltage drives the VCO frequency away from the input frequency so that the VCO frequency must "walk back" on the beat notes—that is, difference frequencies. This phenomenon directly affects the loop filter output, causing overshoot and oscillation. However, the amplitude of the beat frequency at the loop filter output becomes smaller and smaller as time increases. This means that the phase error is decreasing. Finally the loop filter output voltage reaches steady state where the output frequency and phase of the PLL are synchronized to the input signal. When the phase error is initially zero, the error voltage at the loop filter output becomes the steady state DC voltage without beat frequencies.

As shown in Figure 5.2, the APLL macromodel is simulated for initial phase differences of $-\pi$, $-3\pi/4$, $-\pi/2$, $-\pi/4$, and 0 radians between the input signal and feedback signal. In these simulations, the relative phase error is defined as follows:

Relative phase error = Phase difference between two signals $+\pi/2$

$$\theta_e = \theta_{diff} + \frac{\pi}{2} \tag{5.3}$$

When the phase difference is $-\pi$ radians ($\theta_e = -\pi/2$), both APLL systems have beat notes with negative overshoot in the transient response. This means that the phases of the two signals are not synchronized at t = 0. However, the amplitude of the beat frequency decreases with time, finally settling to a steady-state voltage of zero. This steady-state DC voltage does not affect the VCO frequency modulation because the average value is zero. Thus, the output frequency of the VCO remains equal to the free-running frequency, and the relative phase error is zero in steady state.

When the phase difference is $-3\pi/4$ radians ($\theta_e = -\pi/4$), the capture procedure is the same as in the first case. The peak voltage of the negative overshoot becomes smaller, and the settling time decreases. If the phase difference is $-\pi/2$ radians ($\theta_e = 0$), the overshoot is a minimum and very little in the way of oscillations occurs in the error

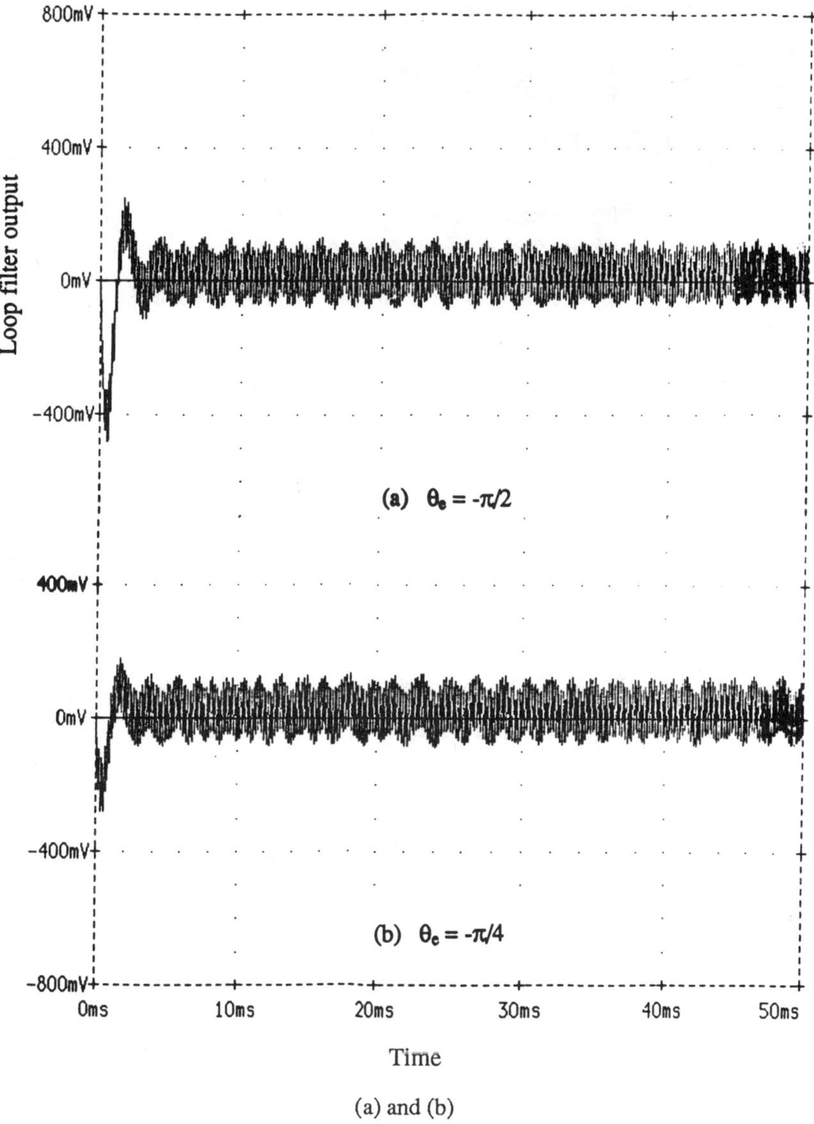

(a) and (b)

Figure 5.2 Transient responses of error voltage of phase inputs. (a) through (e) for System 1 ($\tau_1 = 612\ \mu$s); (f) through (j) for System II ($\tau_2 = 3.6$ ms).

voltage of both systems. This means that the phases of the input and feedback signals are almost synchronized initially. Therefore, the lock-up time is almost instantaneous. When the relative phase error increases from 0 to $\pi/2$, a positive overshoot occurs. The number of oscillation cycles also increases. When the phase difference is zero ($\theta_e = \pi/2$), the overshoot reaches a maximum value, and the lock-up time is very long.

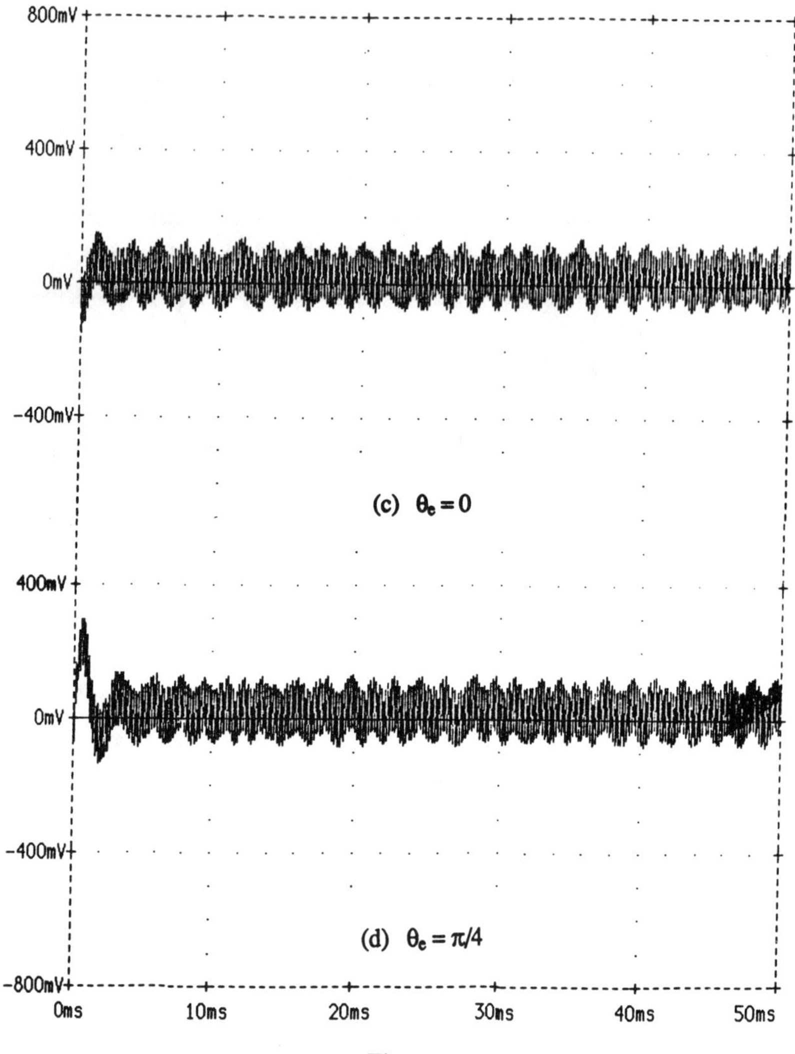

Figure 5.2(c) and (d)

Another important performance index to be verified through these simulations is the interference-rejection. The high-frequency rejection of System II is much better than that of System I because the ripple width in the transient responses of System II is smaller than that of System I. This is due to the lower cutoff frequency of the low-pass filter. When both systems are driven by the same input, the number of oscillation cycles of System II is much greater than that of System I. Also, the lock-up time of System II is longer than that of System I. These performance characteristics of the APLL macromodel are a good match to the characteristics of the low-pass filter cited previously.

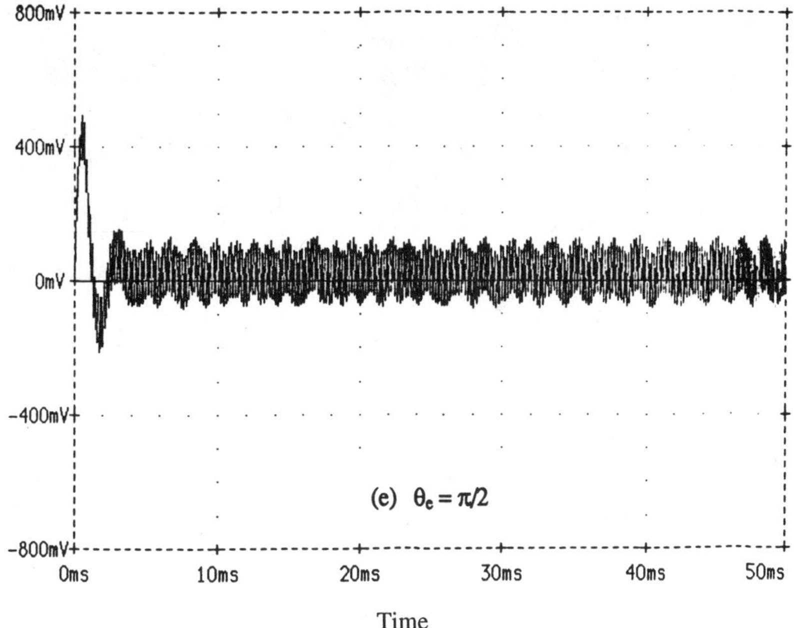

Figure 5.2(e)

Step of frequency input. This type of input occurs when the input frequency is instantaneously changed from one frequency to another, as is done in modem applications. In Figure 5.3, the APLL macromodel is established with the free-running frequency, f_0', being 2.5 KHz. Then either a 2.65 KHz or a 2.35 KHz sine wave is supplied to both example systems as the input signal. These small frequency deviations ($\Delta f = \pm 150$ Hz) are low enough to permit linear operation of the phase detector.

As illustrated in Figures 5.4(a) and 5.5(a), the positive, small-frequency deviations cause the loop filter output to be modulated by the beat note or the difference signal in both systems. When this happens, the beat note frequency at the low-pass filter output becomes a function of time. Because the VCO frequency varies with time, it alternately moves closer to and farther away from the input frequency. Since the filtered error voltage depends on the difference between the VCO frequency and the input frequency, it also alternately increases and decreases the VCO frequency about an average or DC value. Therefore, the beat note becomes asymmetrical and looks like a series of cusps, as in Figure 5.5(a). Note that the portion of the beat note that modulates the VCO closer to the input signal appears more rounded, and the portion that modulates the VCO away from the input signal appears more peaked. Because of this asymmetry, the beat note contains a finite DC voltage, which steadily pushes the VCO frequency toward the input signal. As the VCO drifts toward f_{in}, the beat note frequency rapidly decreases, the asymmetry increases, and the transient rapidly converges to a steady-state DC value corresponding to the lock condition where the VCO frequency is equal to the input frequency. Figures 5.4(b) and 5.5(b) show the frequency spectrum of the triangle wave

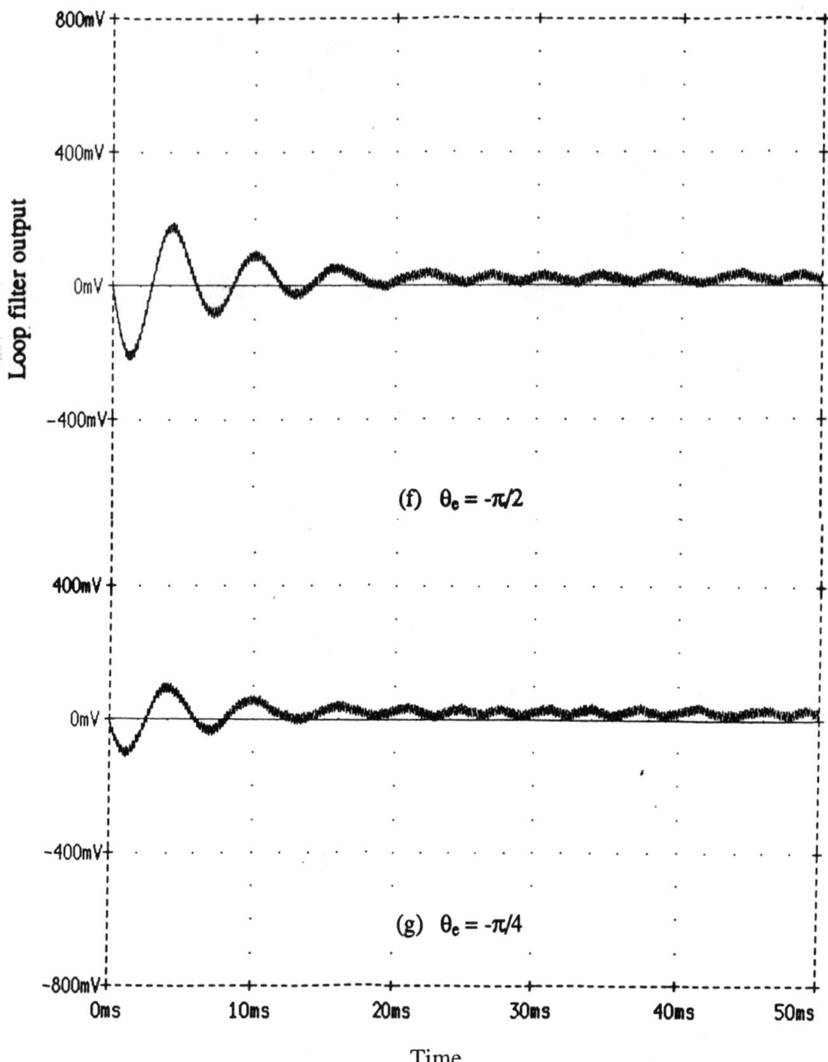

Figure 5.2(f) and (g)

output of the VCO. The dominant Fourier component in both systems occurs at 2.66 KHz (magnitude = 624 mV in System I and 640.5 mV in System II). This frequency is not identical to the input frequency, but very close. The percentage error of the mismatching between the input and the output frequency is

$$\begin{aligned} \%\text{Error of frequency mismatching} &= \frac{2{,}660 - 2{,}650}{2{,}650} \cdot 100(\%) \\ &\approx 0.38\% \end{aligned} \tag{5.4}$$

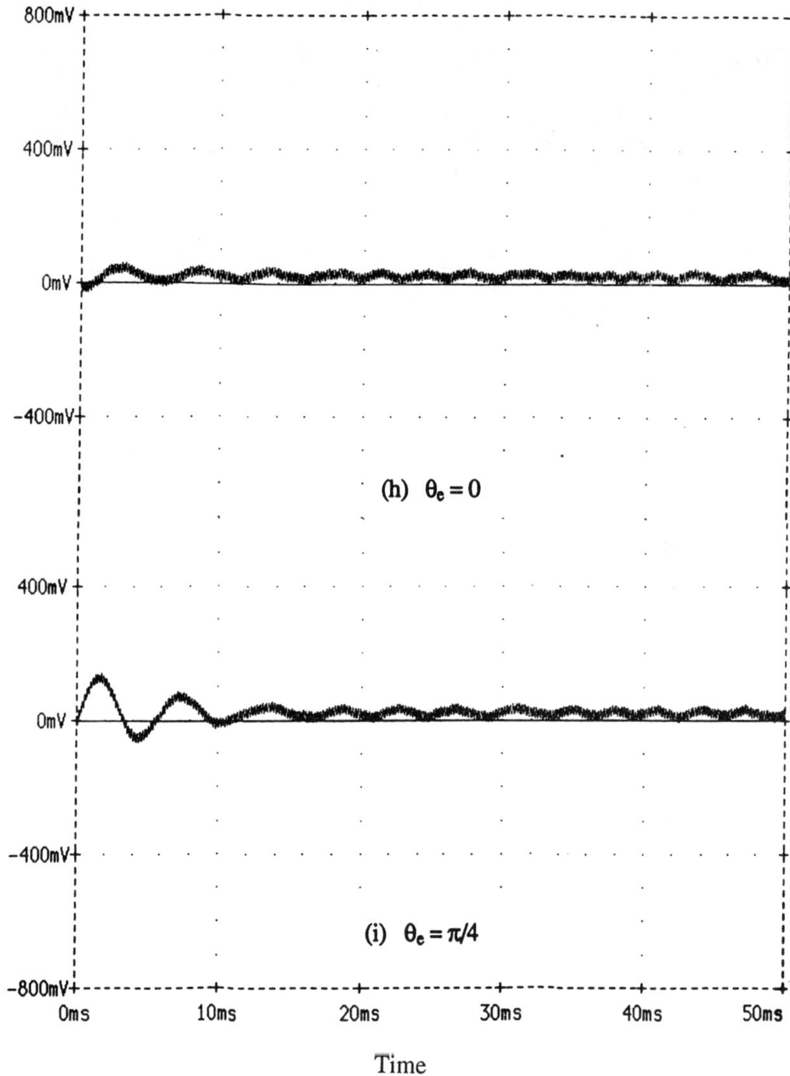

Figure 5.2(h) and (i)

Another interesting feature is that System II, unlike System I, has pronounced double-sided spurious components. The frequency spread between the main and spurious components is usually dependent upon the beat frequency at the loop filter output. The unsymmetrical spurious signal of System I occurs at 2.22 KHz (magnitude = 20 mV) and that of System II at 2.80 KHz (magnitude = 86.02 mV). From the frequency spectrums of both systems, the Signal-to-Noise Ratio (SNR) can be calculated as follows:

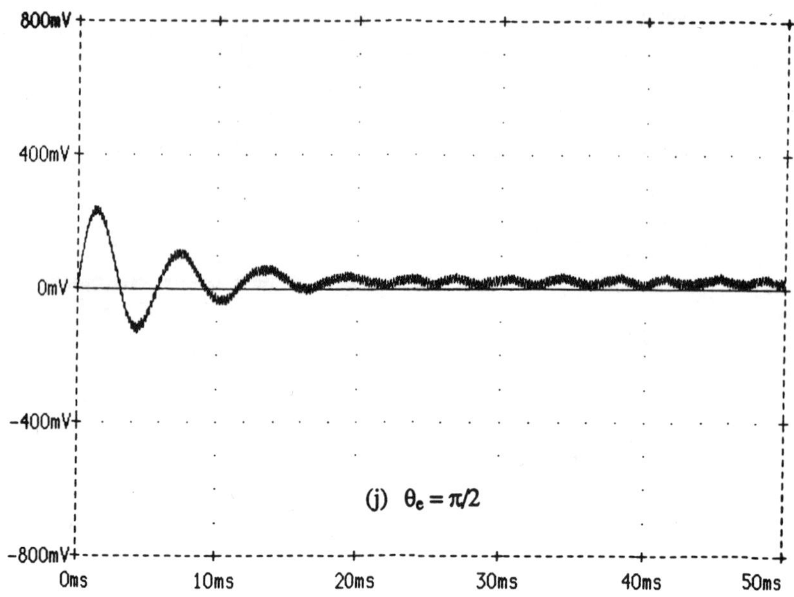

Figure 5.2(j)

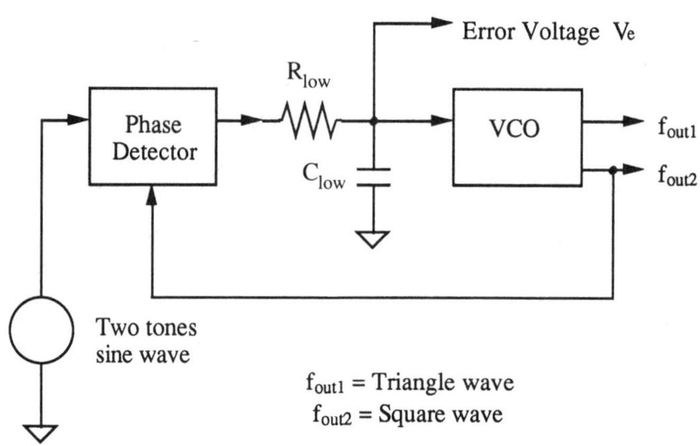

Figure 5.3 Initial setup of the APLL macromodel for a step of frequency input.

$$\text{SNR of the PLL system} = 20 \cdot \log \frac{C_{main}}{C_{spur}}$$

$$\text{SNR of the system I} = 20 \cdot \log \frac{624\text{mV}}{20\text{mV}} \approx 30 \text{ dB} \tag{5.5}$$

$$\text{SNR of the system II} = 20 \cdot \log \frac{640\text{mV}}{86.02\text{mV}} \approx 17 \text{ dB}$$

When the input frequency is 2.35 KHz, the loop filter output is also modulated by the beat note difference signal. This is shown in Figures 5.4(c) and 5.5(c). The signal capture process is the same as that of the positive frequency step case. The only difference is that the steady-state value of the loop filter output converges to a negative voltage instead of a positive value. The frequency spectra of the triangle wave output are shown in Figures 5.4(d) and 5.5(d). Since the main lobe appears at 2.34 KHz, the output frequency of the APLL macromodel closely matches the frequency of the input signal. The percentage error of the mismatch is given by

$$\begin{aligned}\%\text{Error of frequency mismatch} &= \frac{2{,}340 - 2{,}350}{2{,}350} \cdot 100(\%) \\ &\approx -0.4\%\end{aligned} \tag{5.6}$$

Figures 5.4 and 5.5 show important functions of the low-pass filter in the PLL operation. First, it provides a short-term memory for the PLL and ensures a rapid recapture of the signal if the system is briefly thrown out of lock because of an interfering transient. In other words, the low-pass filter constrains the error voltage $V_e(t)$ to be a slowly varying function of time, such that if the PLL temporarily loses lock as a result of a noise or interference transient, the VCO frequency cannot change significantly over a short period of time. Second, the low-pass filter controls the dynamic characteristics such as the interference-rejection, the lock-up time, loop damping factor, and so on. If the passband of the loop filter is low as in System II, the interference-rejection is enhanced but the lock-up time is increased. Actually the lock-up time depends on the initial phase error and the frequency difference between the input and feedback signals, as well as on the overall loop gain and loop filter characteristics. Under certain conditions, however, the lock-up time can be shorter than the period of a beat note, and the loop can lock without an oscillatory error transient. This phenomenon, illustrated in Figure 5.4(c), occurs when the phases of the two signals are synchronized initially.

Another characteristic to be verified using the example APLL macromodel is that the PLL loses its capture capability when a step of frequency larger than the pull-out range is applied. To test this characteristic, the example APLL System I is driven by

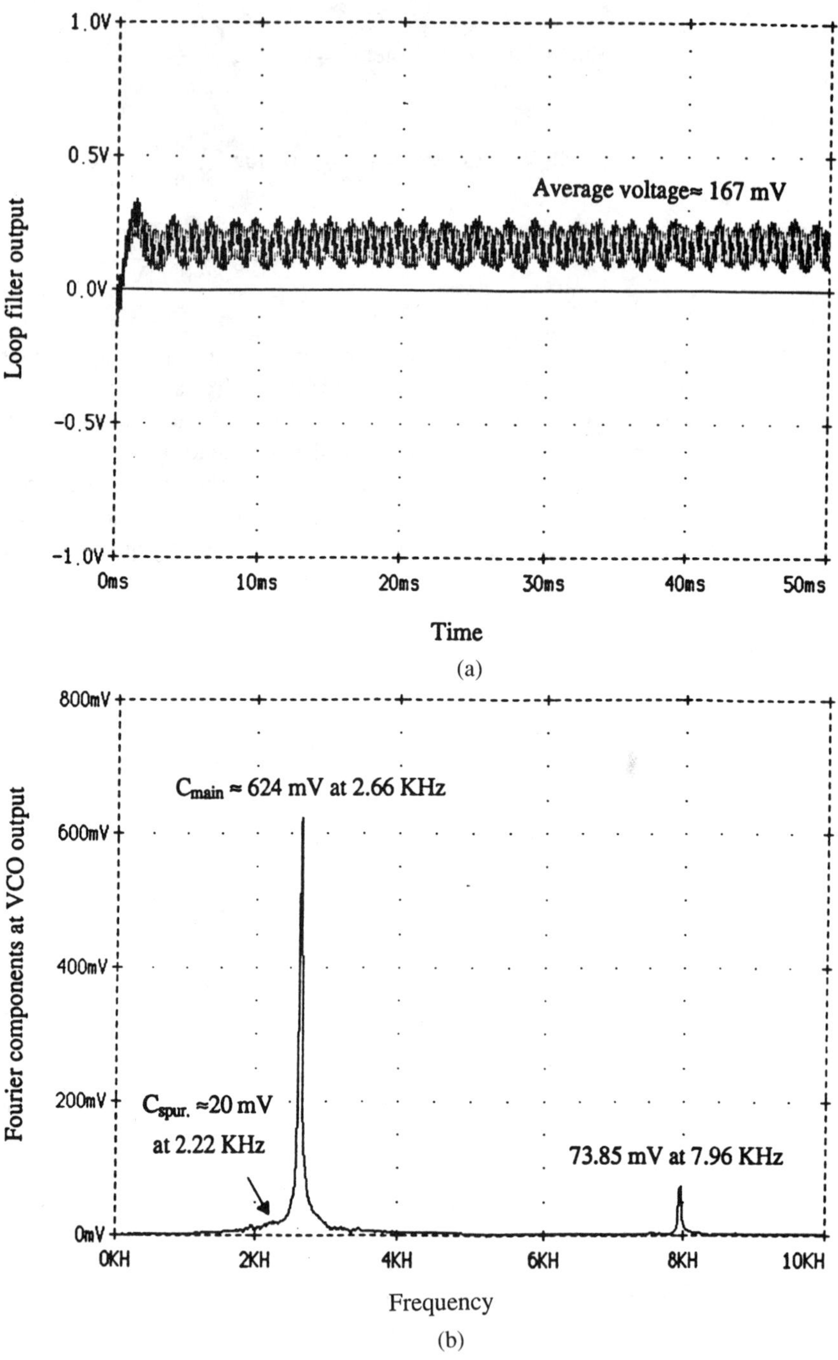

Figure 5.4 Transient response of System I macromodel for a step of frequency input (small-frequency deviation, f_{in} = 2,650 Hz). (a) Loop filter output. (b) Frequency spectrum of f_{out1}. (c) Loop filter output. (d) Frequency spectrum of f_{out1}.

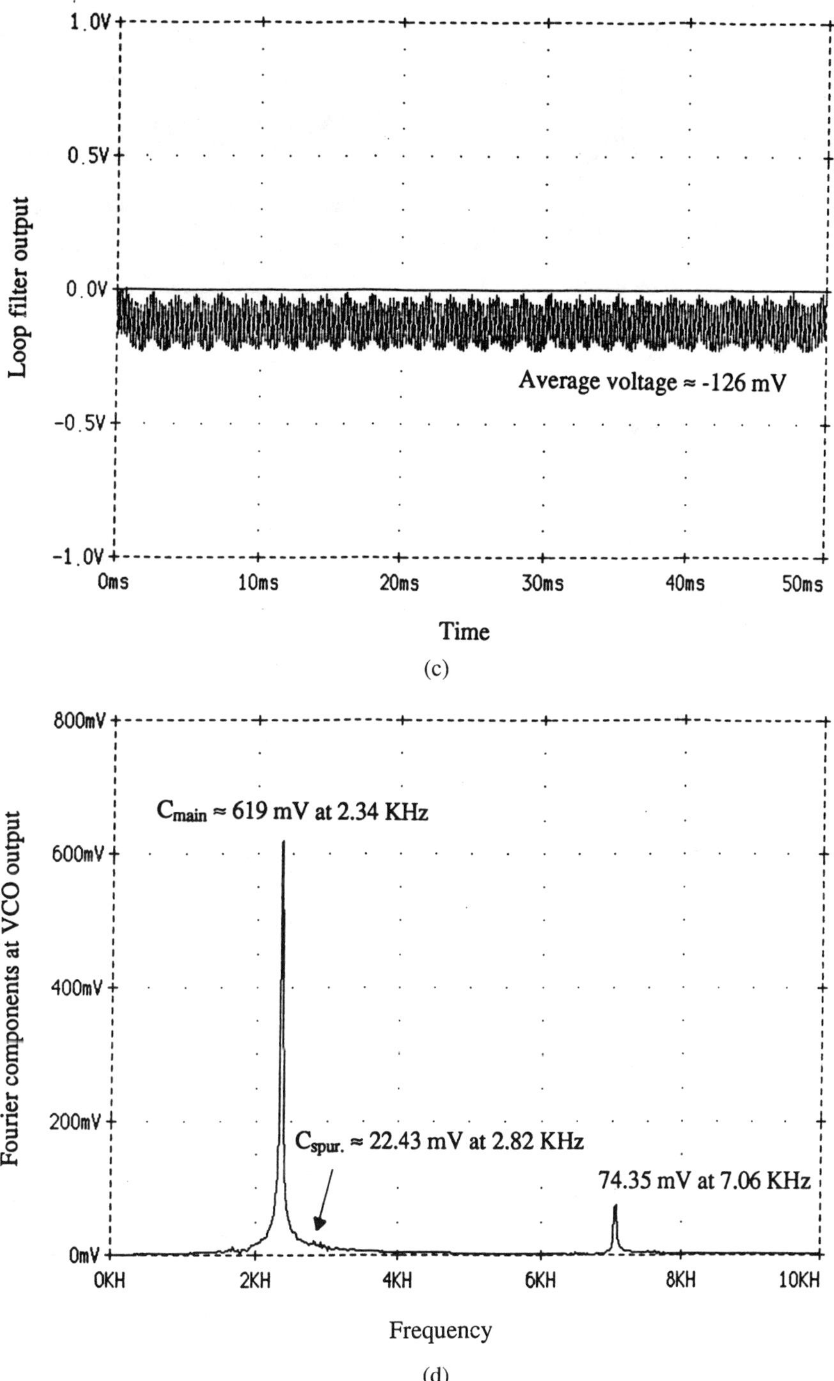

(c)

(d)

Figure 5.4(c) and (d)

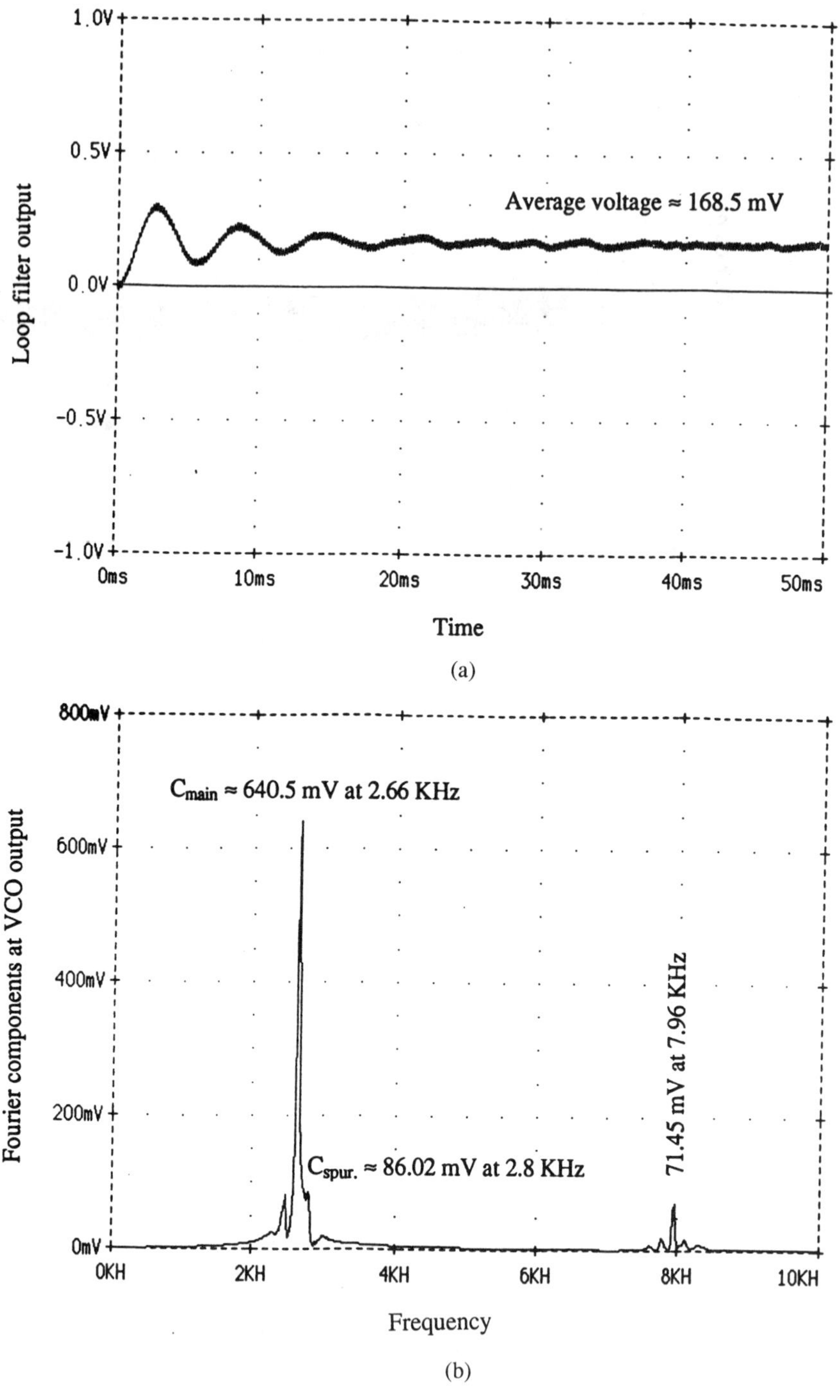

Figure 5.5 Transient response of System II macromodel for a step of frequency input (small-frequency deviation, when f_{in} = 2,650 Hz). (a) Loop filter output. (b) Frequency spectrum of f_{out1} (when f_{in} = 2,350 Hz). (c) Loop filter output. (d) Frequency spectrum of f_{out1}.

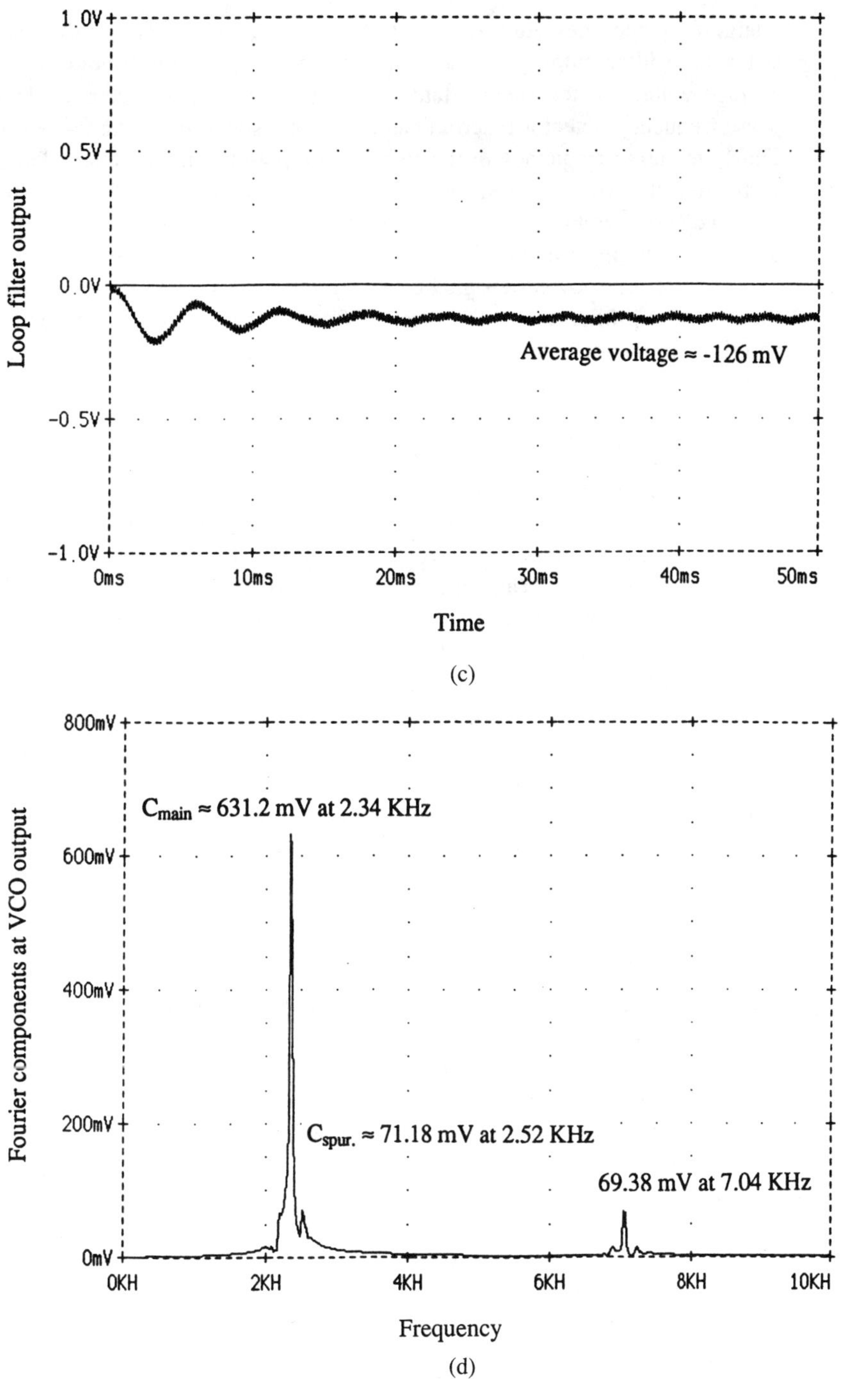

Figure 5.5(c) and (d)

a large-frequency deviation ($\Delta f = \pm 1.5$ KHz). Figure 5.6 shows the transient response at the loop filter output and the frequency spectrum of the triangle wave output. The average voltage at the steady state in both cases is virtually zero. This means that the phase/frequency detector macromodel does not respond to the large-frequency deviation. Thus, the output frequency of the VCO is not modulated. These simulation results again verify that the low-pass filter acts as a short-term memory. Because there is no low-frequency error voltage at the loop filter output, the VCO output frequency remains equal to the free-running frequency.

Figure 5.7 shows the signal capture process of the APLL macromodel. This figure shows the loop filter output for a step of frequency input. Using approximation methods described in a later section of this chapter, the damping factor and natural frequency can be determined from this transient response waveform.

Ramp of frequency input. The performance of the APLL macromodel when excited by a ramp of frequency input is very important, since most of the dynamic characteristics of the PLL system can be determined. Present mathematical or equivalent circuit structure macromodels have limited use because they do not cover both the linear (lock) and the nonlinear (unlock) regions. The APLL macromodel describes both regions and shows the capture process, enabling the transition from the unlock to lock states, or vice versa, to be determined. Also, the tracking characteristics in the linear operation region are evident. Thus, all the key system performance parameters can be determined using a ramp of frequency as the input signal.

Again, SPICE functional primitives do not include an independent source to generate the ramp of frequency. However, this type of independent source is modeled by using the drop-in voltage controlled oscillator macromodel described in Chapter 4. Figure 5.8 shows a model structure for the simulation to find the capture and lock ranges. Since the main transfer function of the VCO-I model is voltage-to-frequency conversion, the ramp of voltage input with constant sweeping rate produces the ramp of frequency output with same rate with a constant signal amplitude. Thus, the signal input of the APLL macromodel is swept linearly through the frequency range of interest.

Another consideration in the simulation is the selection of the proper sweeping rate to estimate correctly the capture and lock ranges. The sweep frequency must be much slower than the free-running frequency, especially when the capture range is below 10% of the free-running frequency. Otherwise, the capture and lock ranges will be functions of the sweep frequency. Thus, it is best to start sweeping as slowly as possible and, if desired, increase the rate until the capture range begins to show an apparent reduction, indicating that the sweep is too fast.

The first verification was performed by sweeping the input frequency at a rate of +31.25 Hz/ms. This simulation was done using a Sun 3/60 workstation running the commercial version 4.03 of PSpice. Approximately four to five hours of simulation time were required. There are several reasons why such a long simulation time was necessary. First, the most time consuming part of the APLL is the VCO block where sharp transitions occur at every half cycle. Two VCO blocks are necessary: one as part of the APLL and the other as the function generator signal source. Second, the sweep rate

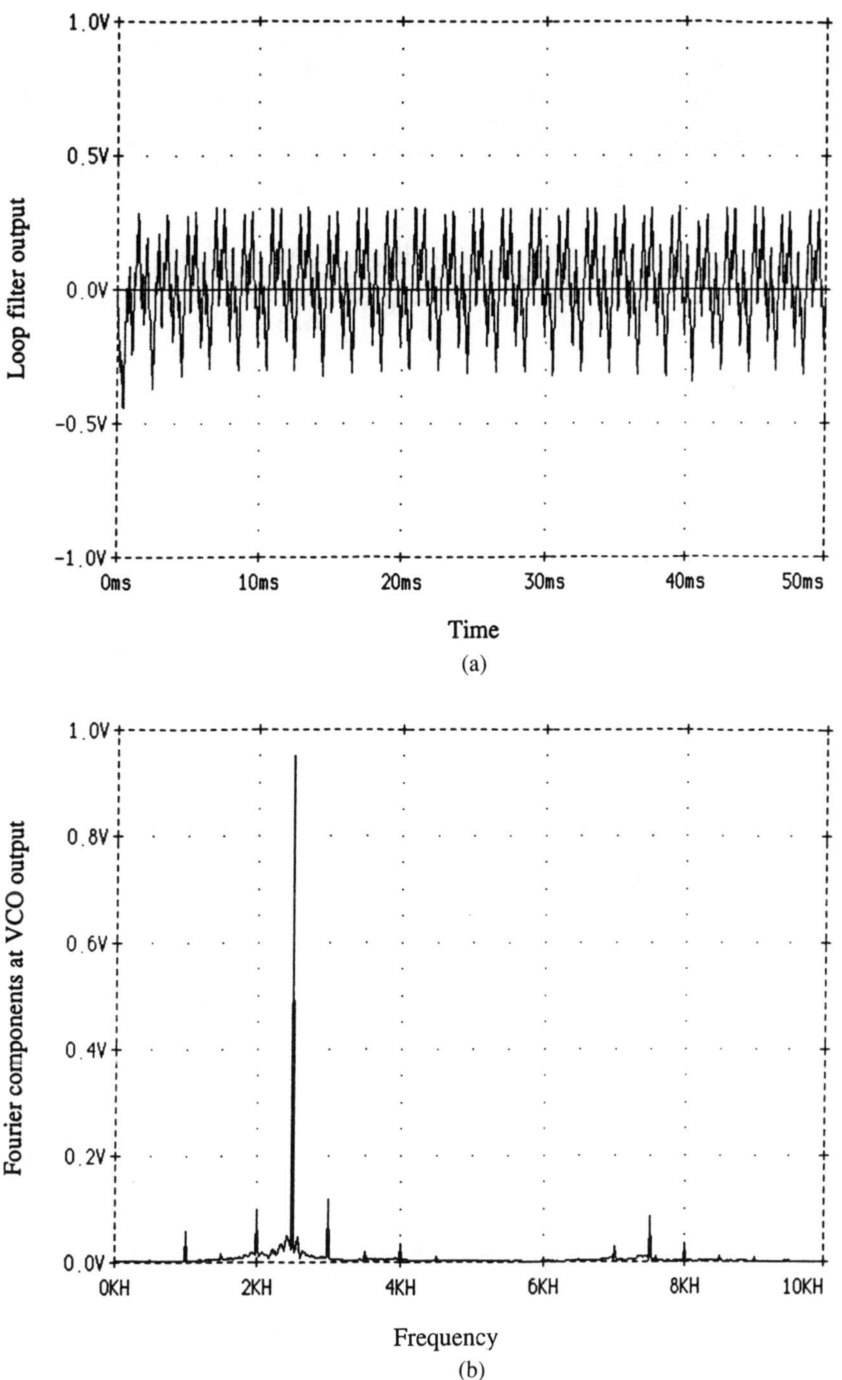

Figure 5.6 Transient response of System I macromodel for a step of frequency input (large-frequency deviation, $f_{in} = 1{,}000$ Hz). (a) Loop filter output. (b) Frequency spectrum of f_{out1}. (c) Loop filter output. (d) Frequency spectrum of f_{out1}.

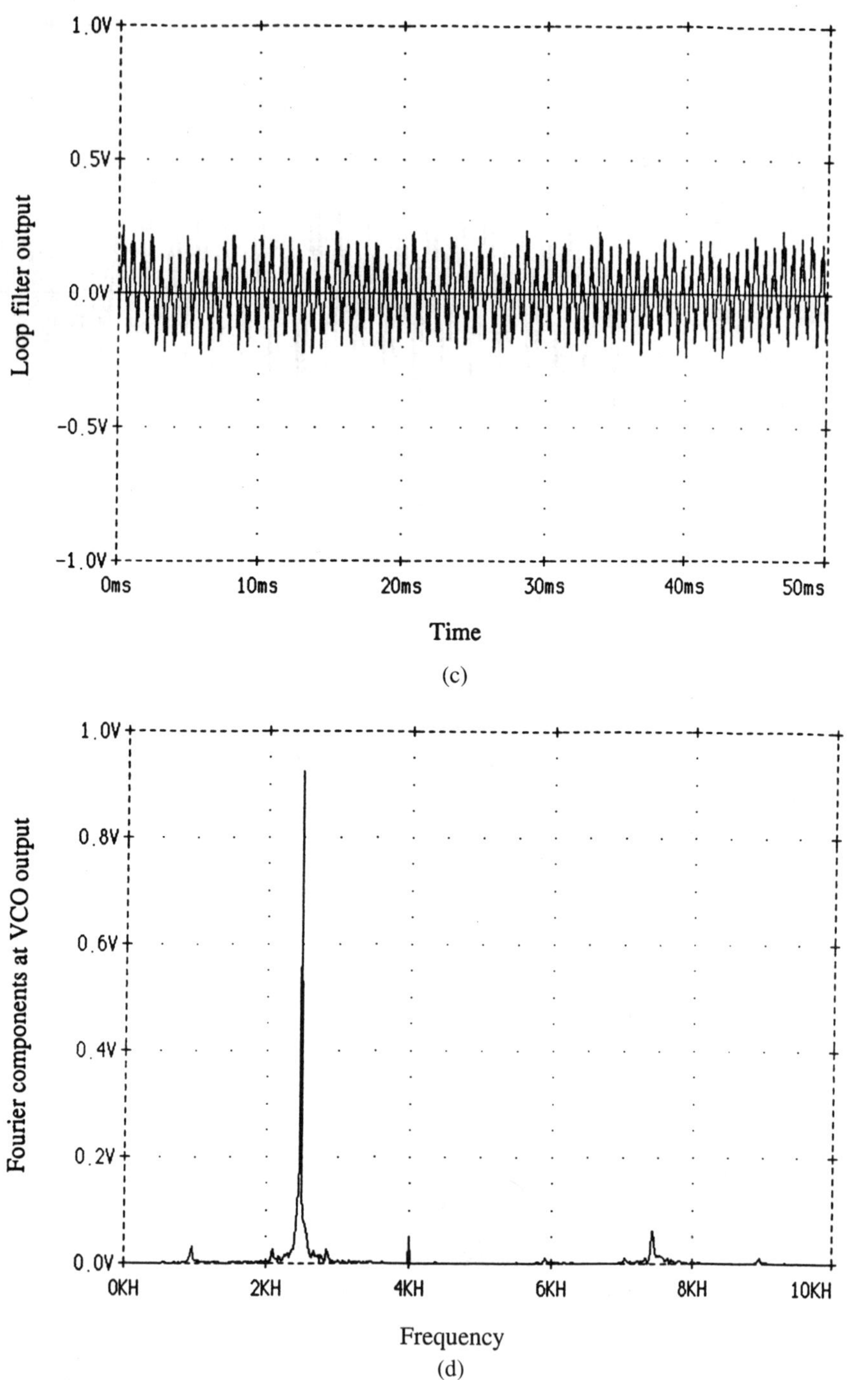

Figure 5.6(c) and (d)

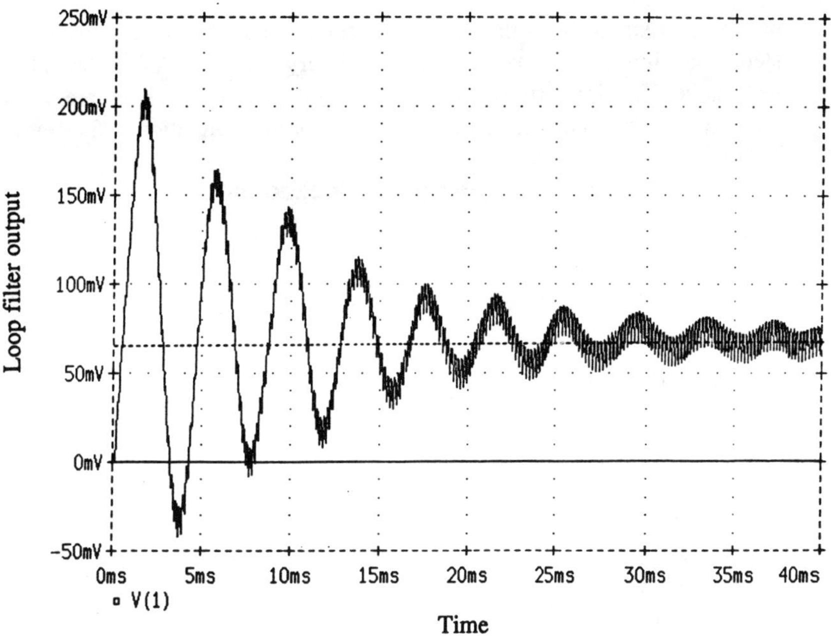

Figure 5.7 The signal capture process.

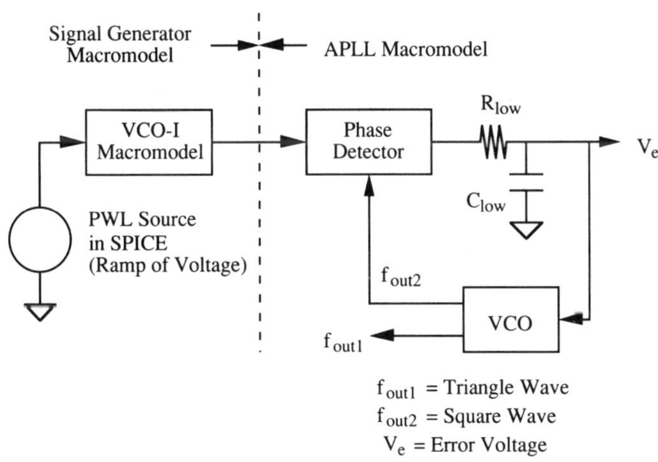

Figure 5.8 Initial setup of the APLL macromodel for the measurement of capture and lock ranges.

of the input signal from the function generator must be slow to achieve good accuracy. Hence the simulation time interval will be very long in order to show the upper or lower lock range. Third, the transient analysis requires longer times compared with other types of analysis. The complete SPICE code used for the simulation of System I is given below.

SPICE code for analog phase locked loop

```
Analog Phase Locked Loop

*Function Generator Model*        ;This macromodel is required since SPICE
                                  ;doesn't have a ramp of frequency source.

VIN 99 0 PWL(0 -0.576 80M 0.864)
REFUN1 99 0 1K
EFUN0 100 0 POLY(1) 99 0 1.44 1
RFUN0 100 0 1
EFUN1 101 0 POLY(2) 100 0 105 0
+ 0 0 0 0 -0.1
RFUNINT 101 102 12K
RFUN2 102 0 1MEG
CFUNINT 102 103 10N               ;V(103) is the triangle wave output
EFUN2 103 0 102 0 100MEG
EFUN3 104 0 POLY(2) 103 0 105 0
+ 0 -10MEG 1.2MEG
RFUN3 104 105 1T
CFUN1 105 0 10E-15
DFUN1 105 106 DIODE3
VFUN1 106 0 DC 10.0
DFUN2 107 105 DIODE3
VFUN2 107 0 DC -10.0
.MODEL DIODE3 D(IS=10U N=0.001)
.IC V(105)=10.0 V(103)=1.2

*Phase Detector Macromodel*

*RIN 1 0 1MEG                     ;This section is not necessary because the input
*E1 2 0 POLY(2) 1 0 3 0 0         ;signal is a square wave. The actual function of
*+ 10MEG -1E-11                   ;this section is for sine-to-square wave conversion.
*R1 2 3 1T
*C1 3 0 10E-15
*D1 3 4 SWITCH
*V1 4 0 DC=10
*D2 5 3 SWITCH
*V2 5 0 DC=-10

E2 6 0 POLY(2) 105 0 17 0         ;V(105) is the square wave output of the function
+ 0 0 0 0 1.0367E-2               ;generator.
R2 6 0 1
```

```
*Loop Filter*

E3 7 0 POLY(1) 6 0 0 1
RFILTER 7 8 3.6K
CFILTER 8 0 170N                    ;V(108) is the loop filter output

*Voltage Controlled Oscillator*

E4 10 0 POLY(1) 8 0 1.44 0.589
R3 10 0 1T
D3 10 11 SWITCH
V3 11 0 DC=0
D4 12 10 SWITCH
V4 12 0 DC=-10
E5 13 0 POLY(2) 10 0 17 0 0 0 0 0 -0.1
RTIME 13 14 12K
CTIME 14 15 10N
R4 14 0 1MEG
E6 15 0 14 0 100MEG                 ;V(15) is the triangle wave output of the VCO
E7 16 0 POLY(2) 15 0 17 0
+ 0 -10MEG 1.2MEG
R5 16 17 1T
C2 17 0 10E-15                      ;V(17) is the square wave output of the VCO
D5 17 18 SWITCH
V5 18 0 DC=10
D6 19 17 SWITCH
V6 19 0 DC=-10
.MODEL SWITCH D(IS=10U N=0.001)
.IC V(17)=10 V(15)=-1.2             ;Initial conditions of PLL Macromodel

*Run Control Section*

.TRAN 50U 80M 0 1E-5 UIC
.OPTIONS LIMPTS=5000 ITL5=0 PIVTOL=1E-30 ITL4=40 RELTOL=0.01 ACCT
.PROBE V(8)
.END
```

Figures 5.9(a) and 5.10(a) illustrate the transient responses of the error voltage (output of the low-pass filter) of Systems I and II, respectively. In both, the APLL macromodel does not respond to the input signal until the input frequency reaches f_{LC} (System I = 2,063 Hz; System II = 2,369 Hz), corresponding to the lower edge or frequency of the capture range. Up to the lower capture frequency, f_{LC}, the VCO output frequency does not change significantly since the average voltage of the beat frequency is virtually zero. At f_{LC}, the macromodel suddenly locks to the input signal, causing a negative jump of the loop error voltage. This means that the nonlinear capture transients are occurring at that frequency. Once the APLL model is locked to the input signal, it tracks small-frequency changes of the input signal by generating the additional phase

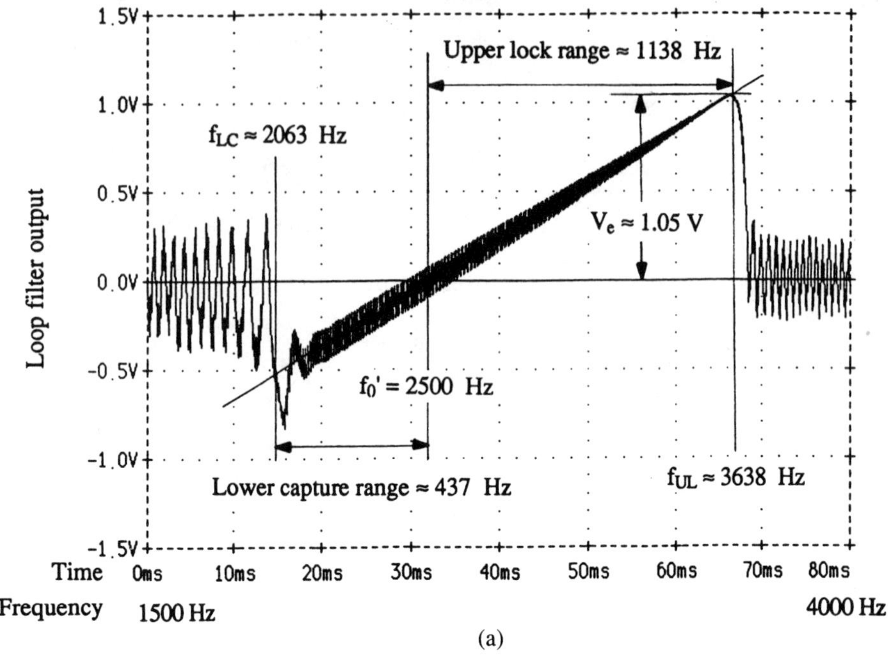

(a)

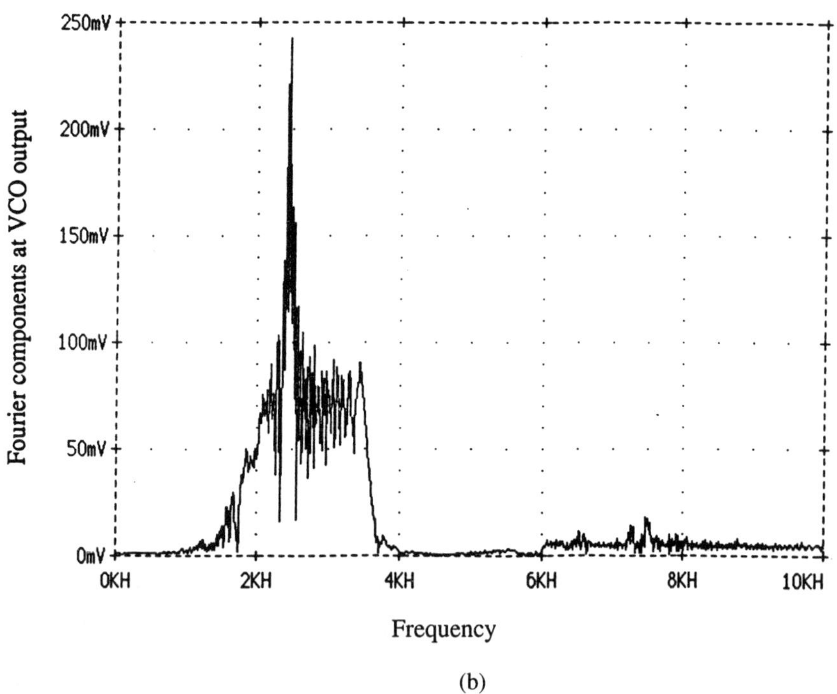

(b)

Figure 5.9 Transient response of the System I macromodel for a ramp of frequency input (when $\Delta f_{in}/\Delta t$ = 31.25 Hz/ms). (a) Loop filter output. (b) Frequency spectrum of f_{out1} (when $\Delta f_{in}/\Delta t = -31.25$ Hz/ms). (c) Loop filter output. (d) Frequency spectrum of f_{out1}.

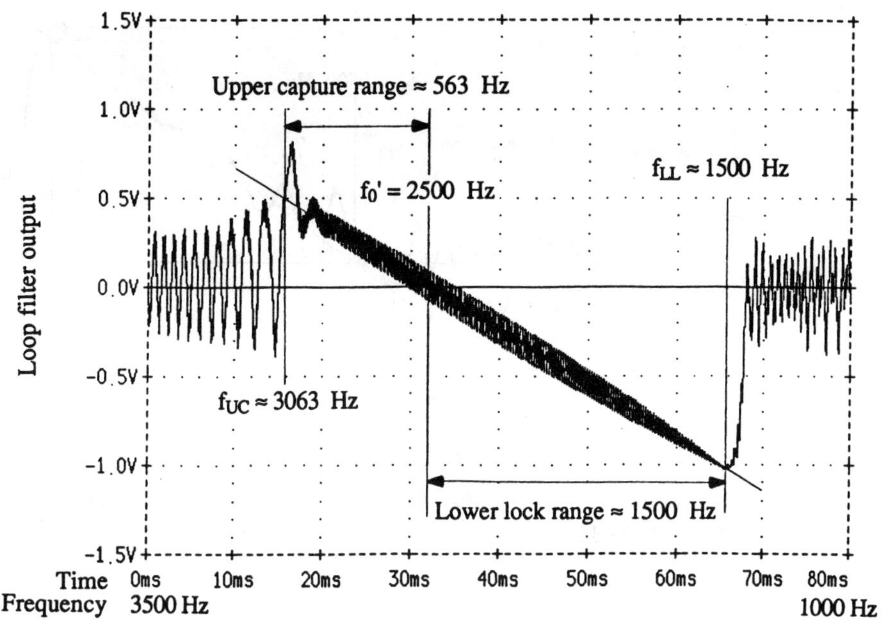

(c)

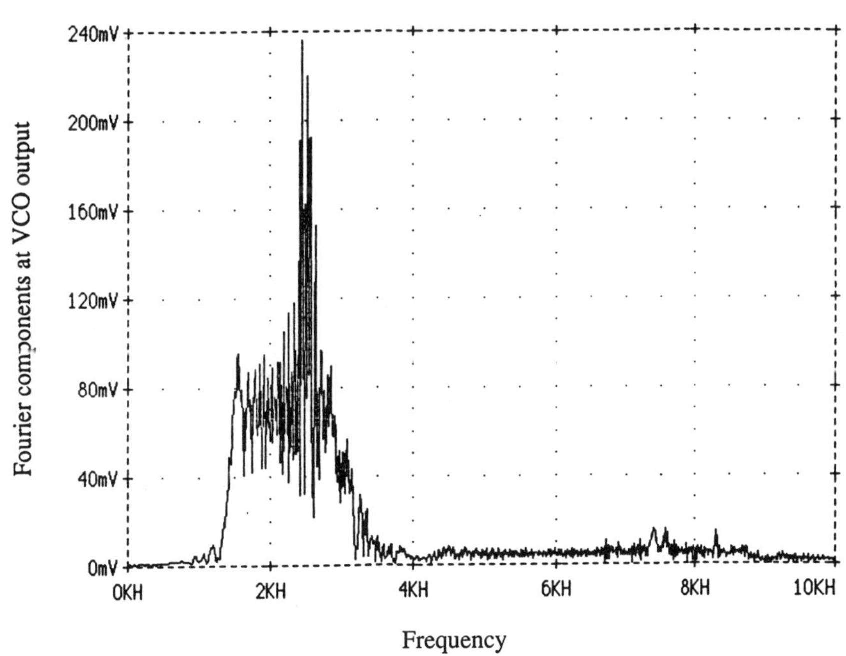

(d)

Figure 5.9(c) and (d)

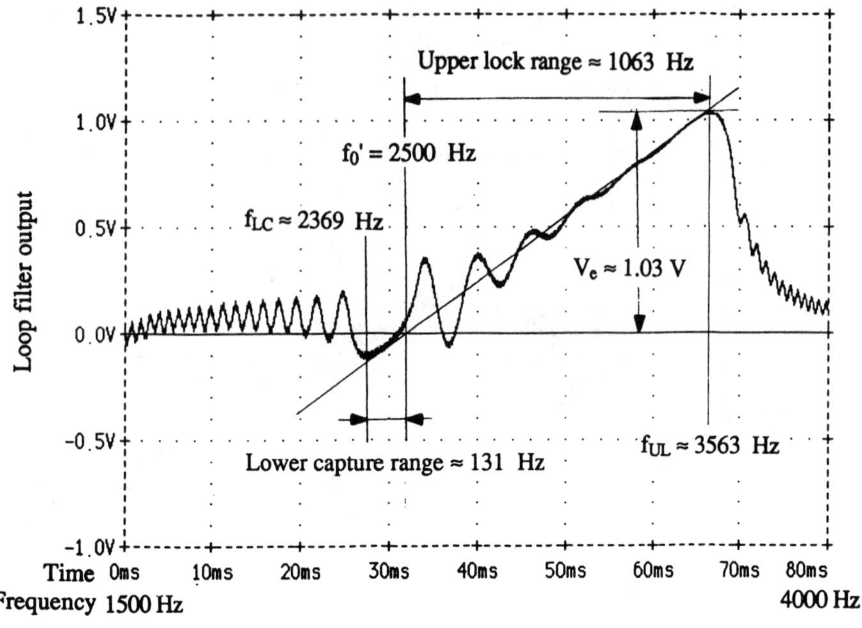

(a)

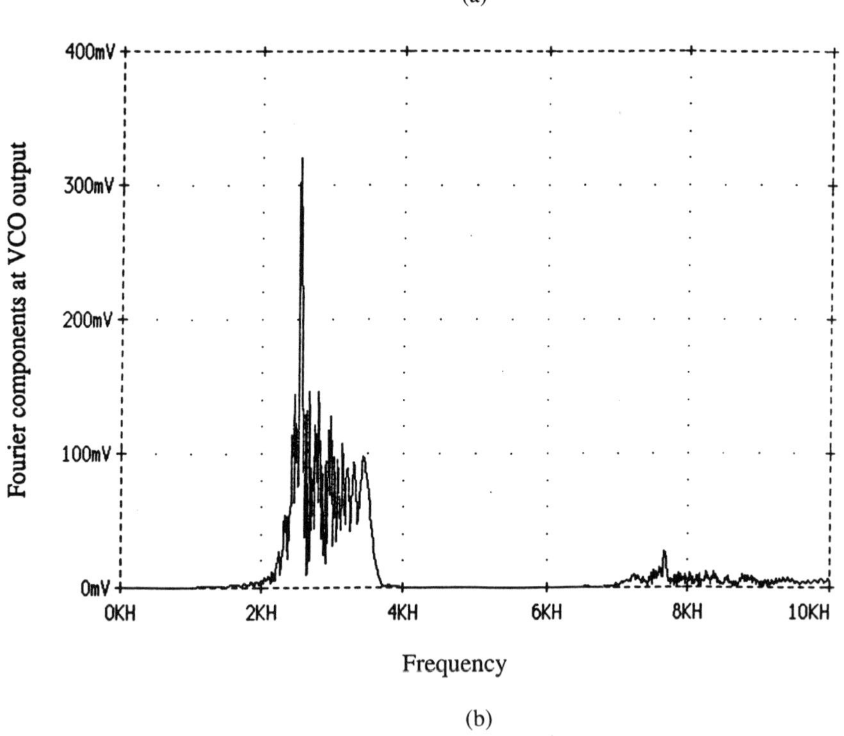

(b)

Figure 5.10 Transient response of the System II macromodel for a ramp of frequency input (when $\Delta f_{in}/\Delta t = 31.25$ Hz/ms). (a) Loop filter output. (b) Frequency spectrum of f_{out1} (when $\Delta f_{in}/\Delta t = -31.25$ Hz/ms). (c) Loop filter output. (d) Frequency spectrum of f_{out1}.

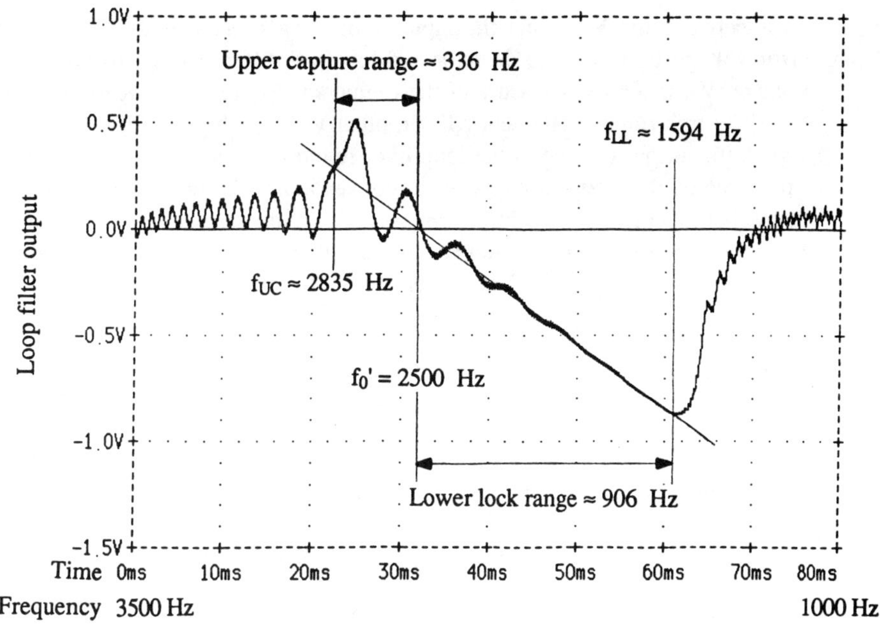

(c)

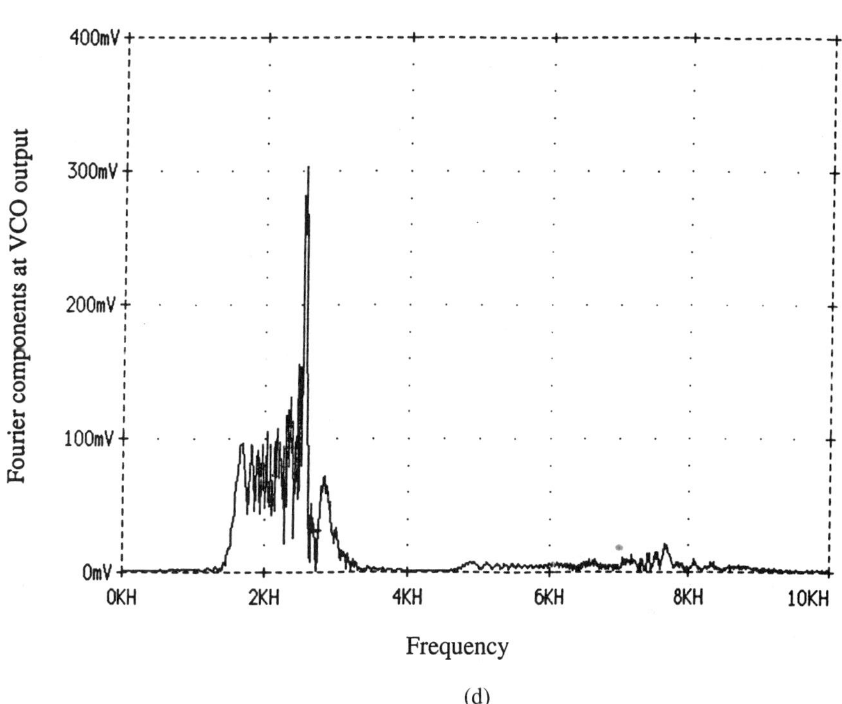

(d)

Figure 5.10(c) and (d)

error between the VCO and the input signal. This signal is then converted to the DC error voltage by the phase detector and low-pass filter. During tracking, the loop error voltage, V_e, is a direct measure of the frequency difference between the input signal and the VCO free-running frequency f'_0. In other words, while the PLL is tracking the input signal, the output voltage of the loop filter functions as a frequency-to-voltage converter. The slope of the frequency-to-voltage conversion is the reciprocal of the VCO voltage-to-frequency conversion gain. From the simulation results, the percentage error of the frequency-to-voltage conversion at the loop filter output and of the voltage-to-frequency conversion gain of the VCO are given by the following equations:

System I:

- Frequency-to-voltage conversion gain in the tracking range:

$$\frac{1}{K_v} = \frac{1.05}{3{,}638 - 2{,}500} = 9.23 \cdot 10^{-4} \text{ (V/Hz)}$$

- Voltage-to-frequency conversion gain of the VCO:

$$K_v = \frac{2\pi}{9.23 \cdot 10^{-4}} \approx 6{,}807 \text{ (rad/s/V)}$$

$$\therefore \ \%\text{Error of } K_v = \frac{6{,}807 - 6{,}422}{6{,}422} \cdot 100(\%) \approx 5.9\ (\%)$$

System II:

- Frequency-to-voltage conversion gain in the tracking range:

$$\frac{1}{K_v} = \frac{1.03}{3{,}563 - 2{,}500} = 9.69 \cdot 10^{-4} \text{ (V/Hz)}$$

- Voltage-to-frequency conversion gain of the VCO:

$$K_v = \frac{2\pi}{9.69 \cdot 10^{-4}} \approx 6{,}484 \text{(rad/s/V)}$$

$$\therefore \ \%\text{Error of } K_v = \frac{6{,}484 - 6{,}422}{6{,}422} \cdot 100(\%) \approx 0.97(\%)$$

When the input frequency reaches f_{UL}, the upper lock range frequency, (System I = 3,638 Hz; System II = 3,563 Hz), the APLL macromodel loses lock. The error voltage suddenly drops and then becomes a beat note. After f_{UL}, the VCO frequency returns to f'_0 since the average DC error voltage is almost zero. Thus, the VCO frequency does not change significantly as a result of the integrator action of the low-pass filter. These characteristics are clearly shown by the frequency spectrum of the triangle wave output of the VCO, as illustrated in Figures 5.9(b) and 5.10(b). In these figures the Fourier spectra between about 2.0 KHz and 3.5 KHz are much larger than those of any other frequency range. This means that the output frequency of the APLL macromodel is tracking the input frequency.

If the input frequency, f_{in}, is slowly swept back toward low frequencies at a −31.25 Hz/ms sweeping rate, the transient responses of Systems I and II are shown in Figures 5.9(c) and 5.10(c). Here the loop recaptures the signal at $f_{in} = f_{UC}$ (System I = 3,063 Hz; System II = 2,836 Hz) corresponding to the upper capture frequency. Then it tracks the input signal to $f_{in} = f_{LL}$ (System I = 1,500 Hz; System II = 1,594 Hz), called the lower lock frequency. The lower side of the frequency spectrum of the triangle wave output is larger than that of the upper side, as seen in Figures 5.9(d) and 5.10(d). Thus, the VCO output is tracking properly the input signal. From these simulation results, the frequency spreads between f_{UC}, f_{LC} and f_{UL}, f_{LL} correspond to the total capture range and the total lock range of the system. These are given by

$$\begin{aligned}
\text{Capture range, } 2 \cdot \Delta f_C &= 2 \cdot \text{ Lock-in range} = f_{UC} - f_{LC} \text{ (Hz)} \\
2 \cdot \Delta f_C \text{ of System I } &= 3{,}063 - 2{,}063 = 1{,}000 \text{ (Hz)} \\
2 \cdot \Delta f_C \text{ of System II } &= 2{,}836 - 2{,}369 = 467 \text{ (Hz)} \\
\text{Lock range, } 2 \cdot \Delta f_L &= 2 \cdot \text{ Tracking range } = f_{UL} - f_{LL} \text{ (Hz)} \\
2 \cdot \Delta f_L \text{ of System I } &= 3{,}638 - 1{,}500 = 2{,}138 \text{ (Hz)} \\
2 \cdot \Delta f_L \text{ of System II } &= 3{,}563 - 1{,}594 = 1{,}969 \text{ (Hz)}
\end{aligned} \tag{5.7}$$

With reference to Figures 5.9 and 5.10, the basic performance characteristics of the APLL macromodel can be summarized as follows:

1. The APLL macromodel shows that when the model is in lock, V_e is primarily a DC voltage, and the loop filter has little effect on the tracking range but does control the interference-rejection properties.
2. Decreasing the low-pass filter bandwidth decreases the capture range and improves the interference-rejection properties of the PLL. This is due to the increased attenuation of the error voltage. Also, the transient response of the loop becomes underdamped because of a smaller loop damping factor.
3. The APLL macromodel exhibits a selective frequency-to-voltage conversion characteristic, centered around the VCO free-running frequency, f'_0.
4. The APLL can acquire lock (capture) with only those signals that fall within the total capture range, centered about f'_0.
5. Once locked, the APLL tracks an input signal over the total lock range, centered about f'_0.
6. The slope of the frequency-to-voltage conversion characteristic in lock is the reciprocal of the VCO voltage-to-frequency conversion gain. Thus, if there is no amplification in the APLL system, one-half of the lock range can be expressed by

$$\frac{\Delta f_L}{2} = \Delta(V_e)_{max} \cdot K_v \tag{5.8}$$

7. The APLL macromodel has stable modes of locked and unlocked states. When the input frequency increases linearly, the APLL macromodel moves from the unlocked to the locked state, and then back to the unlocked state. Thus the new APLL macromodel covers both operating regions as well as the transition between these two regions.

Stability analysis. The condition for phase locking does not ensure loop stability. A PLL can oscillate if at some frequency its open-loop gain is greater than unity and the phase shift exceeds 180 degrees. Thus, the stability analysis of PLL system is another important consideration in designs and simulations.

Linear mathematical PLL models are widely used to analyze the stability and overshoot of a PLL system. However, the mathematical transfer function must be rederived whenever loop filter configuration changes. The linearized PLL macromodel does not have this problem because each functional block is a drop-in type of macromodel. Figure 5.11 shows frequency responses of both APLL systems. The phase and gain margins of each open loop are determined from the simulations results to be:

$$\text{System I}: \quad \text{Phase margin} \approx 34.8 \text{ (degrees)}$$

$$\text{Gain margin} \approx 61.65 \text{ (dB)}$$

$$\text{System II}: \quad \text{Phase Margin} \approx 19 \text{ (degrees)}$$

$$\text{Gain Margin} \approx 46.11 \text{ (dB)}$$

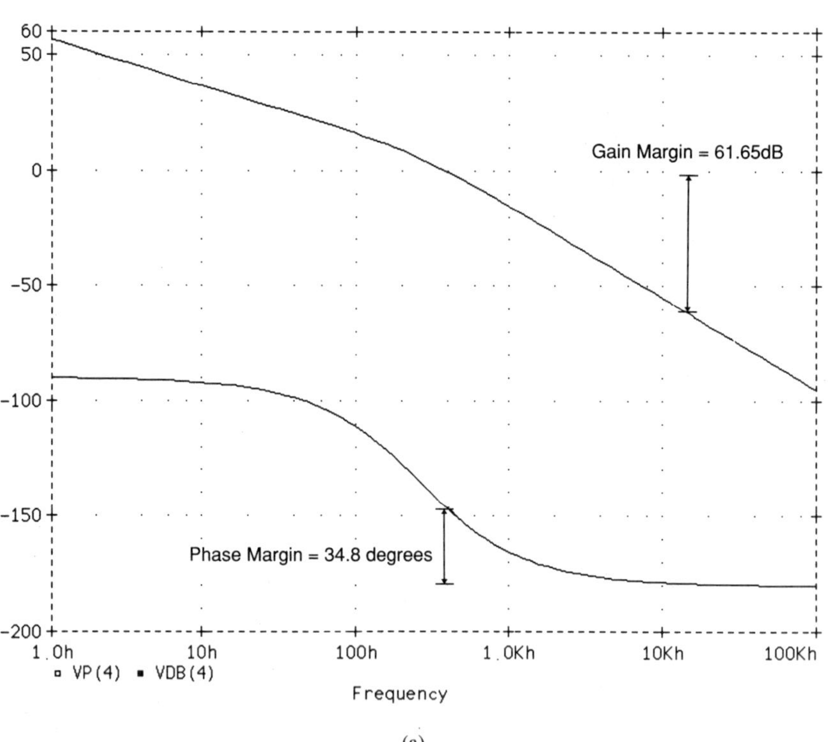

(a)

Figure 5.11 Frequency response of the linearized APLL macromodels. (a) System I. (b) System II.

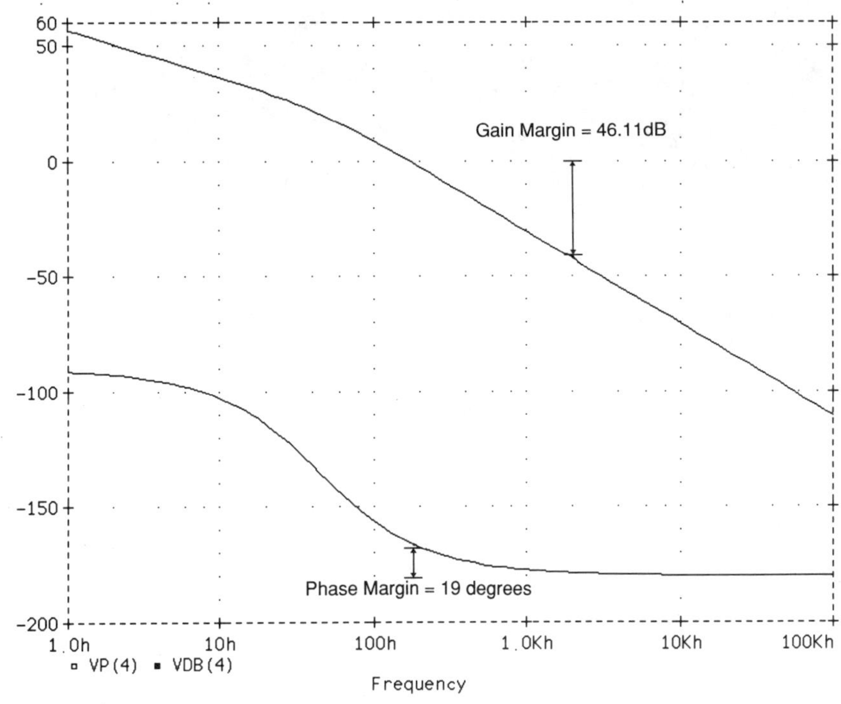

Figure 5.11(b)

As shown in Figure 5.12, the linearized PLL macromodel is also used to measure overshoot of each system. When a step of voltage is applied to the input of the closed loop macromodels, the overshoot of System I is about 35%, while that of System II is 66%.

5.4 ACCURACY OF THE APLL MACROMODEL

In this section, the APLL macromodel is examined more closely to determine its accuracy. The key performance parameters of the APLL macromodel are compared to those of the NE 565F analog phase locked loop system produced by Signetics and other IC manufacturers. The simulated performance parameters of the NE 565F macromodel are compared with experimental results. The macromodel accuracy is expressed as a percentage error according to

$$\%\ \text{Error of macromodel} = \frac{\text{Macromodel—Experimental}}{\text{Experimental}} \cdot 100\% \qquad (5.9)$$

Capture and lock ranges are compared with mathematical expressions developed by the manufacturer. The linear mathematical PLL model also is used to define the accuracy of the loop natural frequency and the loop damping factor.

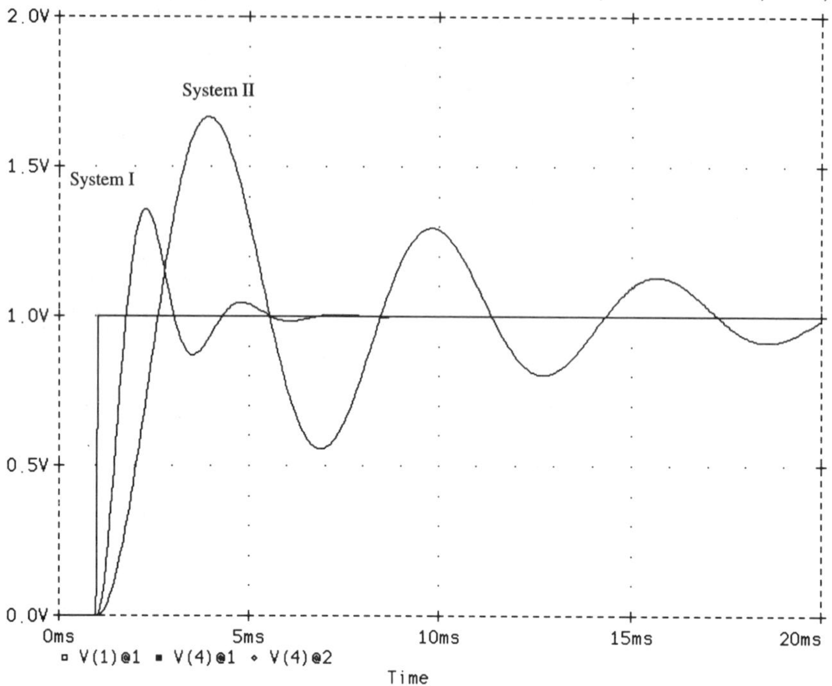

Figure 5.12 Transient response of the linearized APLL macromodels for a step of input voltage

Free-running frequency (f_0'). Only the VCO portion of the NE 565F is used here. Figure 5.13 shows the test circuit used to measure the free-running frequency, f_0'. A digitizing oscilloscope (HP 54110D) is connected to the VCO output of the loop. A capacitor, C_1, is connected between pins 7 and 8 in the test circuit to avoid possible oscillation problems. No input signal is present in this measurement.

With $R_{time} = 12\ K\Omega$, $C_{time} = 10$ nF, $+V = +10$ V, and $-V = -10$ V, the VCO oscillated at an average free-running frequency of $f_0' = 2{,}525$ Hz. Figure 5.14 compares the actual PLL output waveshapes with the simulation results of the macromodel. The percentage error of f_0' is

$$\%\text{Error of } f_0' = \frac{2{,}500 - 2{,}525}{2{,}525} \cdot 100\% \approx -1\% \tag{5.10}$$

VCO conversion gain (K_v). The test circuit in Figure 5.15 was used to measure the VCO conversion gain, K_v. The variable DC voltage at pin 7 is swept and the output frequency is measured. The VCO conversion gain, K_v is equal to the deviation of the VCO angular frequency, $\Delta\omega_0$, caused by a variation of the input voltage, ΔV_e. The VCO output frequency versus the input controlling voltage is illustrated in Figure 5.16. The solid line with asterisks (*) represents a best-fit straight line to the measured data of the VCO section. The dashed line with white squares is from the VCO macromodel.

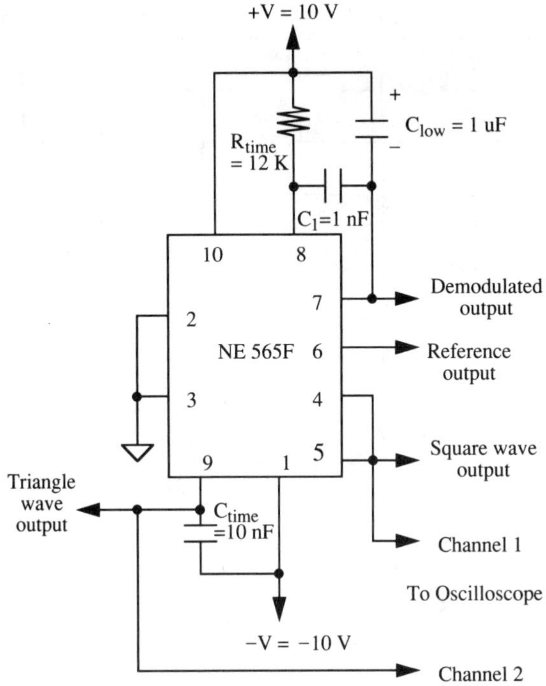

Figure 5.13 Test circuit for the measurement of the free-running frequency, f_0'.

The linear equations of these best-fit lines are given by

$$\text{NE 565F}: \omega_0 = -(2\pi \cdot 1{,}017) \cdot V_{in} + 10{,}414.7 \tag{5.11}$$

$$\text{Macromodel}: \omega_0 = -(2\pi \cdot 1{,}022) \cdot V_{in} + 10{,}473$$

In Equation (5.11), the coefficient of V_{in} is the conversion gain of the VCO. The constant term represents the free-running angular frequency ω_0'. Thus, the percentage error of the conversion gain of the VCO macromodel is determined by

$$\%\ \text{Error of } K_v = \frac{-1{,}022 + 1{,}017}{-1{,}017} \cdot 100\ (\%) \approx 0.5\% \tag{5.12}$$

The percentage error of the free-running frequency in this measurement is given by

$$\%\ \text{Error of } f_0' = \frac{10{,}473 - 10{,}415}{10{,}415} \cdot 100\ (\%) \approx 0.6\% \tag{5.13}$$

Phase detector conversion gain (K_d). After the PLL locks, it will remain in lock if the phase difference is changed by a small amount because of a small variation of the input frequency. This phase difference provides an additional voltage at the loop filter output. The test circuit of Figure 5.17 was used for the measurement of the conversion gain, K_d, of the phase detector. The initial frequency of the input signal must be close to the free-running frequency of the NE 565F so that the PLL system operates in its lock range. After determining that the PLL is locked, the phase difference between the input

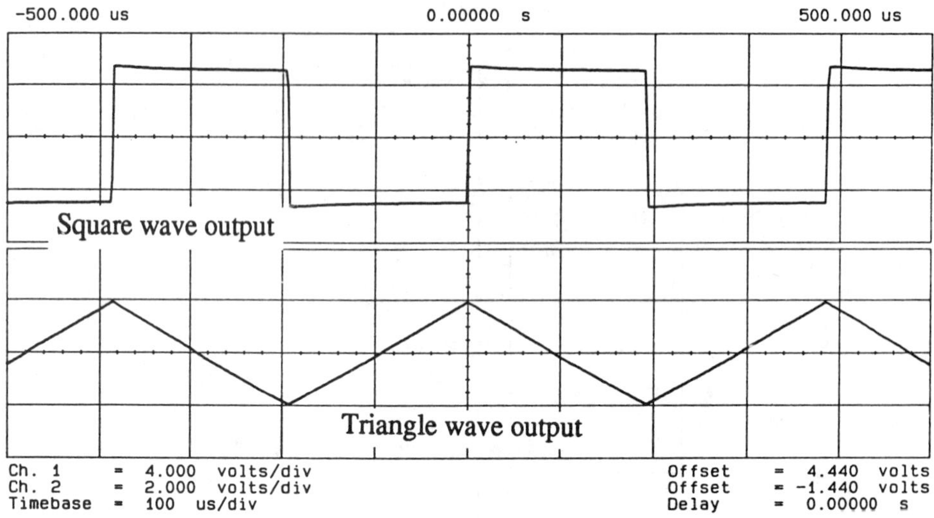

(a)

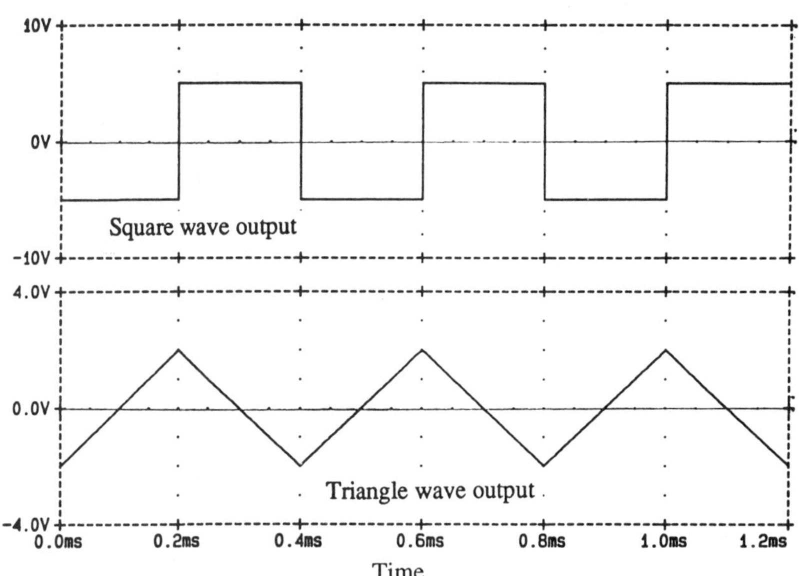

(b)

Figure 5.14 Comparison of waveforms at VCO output. (a) NE 565F phase locked loop. (b) Macromodel of NE 565F.

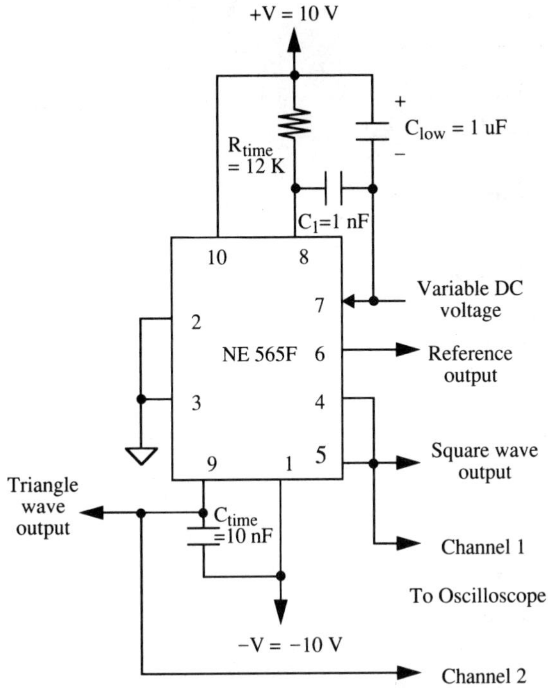

Figure 5.15 Test circuit for the measurement of the VCO conversion gain, K_V.

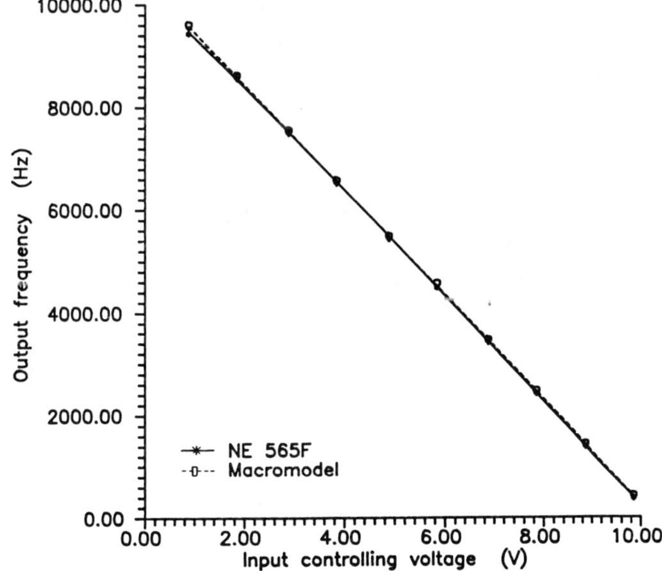

Figure 5.16 Comparison of the VCO conversion gain, K_V, between the NE 565F PLL and its macromodel.

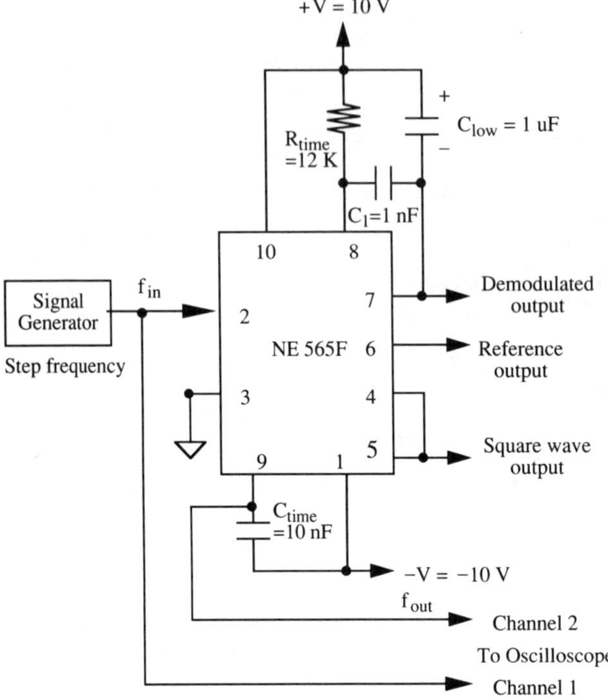

Figure 5.17 Test circuit for the measurement of the conversion gain of the phase detector, K_d, of the NE 565F.

(Channel 1) and the PLL output (Channel 2) is measured with the oscilloscope. The DC voltage at the loop filter output is also measured.

Figure 5.18 is a graphical representation of the comparison between the NE 565F system and its macromodel. The solid line with asterisks (*) is the best-fit line for the experimental data whose linear equation is

$$\text{NE565 F: } V_d = -0.0117183 \cdot \theta_d \text{ (degree)} + 1.02295$$

$$\text{NE565 F: } V_d \approx -0.67 \cdot \theta_d \text{ (radian)} + 1.02295 \tag{5.14}$$

The simulation results represented by the best-fit dashed line produce the equation:

$$\text{Macromodel: } V_d = -0.0111812 \cdot \theta_d \text{ (degree)} + 0.9103581$$

$$\text{Macromodel: } V_d \approx -0.64 \cdot \theta_d \text{ (radian)} + 0.9103581 \tag{5.15}$$

The percentage error of the conversion gain of the phase detector macromodel is given by

$$\%\text{ Error of } K_d = \frac{-0.64 + 0.67}{-0.67} \cdot 100(\%) = -4.6(\%) \tag{5.16}$$

Capture (Δf_C) and lock ranges (Δf_L). Consider the test circuit shown in Figure 5.19. A simple first-order low-pass RC filter is chosen for simplicity, consisting of the on-chip resistor, $R_{low} = 3.6\ K\Omega$, and the external capacitor $C_{low} = 1\ \mu F$. A signal generator is applied to the input (pin 2) of the NE 565F. Both the input signal, f_{in} and

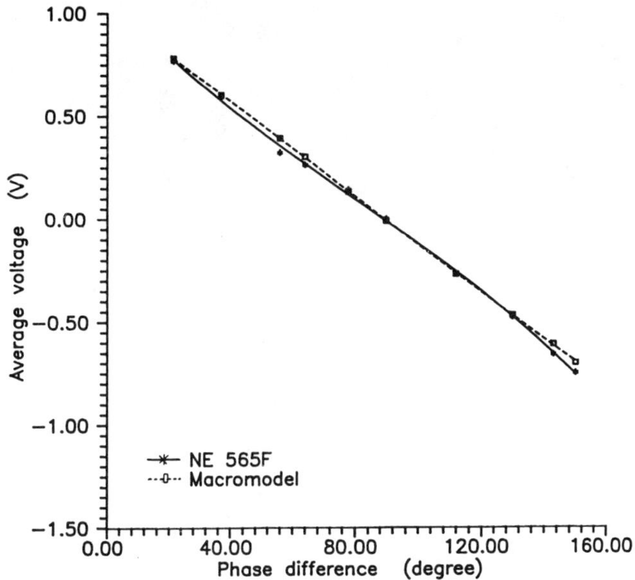

Figure 5.18 Comparison of the phase detector gain, K_d, between the NE 565F PLL and its macromodel.

the VCO output signal, f_{out}, are displayed on an oscilloscope, which is triggered by the VCO output signal.

The frequency of the signal generator is now varied manually until lock is observed. Figure 5.20(a) shows when the input frequency is far away from the free-running frequency, f_0', and the system is unlocked. Figure 5.20(b) shows where f_{in} is at the edge of the capture range. This figure shows that the PLL is trying to lock to the input frequency since the VCO output frequency is barely synchronized to the input. This means that the phase error between the input and the VCO output is at the edge of the dynamic range of the phase detector. As the input frequency approaches f_0', the PLL suddenly locks, as shown in Figure 5.20(c).

Determination of the capture and lock ranges is very simple. The capture range is measured by slowly varying the input frequency and monitoring the upper and lower values of f_{in} where the system locks. In a similar way, the lock range is determined by monitoring the upper and lower values of the input frequency where the system becomes unlocked. However, the value obtained for the capture range was often larger than the value specified by the manufacturer. In effect, the method shown in Figure 5.19 yields a value that is between the capture range and pull-in range. If the input frequency, f_{in}, is swept more slowly, the measured value becomes closer to the pull-in range. If it is swept faster, the measured value comes closer to the lock range. Thus, the sweeping rate of the input signal is critical and must be very slow as the edge of the capture range is approached (sweeping toward the free-running frequency). If swept too fast, the lock-up transient delay will cause an error in reading the band edge. Also, the frequency must

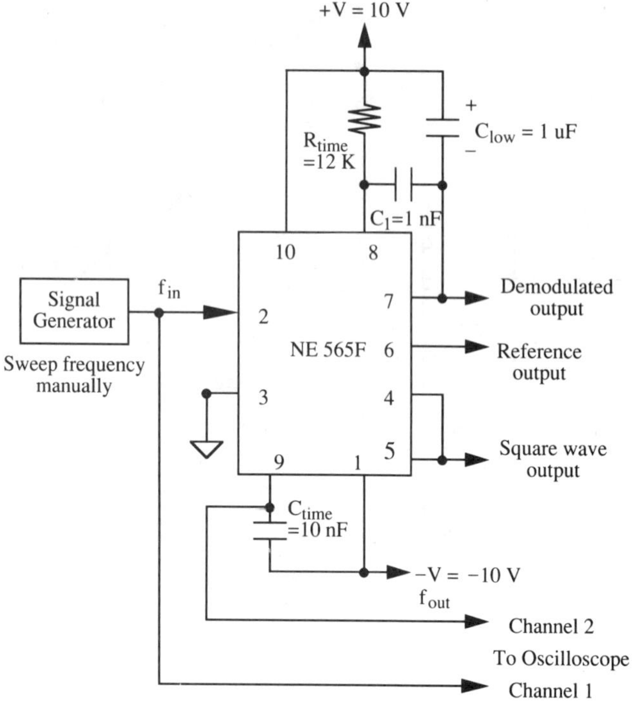

Figure 5.19 Test circuit for the measurement of the capture and lock ranges.

be read from the signal generator rather than the loop VCO because the VCO frequency gyrates widely at the edge of the lock-up range just before and after lock, as was shown in Figure 5.20(b).

Figure 5.21(a) and (b) show the upper and lower capture frequencies, and (c) and (d) are the upper and lower lock frequencies of the NE 565F PLL test system. Recorded frequencies are not precise values, but very close. Thus, this measurement was repeated 10 times to average the errors. Data are listed in Table 5.1. The errors of the capture and lock range of the APLL macromodel are given by

$$\begin{aligned} \text{\% Error of capture range} &= \frac{467 - 455}{455} \cdot 100\ (\%) \approx 2.6\% \\ \text{\% Error of lock range} &= \frac{1{,}969 - 1{,}910}{1{,}910} \cdot 100\ (\%) \approx 3.1\% \end{aligned} \tag{5.17}$$

Another way to determine the accuracy of the capture and lock ranges of the APLL macromodel is to consider the data sheet, which states that the NE 565F PLL can lock to and track the input signal over typically $\pm 60\%$ bandwidth with respect to f_0', the free-running frequency. The lock range and capture range of the NE 565F PLL are estimated as follows:

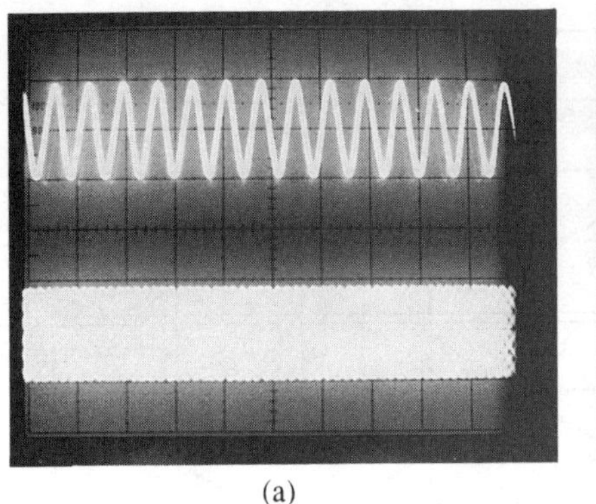
(a)

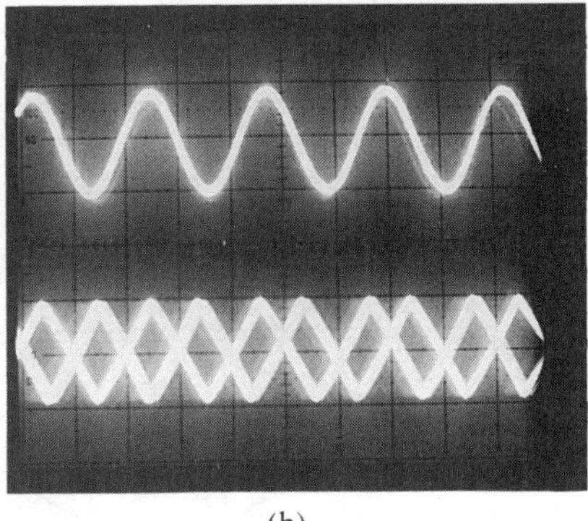
(b)

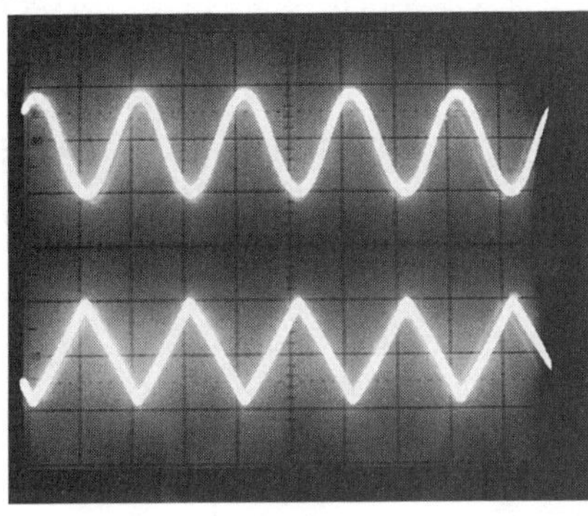
(c)

Figure 5.20 Waveforms of the input and output signals in the test circuit of Figure 5.19. Upper trace: Input. Lower trace: Triangle wave output. X-axis: 0.2 ms/div. Y-axis: 2V/div. (a) PLL unlocked (f_{in} = 7 KHz). (b) PLL is at the edge of the capture frequency (f_{in} = 2,300 Hz). (c) PLL locked (f_{in} = 2,500 Hz).

TABLE 5.1 CAPTURE AND LOCK RANGE OF THE NE 565F PLL

Measurement	Upper Capture Frequency f_{UC}	Lower Capture Frequency f_{LC}	Upper Lock Frequency f_{UL}	Lower Lock Frequency f_{LL}	Unit
1	2,799	2,330	3,528	1,596	Hz
2	2,768	2,330	3,553	1,585	Hz
3	2,799	2,309	3,568	1,585	Hz
4	2,783	2,330	3,553	1,596	Hz
5	2,815	2,386	3,553	1,590	Hz
6	2,815	2,397	3,553	1,585	Hz
7	2,815	2,330	3,578	1,585	Hz
8	2,815	2,330	3,528	1,590	Hz
9	2,815	2,374	3,553	1,596	Hz
10	2,799	2,352	3,528	1,590	Hz
Average	2,802	2,347	3,500	1,590	Hz
	Capture range		**Lock range**		
	NE 565F PLL	Macromodel	NE 565F PLL	Macromodel	
	455	467	1,910	1,969	Hz

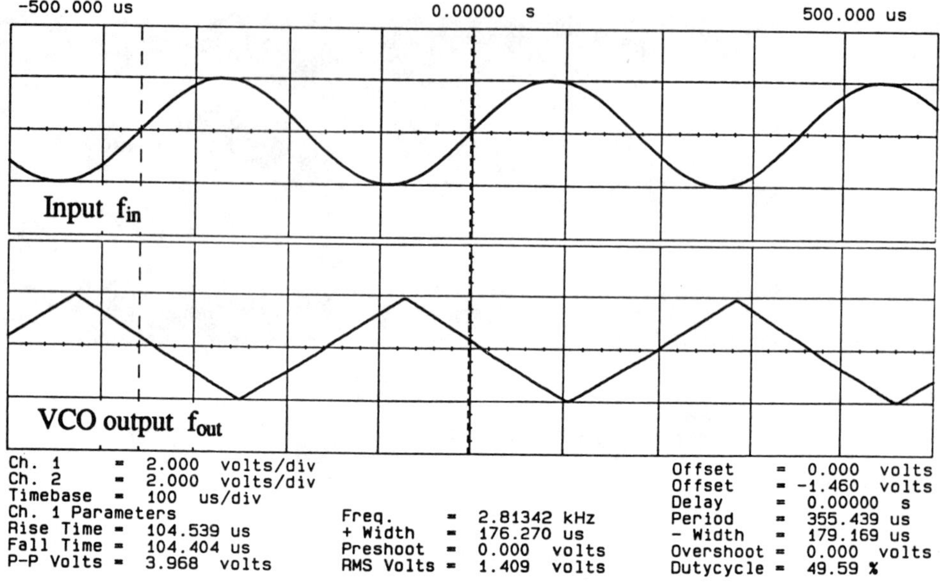

(a)

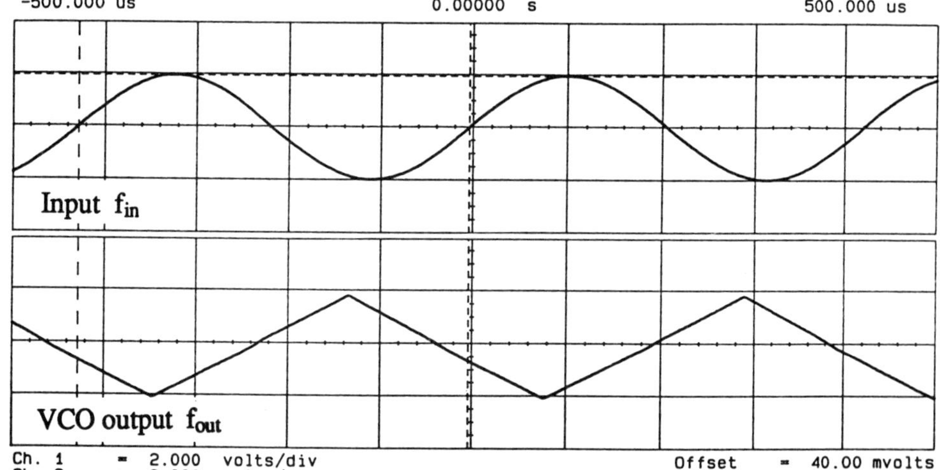

(b)

Figure 5.21 Experimental data of the capture and lock frequencies. (a) Upper capture frequency. (b) Lower capture frequency. (c) Upper lock frequency. (d) Lower lock frequency.

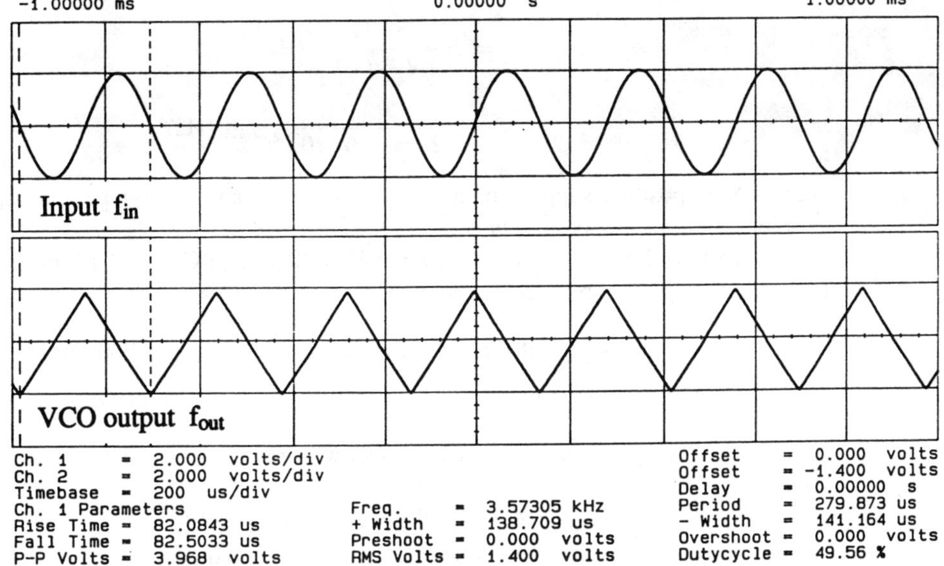

(c)

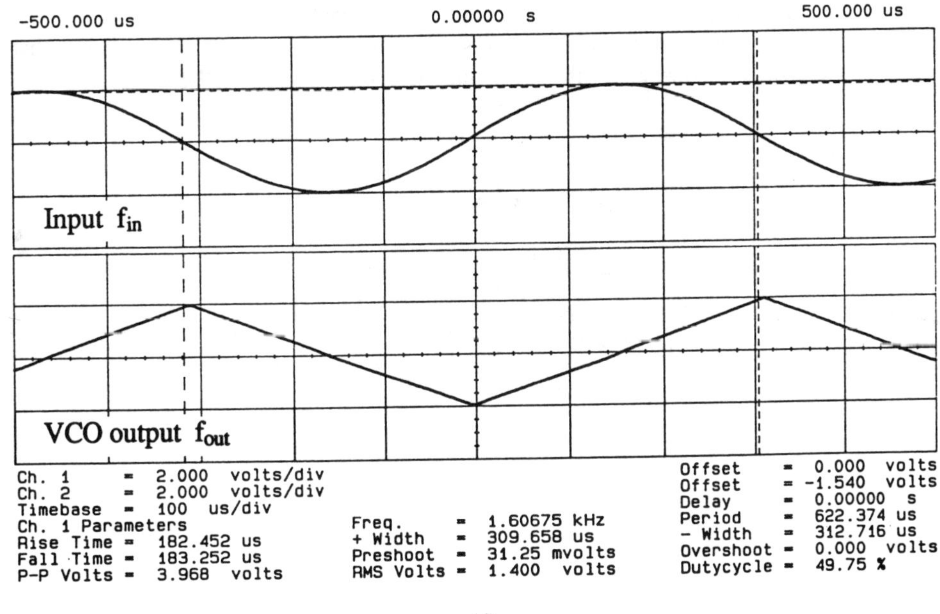

(d)

Figure 5.21(c) and (d)

$$f_{\text{tracking}} = \pm\frac{8 \cdot f_0'}{+V - (-V)}\ \text{(Hz)} \tag{5.18}$$

$$f_{\text{lock-in}} = \pm\frac{f_{\text{tracking}}}{2 \cdot \pi \cdot 3.6 \cdot 10^3 \cdot C_{\text{low}}}\ \text{(Hz)}$$

where +V is positive supply voltage, −V is negative supply voltage, and C_{low} is the capacitance in the low-pass filter.

Since $f_0' = 2.5$ KHz, $+V - (-V) = 2\ V = 20$ V, and $C_{low} = 1\ \mu F$ in Figure 5.19, the lock and capture range are given by

$$\text{Lock range of NE 565F system} = 2 \cdot f_{\text{tracking}} = 2{,}000\ \text{(Hz)}$$
$$\text{Capture range of NE 565F system} = 2 \cdot f_{\text{lock-in}} = 421\ \text{(Hz)} \tag{5.19}$$

Thus, the percentage error of the macromodel with respect to the data sheet is

$$\%\ \text{Error of lock range} \approx -1.6\% \tag{5.20}$$
$$\%\ \text{Error of capture range} \approx 10.9\%$$

Natural frequency (ω_n) and damping factor (ζ). The equivalent circuit structure APLL macromodel of Figure 5.22 was used for the measurement of the loop natural frequency, ω_n, and the loop damping factor, ζ.

To measure these parameters from simulation results, a disturbance that forces the system to settle at a different stable state was applied to the APLL macromodel. This

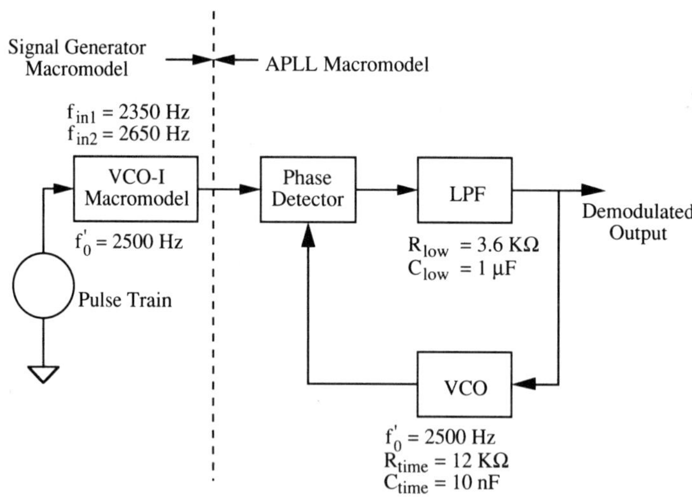

Figure 5.22 Equivalent circuit macromodel for the measurement of the natural frequency and the damping factor.

was easily done by modulating the output frequency of the VCO-I with a square wave input signal. This type of signal is modeled by using a PULSE independent voltage source in SPICE. The amplitude of the square wave is 0.0864 V to produce a small-frequency deviation of $\Delta f = \pm 150$ Hz. Thus, the output frequency of the VCO-I is two tones of 2,650 Hz and 2,350 Hz. Within this frequency range, the APLL model stays in its lock state, since its capture and lock ranges are 455 Hz and 1,910 Hz, respectively.

When the output frequency of the signal generator is abruptly changed from one level to another because of the voltage level shift at the VCO-I input, a phase error, θ_e, results. The output signal of the loop filter can be considered a measure of the average phase error. Thus the transient response of the APLL macromodel can be easily analyzed by plotting the loop filter output signal, as shown in Figure 5.23.

The loop filter output signal shows a damped sinusoidal oscillation on each transition of the 30 ms repetition rate, and then it settles at a stable DC level. Now ζ and ω_n can be calculated for the waveform of the loop filter output signal; ζ can be calculated from the ratio of the amplitudes of two subsequent half-waves, A_1 and A_2. This

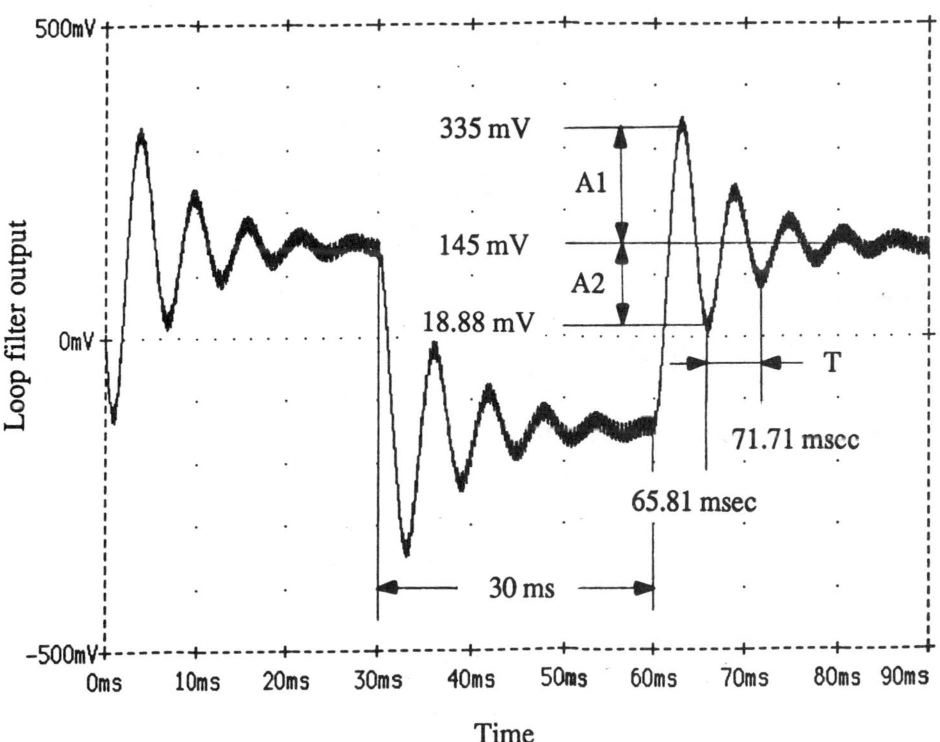

Figure 5.23 Transient response at the loop filter output of Figure 5.22.

relationship is given by

$$\zeta = \frac{\ln \frac{A_1}{A_2}}{\sqrt{\pi^2 + \left(\ln \frac{A_1}{A_2}\right)^2}} \tag{5.21}$$

The natural frequency ω_n is calculated from the period T of one oscillation in Figure 5.23 according to

$$\omega_n = \frac{2\pi}{T \cdot \sqrt{1 - \zeta^2}} \tag{5.22}$$

From the macromodel simulation results, the loop damping factor and the natural frequency are given by

$$\zeta = \frac{\ln \frac{190\ \text{mV}}{126\ \text{mV}}}{\sqrt{\pi^2 + \left(\ln \frac{190\ \text{mV}}{126\ \text{mV}}\right)^2}} \approx 0.1296 \tag{5.23}$$

$$\omega_n = \frac{2\pi}{5.9\text{msec} \cdot \sqrt{1 - 0.1296^2}} \approx 1{,}073.8\ (\text{rad/s}) \approx 170.9\,(\text{Hz})$$

The mathematical linear model of the NE 565F PLL system used in this verification is

$$\frac{\theta_{out}(s)}{\theta_{in}(s)} = \frac{K_d \cdot K_v \cdot F(s)/s}{1 + K_d \cdot K_v \cdot F(s)/s} \tag{5.24}$$

where F(s) is the transfer function of the low-pass filter in the s-domain. The F(s) is given by

$$F(s) = \frac{1}{s\tau + 1} \tag{5.25}$$

where τ = 3.6 msec. Thus,

$$\frac{\theta_{out}(s)}{\theta_{in}(s)} = \frac{K_d \cdot K_v}{s^2 + \frac{1}{\tau}s + \frac{K_d \cdot K_v}{\tau}} = \frac{\omega_n^2}{s^2 + 2\zeta\omega_n s + \omega_n^2} \tag{5.26}$$

From Equation (5.26), the natural frequency and loop damping factor are given by

$$\omega_n = \sqrt{\frac{K_d \cdot K_v}{\tau}} = \sqrt{\frac{0.67 \cdot 2\pi \cdot 1017}{3.6 \cdot 10^{-3}}} \approx 1{,}090.53\ (\text{rad/s}) \approx 173.6\ (\text{Hz}) \tag{5.27}$$

$$\zeta = \frac{1}{2 \cdot \sqrt{K_d \cdot K_v \cdot \tau}} \approx 0.1274$$

Thus, the percentage errors of the macromodel are given by

$$\%\ \text{Error of}\ \zeta = \frac{0.1296 - 0.1274}{0.1274} \cdot 100\ (\%) \approx 1.7\% \tag{5.28}$$

$$\%\ \text{Error of}\ \omega_n = \frac{170.9 - 173.6}{173.6} \cdot 100\ (\%) \approx -1.6\%$$

5.5 APPLICATIONS OF THE APLL MACROMODEL

FSK Demodulator. The APLL macromodel was configured for a Frequency Shift Key (FSK) demodulation application. FSK refers to digital data transmission by means of a carrier that is shifted between two preset frequencies. This frequency shift is usually accomplished by driving a VCO with binary data so that the two resulting frequencies correspond to the logic "0" and "1" states, commonly called *space* and *mark* frequencies. The NE 565F phase locked loop is frequently used as a receiving converter to demodulate FSK tones and to provide a shifting DC voltage to initiate mark or space frequencies. A simple schematic using the NE 565F PLL to demodulate FSK frequencies of 2,250 Hz and 2,750 Hz is shown in Figure 5.24(a). When the signal appears at the input, the loop locks to the input frequency and tracks it between the two frequencies, with corresponding DC shifts at the loop filter output.

The PLL output is filtered with a third-order low-pass filter to remove the higher order sum frequency components. The order of the filter chosen depends on the degree of uncertainty allowable in the comparator output pulse. In this simulation, the band edge of the ladder filter was chosen to be approximately 2 KHz (approximately halfway between

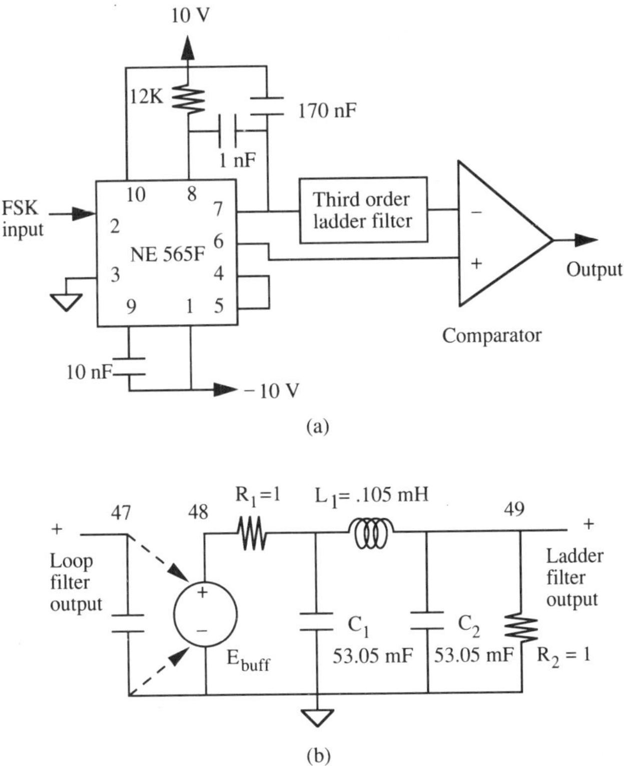

Figure 5.24 FSK demodulator. (a) Functional block diagram. (b) Third-order low-pass filter.

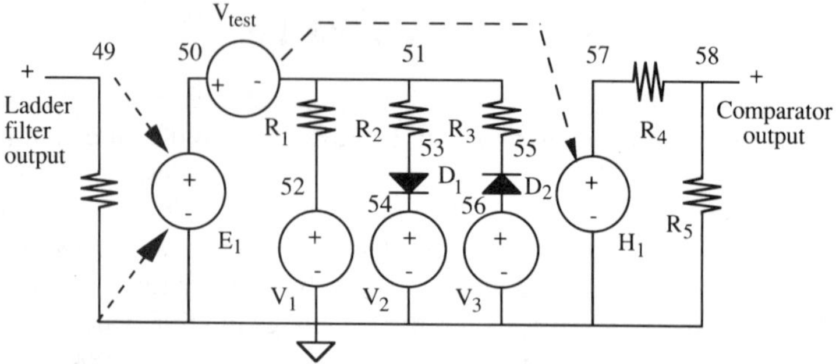

Figure 5.25 Comparator macromodel equivalent circuit.

the maximum keying rate of 500 Hz and twice the carrier frequency). The filtered output signal can now be made logic-compatible by connecting a voltage comparator. An equivalent circuit macromodel of the voltage comparator used in this simulation is shown in Figure 5.25. The voltage transition at the comparator output occurs between $-10\mu V$ and $+10\mu V$ of the input.

SPICE code for Figure 5.25

```
E1 50 0 49 0 1
VTEST 50 51 DC=0 AC=0
R1 51 52 -2E-6
V1 52 0 2.5E-8
R2 51 53 2E-6
D1 53 54 SWITCH2
V2 54 0 -1E-5
R3 51 55 2E-6
D2 56 55 SWITCH2
V3 56 0 1E-5
H1 57 0 POLY(1) VTEST 0 1
R4 57 58 1K
R5 58 0 1G
.MODEL SWITCH2 D(N=1E-6)
```

Figure 5.26 shows the simulation results of the FSK macromodel of Figure 5.24(a). The upper trace of each figure is the FSK input signal, the middle trace is the output of the third-order ladder filter, and the lower trace represents the output of the comparator model. When the keying rate of the input is 100 and 200 bits/s in Figures 5.26(a) and (b), the comparator model produces the correct outputs of logic "1" and "0," with some time delay. When the input signal is 500 bits/s, the logic output becomes stable after about 2.5 ms [Figure 5.26(c)], but the signal delay becomes longer. Figure 5.26(d) shows simulation results when the input signal is 1,000 bits/s. The logic output of the comparator model is not correct because the output of the comparator is a logic "0" when the input is the mark frequency, and vice versa. Thus, the FSK demodulator used in this simulation fails to operate correctly with this high-input baud rate.

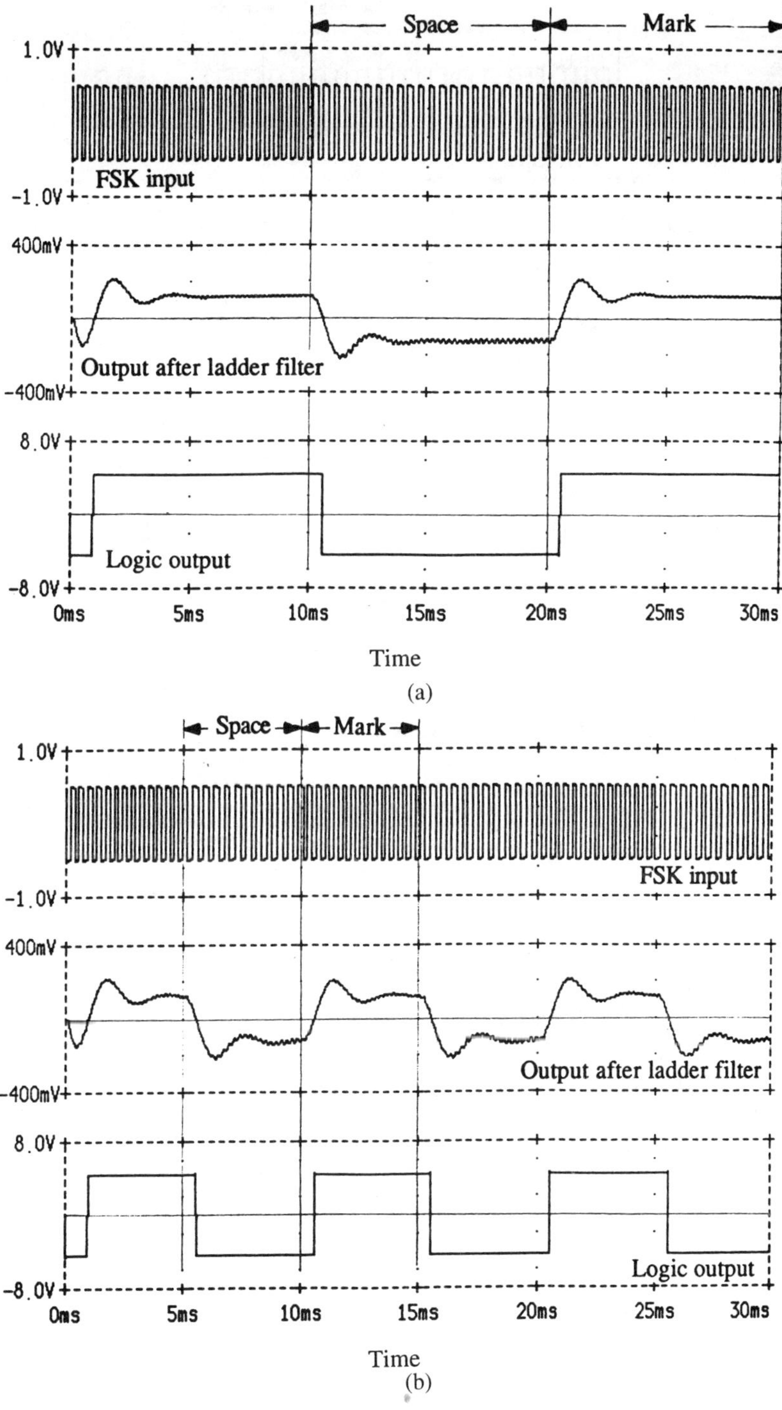

Figure 5.26 Simulation results of the FSK demodulator macromodel. (a) Input = 100 bits/s. (b) Input = 200 bits/s. (c) Input = 500 bits/s. (d) Input = 1,000 bits/s.

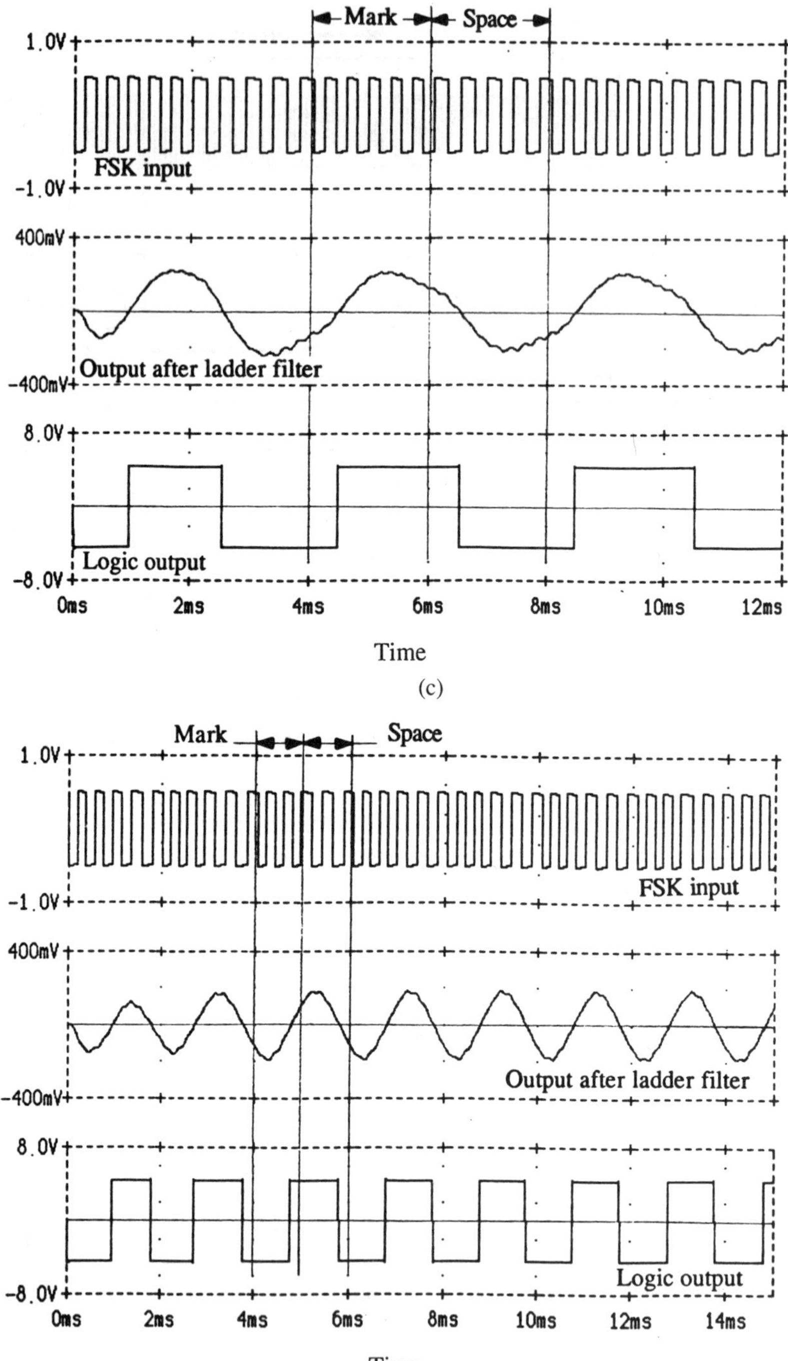

Figure 5.26(c) and (d)

5.6 SUMMARY

The APLL macromodel was verified by examining its performances for various input signals such as a step of phase, a step of frequency, and a ramp of frequency. From these simulations, the key performance parameters of the phase locked loop were determined without tedious mathematical calculations. The loop performance parameters of the macromodel were controlled to a target accuracy of $\pm 5\%$. The APLL macromodel was shown to overcome the serious limitations of other PLL macromodels such as flexibility, generality, and practicality.

The results from the FSK demodulator simulations show that the PLL macromodel can be used to verify many PLL applications such as AM, FM, PM, and waveform generation without either hardware realizations or mathematic solutions. Also the PLL model can be used in its own redesign to improve its performance parameters. Furthermore, the limitations of the input signal to achieve proper system performance can be determined with this APLL macromodel.

REFERENCES

1. Signetics Corporation, *Analog Data Manual*, Sunnyvale, California, June 1983, pp. 11-12 to 11-16; J. Alvin Connelly, *Phase Locked Loops Design and Application Manual*, Signetics Corporation, 1979, pp. 265–342; National Semiconductor Corporation, *Linear Integrated Circuits*, Santa Clara, California, 1975, pp. 5–59 to 5–63.

FURTHER READINGS

D. Kesner, "Take the Guesswork out of Phase Locked Loop Design," *EDN*, January 5, 1973, pp. 54–60.

H. Cohen, "How Phase-Locked Loops Work," *Popular Electronics*, February 1975, pp. 32–34.

A. B. Przedpelski, "PLL Primer Part I," *EDN*, March–April 1983, pp. 18–28.

A. B. Przedpelski, "PLL Primer Part II," *EDN*, May–June 1983, pp. 12–28.

A. B. Przedpelski, "PLL Primer Part III," *EDN* , July–August 1983, pp. 48–96.

P. Hasan, "Nonlinear PLL Analysis," *Electronic Engineering*, Vol. 56, No. 686, February 1984, pp. 30–32.

E. E. L. Mitchell, "Phase Locked Loop Technique: Interactive Simulation with the Advanced Continuous Simulation Language (ACSL)," *Proceedings of the 1978 Summer Simulation Conference*, July 1978, pp. 444–451.

B. Priestley and B. Langley, "Nonlinear Analysis of Phase Locked Loop," *Electronic Engineering*, April 1983, p. 27.

B. Labitt, "Phase Locked Loop Simulation," *Proceedings of the Twenty-fifth Midwest Symposium on Circuits and Systems*, 1982, pp. 200–206.

S. Yousef, I. E. Dokknay, and S. A. Monaem, "A Computer Program DESPIL for Design of Phase Locked Loop PLL," *Proceedings of the IEEE 1983 National Aerospace and Electronics Conference*, Vol. 1, pp. 362–365.

S. Lubs, "Simple Program Finds Parts Values for Phase-Locked Loop," *Electronic Design*, Vol. 32, No. 22, October 31, 1984, pp. 319–320.

S. Zubiel, "Predict PLL Transient Response with Computer Simulation," *RF Design*, Vol. 8, No. 5, May 1985, pp. 35–38.

R. Rippy, "BASIC Program Eases Analysis of Phase Locked Loops," *EDN*, Vol. 31, No. 13, June 26, 1986, pp. 157–162.

J. Kovats, "A Simulation Program for High-Order Nonlinear PLLs," *Simulation of Control Systems*, Selected papers from IFAC symposium, edited by I. T. Kopacek and F. Breitenecker, 1987, pp. 423–426.

P. Champagne, "BASIC Program Simulates Digital Phase Locked Loops," *Microwave and RF*, Vol. 27, No. 1, January 1988, pp. 117–120.

S. J. Garrett and V. K. Jain, "Macromodel of a Phase Lock Loop Integrated Circuit," *IEEE Proceedings of Southeastcon 1979*, April 1979, pp. 438–441.

J. A. Connelly and G. E. Prescott, "Phase-Locked Loop Includes Lock Indicator," *Electronics*, Vol. 47, No. 18, September 5, 1974, pp. 112–113.

J. A. Connelly and G. E. Prescott, "Active Filter Improves Tracking and Capture Ranges of PLL," *Electronic Design*, Vol. 23, No. 7, April 1, 1975, pp. 128–130.

J. A. Connelly and D. Martin, "Tape and Slide Projector Synchronizer Uses PLLs," *Electronics*, Vol. 54, No. 12, June 16, 1981, p. 169.

J. A. Connelly, "A General Description of the HA-2800/2805 High Frequency Phase Locked Loops," Harris Semiconductor Corporation, Application Note 603, October 1972.

J. A. Connelly, "FM Demodulation and Frequency Synthesis," Harris Semiconductor Corporation, Application Note 604, October 1972.

J. A. Connelly, "The HA-2820/2825 Low Frequency Phase Locked Loop," Harris Semiconductor Corpation, Application Note 605, January 1973.

J. A. Connelly, "Interfacing the High Frequency Phase Locked Loop," Harris Semiconductor Corporation, Application Note 606, January 1973.

J. A. Connelly, "Take the Haze Out of the Phase Locked Loop," Harris Semiconductor Data Channel, August 1972.

6

Automatic Generation of Macromodels

A computer software program for the automatic generation of macromodels is given. The program produces a SPICE input file for some of the various macromodels presented in this book. The program is menu-driven for convenience and ease of operation. When the user gives answers to the desired macromodel performance specifications, the program produces an output file whose format is that of either a direct input file or a subcircuit file. The elements included in the macromodel are SPICE primitives. Therefore, the macromodel produced by the program can be used by the large SPICE family of circuit simulators.

6.1 INTRODUCTION

Overview. MACRO is a computer program for the automatic generation of macromodels of analog and sampled data circuits and systems. It is written in QuickBASIC. Many kinds of analog circuits can be modeled quickly as equivalent circuit structure macromodels using this program. All elements in the macromodel are SPICE standard primitives so that the resulting macromodels can be used by any type of the SPICE family of circuit simulators. A menu allows the user to select any of the following circuits or systems for which it will automatically generate a macromodel: analog phase locked loop (APLL), continuous-time (s-) domain filter transfer function, peak detector, sample and hold circuit, Schmitt trigger/bistable network, noise sources (voltage and current), piecewise linear circuit, discrete-time (z-) domain transfer function, and operational amplifier. A listing of each macromodel program is given to allow the interested user the ability to extend the menu to other macromodels or to add modifications to the programs supplied. Original extensions and modifications will need to be compiled using QuickBASIC or a similar compiler.

System requirements. MACRO will run on any IBM-PC, XT, or AT with at least 512 kilobytes of memory and a MS-DOS (3.0 or later) operating system. A fixed disk or floating-point processor is not required. Either a monochrome or a color graphics display can be used. No special type of printer or special printer features are needed. MACRO will also run on IBM-compatible systems, as long as they meet these requirements. Note that MACRO is delivered on two 360K diskettes.

Installation. The following files are stored on the accompanying diskettes:

Disk 1

```
AMP.EXE, APLL.EXE, CFFM.EXE, MACRO.EXE, NOISE.EXE
PEAK.EXE, PWLM.EXE, README.DOC, SAMPLE.EXE
```

Disk 2

```
SCHMITT.EXE, ZFFM.EXE, AMP.BAS, APLL.BAS, CFFM.BAS,
MACRO.BAS, NOISE.BAS, PEAK.BAS, PWLM.BAS, SAMPLE.BAS,
SCHMITT.BAS, ZFFM.BAS
```

No special installation procedure is required. If your system has a hard disk, just copy these files onto your fixed disk. If not, just use the companion diskettes.

Running MACRO. MACRO can be run from the hard disk by typing

```
C:\>MACRO          — MACRO is installed in the root directory
```

When the floppy diskettes are used, insert Disk 1 and then type MACRO according to

```
C:\>A:MACRO
```

If you wish to create a macromodel for either the Schmitt Trigger/Bistable Network (menu option 5) or the Discrete-Time (Z-Domain) Transfer Function model (menu option 8), insert Disk 2 before selecting either of these options since the executable programs for these two models are on the second disk. All other executable programs are on the first disk. The second disk contains all the .BAS source codes which can be modified and recompiled for creating customized macromodels.

Aborting a run. If for some reason you wish to abort MACRO while it is running, simultaneously press the < Ctrl > < C > keys. MACRO will "clean up" wherever it is in the routine and then return you to the main menu.

6.2 DATA REQUIRED AND OUTPUT FILE PRODUCED

Input data. The input data **MACRO** uses are the electrical parameters and transfer characteristics of the circuits or system to be modeled.

Output data. The output of **MACRO** is a circuit description file representing an equivalent circuit macromodel. The format of the circuit description file is either a SPICE

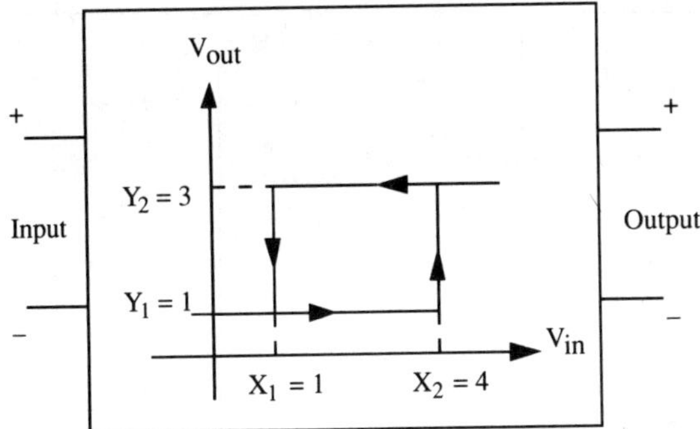

Figure 6.1 An example bistable transfer characteristic for MACRO.

input file or a SPICE subcircuit file. Since the output file is in standard ASCII format, the output file can be used by almost any kind of standard word processor or editor operated in DOS. Note that the user must append the desired form of input signal and specify which options (.AC, .DC, .TRAN, etc.) to the output data file generated by **MACRO**.

Example. Consider a bistable network whose transfer characteristic is shown in Figure 6.1. To model it, type **MACRO**. After a short time, the screen should clear and display the **MACRO** main menu. Figure 6.2 illustrates the procedure for the creation of the bistable macromodel.

6.3 COMPUTER SOFTWARE

6.3.1 Batch File: MACRO.BAS

```
10 REM -- Automatic Generation of Macromodel for Analog Circuits
   & Systems --
20 CLS
30 CLEAR
40 T$ = "SPICE Macromodels of Analog Circuits & Systems"
50 O$ = "Programmed by J.A. Connelly and Pyung Choi"
60 Q$ = "January 15, 1990"
70 L = LEN(T$): LA = LEN(O$): LB = LEN(Q$)
80 PRINT TAB(INT(80 - L) / 2 + .5); T$
90 PRINT : PRINT TAB(INT(80 - LA) / 2 + .5); O$
100 PRINT TAB(INT(80 - LB) / 2 + .5); Q$
110 PRINT : PRINT :
120 PRINT "Menu"
130 PRINT TAB(5); " 1. Analog Phase Locked Loop"
140 PRINT TAB(5); " 2. Continuous-Time (S-) Domain Transfer Function"
150 PRINT TAB(5); " 3. Peak Detector"
160 PRINT TAB(5); " 4. Sample-and-Hold Circuit"
170 PRINT TAB(5); " 5. Schmitt Trigger/Bistable Network"
180 PRINT TAB(5); " 6. Noise Sources (Voltage & Current)"
190 PRINT TAB(5); " 7. Piecewise Linear Circuit or System"
```

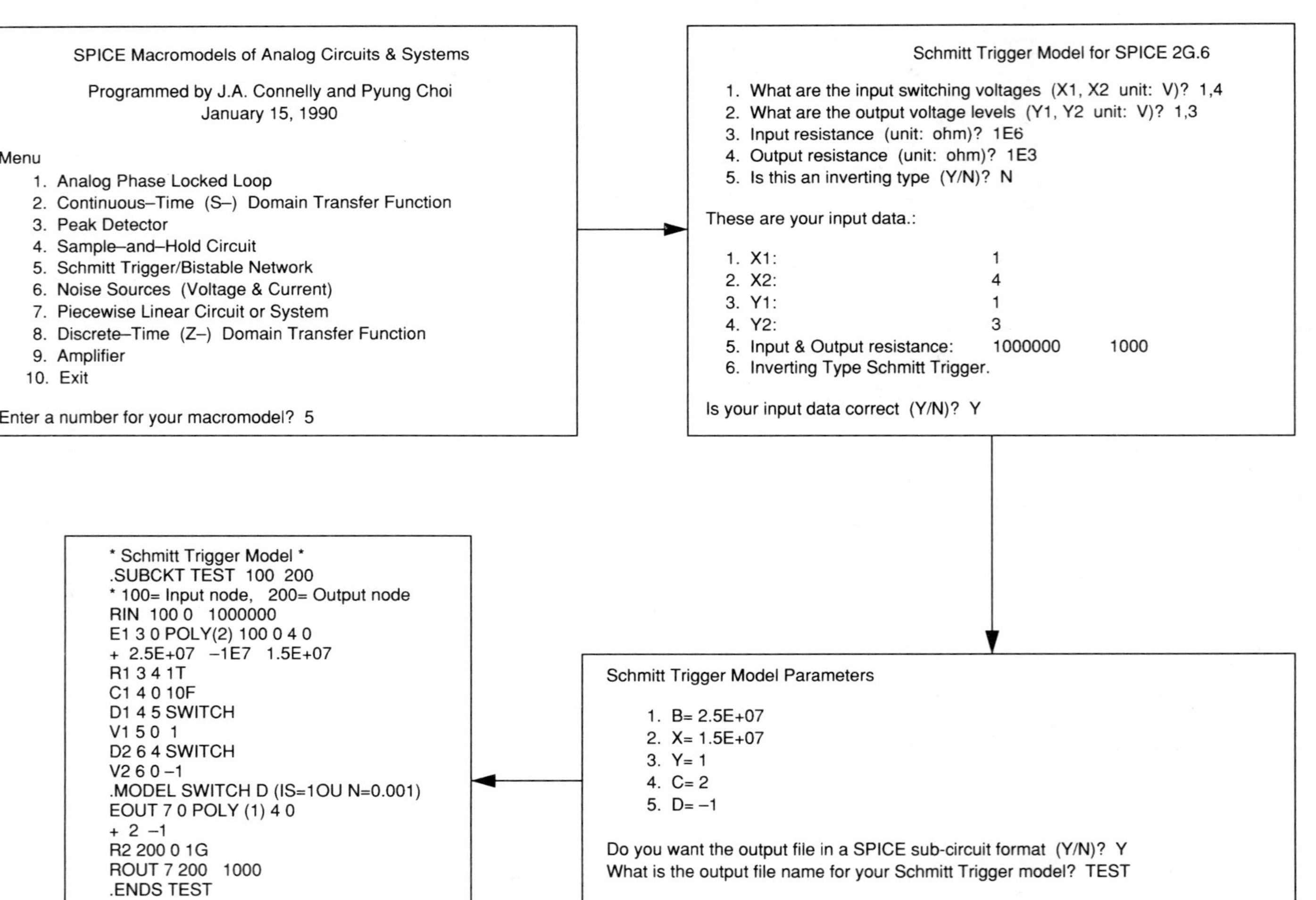

SPICE Macromodels of Analog Circuits & Systems

Programmed by J.A. Connelly and Pyung Choi
January 15, 1990

Menu
1. Analog Phase Locked Loop
2. Continuous–Time (S–) Domain Transfer Function
3. Peak Detector
4. Sample–and–Hold Circuit
5. Schmitt Trigger/Bistable Network
6. Noise Sources (Voltage & Current)
7. Piecewise Linear Circuit or System
8. Discrete–Time (Z–) Domain Transfer Function
9. Amplifier
10. Exit

Enter a number for your macromodel? 5

Schmitt Trigger Model for SPICE 2G.6

1. What are the input switching voltages (X1, X2 unit: V)? 1,4
2. What are the output voltage levels (Y1, Y2 unit: V)? 1,3
3. Input resistance (unit: ohm)? 1E6
4. Output resistance (unit: ohm)? 1E3
5. Is this an inverting type (Y/N)? N

These are your input data.:

1. X1: 1
2. X2: 4
3. Y1: 1
4. Y2: 3
5. Input & Output resistance: 1000000 1000
6. Inverting Type Schmitt Trigger.

Is your input data correct (Y/N)? Y

Schmitt Trigger Model Parameters

1. B= 2.5E+07
2. X= 1.5E+07
3. Y= 1
4. C= 2
5. D= −1

Do you want the output file in a SPICE sub-circuit format (Y/N)? Y
What is the output file name for your Schmitt Trigger model? TEST

```
* Schmitt Trigger Model *
.SUBCKT TEST  100  200
* 100= Input node,   200= Output node
RIN  100 0   1000000
E1 3 0 POLY(2) 100 0 4 0
+  2.5E+07   −1E7   1.5E+07
R1 3 4 1T
C1 4 0 10F
D1 4 5 SWITCH
V1 5 0  1
D2 6 4 SWITCH
V2 6 0 −1
.MODEL SWITCH D (IS=1OU N=0.001)
EOUT 7 0 POLY (1) 4 0
+  2  −1
R2 200 0 1G
ROUT 7 200   1000
.ENDS TEST
```

Figure 6.2 Procedure for the creation of the Schmitt trigger macromodel shown in Figure 6.1.

```
200 PRINT TAB(5); " 8. Discrete-Time (Z-) Domain Transfer Function"
205 PRINT TAB(5); " 9. Amplifier"
210 PRINT TAB(5); "10. Exit"
220 PRINT : PRINT :
230 INPUT "Enter a number for your macromodel"; XY
240 IF XY = 6 GOTO 350 ELSE IF XY = 5 GOTO 330 ELSE IF XY = 4 GOTO 310
250 IF XY = 3 GOTO 290 ELSE IF XY = 7 GOTO 370 ELSE IF XY = 10 GOTO 470
260 IF XY = 2 GOTO 390 ELSE IF XY = 8 GOTO 410 ELSE IF XY = 9 GOTO 425
270 SHELL "APLL.EXE"
280 GOTO 430
290 SHELL "PEAK.EXE"
300 GOTO 430
310 SHELL "SAMPLE.EXE"
320 GOTO 430
330 SHELL "SCHMITT.EXE"
340 GOTO 430
350 SHELL "NOISE.EXE"
360 GOTO 430
370 SHELL "PWLM.EXE"
380 GOTO 430
390 SHELL "CFFM.EXE"
400 GOTO 430
410 SHELL "ZFFM.EXE"
420 GOTO 430
425 SHELL "AMP.EXE"
427 GOTO 430
430 PRINT :
450 INPUT "Would you like to try again for another macromodel (Y/N)";
    R$
460 IF R$ = "Y" GOTO 20 ELSE IF R$ = "y" GOTO 20
470 END
```

6.3.2 Analog Phase Locked Loop Macromodel: APPL.BAS

```
10 REM ---- Analog Phase Locked Loop Macromodel ---
20 DIM X(50)
30 CLS
40 T$="Analog Phase Locked Loop Macromodel (Time Domain) for SPICE
   2G.6"
50 L=LEN(T$)
60 PRINT TAB(INT(80-L)/2+.5); T$
70 LET X(8)=0
80 PRINT:PRINT:
90 INPUT " 1. What is the input resistance (unit: Ω)"; X(12)
100 INPUT " 2. What is the conversion gain of the phase detector
    (unit: V/rad)" ; X(1)
110 INPUT " 3. What is the conversion gain of the VCO
    (unit: rad/s/V)" ; X(2)
```

```
120 PRINT " 4. What is the peak-to-peak voltage of the triangle wave
    output";
130 INPUT " (unit: Vp-p)"; X(3)
140 INPUT " 5. What is the amplitude of the square wave (unit: V)";
    X(4)
150 INPUT " 6. What is the free-running frequency (unit: Hz)"; X(5)
160 INPUT " 7. What is the timing resistance (unit: Ω)"; X(10)
170 INPUT " 8. What is the timing capacitance (unit: farad)"; X(11)
180 INPUT " 9. What is the threshold voltage of the VCO input
    (unit: V)"; X(13)
190 PRINT "10. What type of low-pass filter is used for the loop
    filter ?"
200 PRINT:PRINT TAB(7) "1. Passive simple RC low-pass filter"
210 PRINT TAB(7) "2. Passive lag-lead low-pass filter"
220 PRINT TAB(7) "3. Active RC filter"
230 PRINT TAB(7) "4. Ideal integrator"
240 INPUT "       Enter number (1,2,3, or 4)"; X(6)
250 PRINT:
260 IF X(6)=4 GOTO 270 ELSE IF X(6)=3 GOTO 290 ELSE IF X(6)=2 GOTO 290
270 INPUT "7. What are values of R1, C1 (units: Ω and F)"; X(7),X(9)
280 GOTO 300
290 INPUT "7. What are values of R1,R2, C1 (units: Ω and F)";X(7),X(8),
    X(9)
300 CLS
310 PRINT "These are your input data:"
320 PRINT:
330 PRINT TAB(3) " 1. Input resistance:"; TAB(43) X(12); TAB(55) "Ω"
340 PRINT TAB(3) " 2. Kd of the phase detector:";TAB(43) X(1);TAB(55)
    "V/rad"
350 PRINT TAB(3) " 3. Kvco of the VCO:";TAB(43) X(2);TAB(55) "rad/s/V"
360 PRINT TAB(3) " 4. Vp-p of the triangle wave:"; TAB(43) X(3);
    TAB(55) "Vp-p"
370 PRINT TAB(3) " 5. Vam of the square wave:"; TAB(43) X(4);TAB(55)
    "V"
380 PRINT TAB(3) " 6. fo':";TAB(43) X(5);TAB(55) "Hz"
390 PRINT TAB(3) " 7. Timing resistor:"; TAB(43) X(10); TAB(55) "Ω"
400 PRINT TAB(3) " 8. Timing capacitor:";TAB(43) X(11); TAB(55) "farad"
410 PRINT TAB(3) " 9. Vt of the VCO input:"TAB(43) X(13); TAB(55) "V"
420 IF X(6)=4 GOTO 450 ELSE IF X(6)=3 GOTO 510 ELSE IF X(6)=2 GOTO 490
430 PRINT TAB(3) "10. Passive simple RC low-pass filter :"
440 GOTO 460
450 PRINT TAB(3) "10. Ideal integrator :"
460 PRINT:PRINT TAB(10) ". R1="; SPC(1) X(7); TAB(55) "Ω"
470 PRINT TAB(10) ". C1="; SPC(1) X(9); TAB(55) "farad"
480 GOTO 550
490 PRINT TAB(3) "10. Passive lag-lead low-pass filter:"
500 GOTO 520
510 PRINT TAB(3) "10. Active filter:"
```

```
520 PRINT:PRINT TAB(10) ". R1=";SPC(1) X(7); TAB(55) "Ω"
530 PRINT TAB(10) ". R2=";SPC(1) X(8); TAB(55) "Ω"
540 PRINT TAB(10) ". C1=";SPC(1) X(9); TAB(55) "farad"
550 PRINT:PRINT:
560 INPUT "Is your input data correct (Y/N)"; D$
570 IF D$="Y" GOTO 610 ELSE IF D$="y" GOTO 610
580 INPUT "Do you want to try again (Y/N)"; Z$
590 IF Z$="Y" GOTO 30 ELSE IF Z$="y" GOTO 30
600 GOTO 1470
610 LET PI=3.141592654#
620 LET B=X(1)*PI/(20*X(4))
630 LET E=2*X(3)*X(10)*X(11)*X(5)
640 LET F=2*X(2)*X(3)*X(10)*X(11)/(2*PI)
650 LET H=-1/X(4)
660 LET L=X(4)
670 LET K=X(3)*1E+07/(2*X(4))
680 CLS
690 PRINT "Analog Phase Locked Loop Macromodel Parameters"
700 PRINT:PRINT TAB(5)" 1. A=" X(12)
710 PRINT TAB(5) " 2. B=" B
720 IF X(8) < > 0 THEN PRINT TAB(5) " 3. C1=";X(7);SPC(5) "C2=";X(8)
730 IF X(8)=0 THEN PRINT TAB(5) " 3. C=" X(7)
740 PRINT TAB(5) " 4. D=" X(9)
750 PRINT TAB(5) " 5. E=" E
760 PRINT TAB(5) " 6. F=" F
770 PRINT TAB(5) " 7. G=" X(13)
780 PRINT TAB(5) " 8. H=" H
790 PRINT TAB(5) " 9. I=" X(10)
800 PRINT TAB(5) "10. J=" X(11)
810 PRINT TAB(5) "11. K=" K
820 PRINT TAB(5) "12. L=" L
830 PRINT:PRINT:
840 INPUT "Do you want output file in a SPICE sub-circuit format
    (Y/N)"; P$
850 IF P$="N" GOTO 1470 ELSE IF P$="n" GOTO 1470
860 INPUT "What is your output file name for the APLL macromodel";
    CHOI$
870 REM -- Output File for Analog Phase Locked Loop Macromodel --
880 OPEN CHOI$ FOR OUTPUT AS #1
890 PRINT#1, "* Analog Phase Locked Loop Macromodel *"
900 PRINT#1, "* 101=Input node,          107=Phase detector output,"
910 PRINT#1, "* 108=Loop filter output, 115=Frequency output
    (Triangle wave),"
920 PRINT#1, "* 117= Frequency output (Square wave)"
930 PRINT#1, "  "
940 PRINT#1, "RIN 101"; SPC(1) X(12)
950 PRINT#1, "E1 102 0 POLY(2) 101 0 103 0 0 10MEG -1E-11"
960 PRINT#1, "R1 102 103 1T"
```

```
970 PRINT#1, "C1 103 0 1F"
980 PRINT#1, "D1 103 104 SWITCH"
990 PRINT#1, "V1 104 0 DC=10"
1000 PRINT#1, "D2 105 103 SWITCH"
1010 PRINT#1, "V2 105 0 DC=-10"
1020 PRINT#1, "E2 106 0 POLY(2) 103 0 117 0 0 0 0 0"; SPC(1) B
1030 PRINT#1, "R2 106 0 1"
1040 PRINT#1, "  "
1050 PRINT#1, "E3 107 0 106 0 1"
1060 IF X(6)=4 GOTO 1200 ELSE IF X(6)=3 GOTO 1140 ELSE IF X(6)=2 GOTO
     1100
1070 PRINT#1, "RFILTER 107 108";SPC(1) X(7)
1080 PRINT#1, "CFILTER 108 0";SPC(1) X(9)
1090 GOTO 1240
1100 PRINT#1, "RFILTER1 107 108";SPC(1) X(7)
1110 PRINT#1, "RFILTER2 108 999";SPC(1) X(8)
1120 PRINT#1, "CFILTER 999 0";SPC(1) X(9)
1130 GOTO 1240
1140 PRINT#1, "RFILTER1 107 998";SPC(1) X(7)
1150 PRINT#1, "RFILTER2 998 999";SPC(1) X(8)
1160 PRINT#1, "CFILTER 999 108";SPC(1) X(9)
1170 PRINT#1, "ROP 998 0 100MEG"
1180 PRINT#1, "EOP 108 0 998 0 -100MEG"
1190 GOTO 1240
1200 PRINT#1, "RFILTER 107 999";SPC(1) X(7)
1210 PRINT#1, "CFILTER 999 108";SPC(1) X(9)
1220 PRINT#1, "ROP 999 0 100MEG"
1230 PRINT#1, "EOP 108 0 999 0 -100MEG"
1240 PRINT#1, "  "
1250 PRINT#1, "E4 109 0 POLY(1) 108 0";SPC(1) E;SPC(1) F
1260 PRINT#1, "R3 109 110 1T"
1270 PRINT#1, "D3 110 111 SWITCH"
1280 PRINT#1, "V3 111 0 DC="; X(13)
1290 PRINT#1, "D4 112 110 SWITCH"
1300 PRINT#1, "V4 112 0 DC="; -X(13)
1310 PRINT#1, "E5 113 0 POLY(2) 110 0 117 0 0 0 0 0"; SPC(1) H
1320 PRINT#1, "RTIME 113 114";SPC(1) X(10)
1330 PRINT#1, "CTIME 114 115";SPC(1) X(11)
1340 PRINT#1, "R4 114 0 1MEG"
1350 PRINT#1, "E6 115 0 114 0 100MEG"
1360 PRINT#1, "E7 116 0 POLY(2) 115 0 117 0 0 -10MEG";SPC(1) K
1370 PRINT#1, "R5 116 117 1T"
1380 PRINT#1, "C2 117 0 1F"
1390 PRINT#1, "D5 117 118 SWITCH"
1400 PRINT#1, "V5 118 0 DC="; L
1410 PRINT#1, "D6 119 117 SWITCH"
1420 PRINT#1, "V6 119 0 DC="; -L
1430 PRINT#1, ".MODEL SWITCH D(IS=10U N=0.001)"
```

```
1440 CLOSE #1
1450 PRINT:
1460 PRINT "Your output file (name:"; SPC(1) CHOI$;") is successfully
     created."
1470 END
```

6.3.3 Macromodels for Continuous-Time Domain Transfer Function: CFFM.BAS

```
10 REM *- Macromodel Extractor Program for Continuous Filtering
   Function -*
20 DIM DE(100): DIM DF(100): DIM NU(100): DIM NV(100)
50 CLS
60 T$="Macromodel for Continuous Time domain Transfer Function for
   SPICE 2G.6."
90 P=LEN(T$)
100 PRINT TAB(INT(80-P)/2+.5); T$
130 PRINT:PRINT:INPUT " 1. What is the highest degree of your transfer
    function"; L
140 PRINT
150 LET M=L: LET HO=L: LET MO=L
160 LET LA=L+1
170 LET MA=M+1
180 PRINT " 2. Enter the coefficients of denominator polynomial,
    highest power first."
190 FOR LB=1 TO LA
200 PRINT TAB(7) "Coefficient of s**"; SPC(1) HO;
210 INPUT DE(LB)
220 LET HO=HO-1
230 NEXT
240 PRINT
250 PRINT " 3. Enter  the coefficients of numerator polynomial,
    highest power first."
260 FOR MB-1 TO MA
270 PRINT TAB(7) "Coefficient of s**"; SPC(1) MO;
280 INPUT NU(MB)
290 LET MO=MO-1
300 NEXT
310 PRINT:INPUT " 4. Input resistance (unit: Ω)"; X
320 INPUT " 5. Output resistance (unit: Ω)"; Y
321 CLS : PRINT "These are your input data for the denominator
    polynomial:":PRINT
322 HO = L
323 FOR LB = 1 TO LA
324 PRINT TAB(7); "COEFFICIENT OF S**"; HO; ":";
325 PRINT" "; DE(LB)
326 LET HO = HO -1
327 NEXT
```

```
328 PRINT : PRINT "These are your input data for the numerator
    polynomial:":PRINT
329 MO = L
330 FOR MB = 1 TO MA
331 PRINT TAB(7); "Coefficient of s**"; MO; ":";
332 PRINT " "; NU(MB)
333 LET MO = MO - 1
334 NEXT
336 PRINT : PRINT "INPUT RESISTANCE (Unit Ohm):"; X
337 PRINT "OUTPUT RESISTANCE (Unit Ohm):"; Y:INPUT "Is your input data
    correct"; YN$
338 IF YN$ = "N" OR YN$ = "n" THEN 50
339 REM *** Calculation of Feedback Coefficients***
340 FOR LB=1 TO LA
350 LET DF(LB)=DE(LB)/DE(1)
360 NEXT
370 REM *** Calculation of Forward Gain Coefficients ***
380 FOR MB=1 TO MA
390 LET NV(MB)=NU(MB)/DE(1)
400 NEXT
420 REM *** SPICE 2G6 Output File as Subcircuit Format ***
430 PRINT: INPUT " Enter the output file name"; CHOI$
440 OPEN CHOI$ FOR OUTPUT AS #1
450 PRINT#1, "**-- Continuous Filtering Function Macromodel --**"
460 PRINT#1, ".SUBCKT"; SPC(1) CHOI$; SPC(1) "10  20  30"
470 PRINT#1, "* Node 10= Input, Node 30= Output, Node 20= GND"
480 PRINT#1, "RIN 10 20";SPC(1) X
490 PRINT#1, "EIN 100 0 POLY (";LA;") 10 20"
500 FOR LB=1 TO L
510 PRINT#1, "+"; SPC(2) 100+LB; "0"
520 NEXT
530 PRINT#1, "+   0"; SPC(2) DF(1)
540 FOR LB=2 TO LA
550 LET DF(LB)=-1*DF(LB)
560 PRINT#1, "+" SPC(2) DF(LB)
570 NEXT
580 PRINT#1, "REI 100 0 1"
590 FOR LB=1 TO L
600 LET LK=100+LB
610 PRINT#1, "G";-LB; SPC(1)"0"; LK; LK-1; "0 1"
620 PRINT#1, "R";-LB; LK; "0 1T"
630 PRINT#1, "C";-LB; LK; "0 1"
640 NEXT
650 PRINT#1, "ESO";LK+1;"0 POLY("; MA;")"
660 FOR MB=1 TO MA
670 PRINT#1, "+"; SPC(2) MB+99;"0"
680 NEXT
690 PRINT#1, "+   0"
```

```
700 FOR MB=1 TO MA
710 PRINT#1, "+"; SPC(2) NV(MB)
720 NEXT
730 PRINT#1, "ROT";LK+1;"30" SPC(1) Y
740 PRINT#1, "RDY 30 0 1T"
750 PRINT#1, ".ENDS"; SPC(1) CHOI$
760 CLOSE#1
761 PRINT:PRINT:
770 PRINT "Your output file (name:"; SPC(1) CHOI$;") is successfully
    created."
810 END
```

6.3.4 Peak Detector Macromodel: PEAK.BAS

```
10 REM ---- Peak Detector Model ---
20 DIM X(50)
30 CLS
40 T$="Peak Detector Circuit Model for SPICE 2G.6"
50 L=LEN(T$)
60 PRINT TAB(INT(80-L)/2+.5); T$
70 PRINT:PRINT:
80 INPUT " 1. What is the shortest period of the input signal
   (unit: sec)" ; X(1)
90 INPUT " 2. Input resistance (unit: Ω)"; X(2)
91 INPUT " 3. Output resistance (unit: Ω)";X(3)
100 PRINT:PRINT:
110 PRINT "These are your input data.:"
120 PRINT:PRINT TAB(3) "1. Shortest period of input signal:"; SPC(1)
    X(1)
130 PRINT TAB(3) "2. Input resistance:" SPC(16) X(2)
131 PRINT TAB(3) "3. Output resistance:" ;SPC(15) X(3)
140 PRINT:PRINT:
150 INPUT "Is your input data correct (Y/N)"; D$
160 IF D$="Y" GOTO 200 ELSE IF D$="y" GOTO 200
170 INPUT "Do you want to try again (Y/N)"; Z$
180 IF Z$="Y" GOTO 30 ELSE IF Z$="y" GOTO 30
190 GOTO 510
200 LET TIN=100*X(1)
210 LET C1=.000001
220 LET R1=TIN/C1
230 CLS
240 PRINT "Peak Detector Model Parameters"
250 PRINT:PRINT TAB(5)"1. A="; SPC(2) X(2)
260 PRINT TAB(5) "2. B="; SPC(2) C1
270 PRINT TAB(5) "3. C="; SPC(2) R1
280 PRINT TAB(5) "4. D="; SPC(2) X(3)
290 PRINT:PRINT:
```

```
300 INPUT "Do you want the output file in a SPICE sub-circuit format
    (Y/N)"; P$
310 IF P$="N" GOTO 510 ELSE IF P$="n" GOTO 510
320 INPUT "What is the output file name for your Peak Detector model";
    CHOI$
330 REM -- Output File for Peak Detector Model --
340 OPEN CHOI$ FOR OUTPUT AS #1
350 PRINT#1, "* Peak Detector Model *"
360 PRINT#1, "  "
370 PRINT#1, ".SUBCKT"; SPC(1) CHOI$; SPC(1) "100 200"
380 PRINT#1, "* 100= Input node,  200= Output node"
390 PRINT#1, "RIN 100 0"; SPC(1) X(2)
400 PRINT#1, "E1 3 0 100 0 1"
410 PRINT#1, "D1 3 4 DIODE"
420 PRINT#1, "R1 4 0"; SPC(1) R1
430 PRINT#1, "C1 4 0"; SPC(1) C1
440 PRINT#1, ".MODEL DIODE D (N=0.001)"
450 PRINT#1, "EOUT 5 0 4 0 1"
460 PRINT#1, "R2 200 0 1G"
470 PRINT#1, "ROUT 5 200";SPC(1) X(3)
480 PRINT#1, ".ENDS"; SPC(1) CHOI$
490 CLOSE #1
500 PRINT "Output file (name:"; CHOI$;") is successfully created."
510 END
```

6.3.5 Sample-and-Hold Circuit Macromodel: SAMPLE.BAS

```
10 REM ---- Sample-and-Hold Circuit Model ---
20 DIM X(50)
30 CLS
40 T$="Sample-and-Hold Circuit Model for SPICE 2G.6"
70 L=LEN(T$)
80 PRINT TAB(INT(80-L)/2+.5); T$
110 PRINT:PRINT:
120 INPUT "1. What is the period of the input signal (unit: sec)" ;
    X(1)
130 INPUT "2. What is the sampling period (unit: sec)"; X(2)
131 INPUT "3. What is the input resistance (unit: Ω)"; X(3)
132 INPUT "4. What is the output resistance (unit: Ω)"; X(4)
140 PRINT:PRINT:
150 PRINT "These are your input data.:"
160 PRINT:PRINT TAB(3) "1. Period of input signal:"; SPC(5) X(1)
161 PRINT TAB(3) "2. Sampling period:" SPC(12) X(2)
170 PRINT TAB(3) "3. Input & Output resistances:" SPC(2) X(3);SPC(10)
    X(4)
180 PRINT:PRINT:
190 INPUT "Is your input data correct (Y/N)"; D$
200 IF D$="Y" GOTO 240 ELSE IF D$="y" GOTO 240
```

```
210 INPUT "Do you want to try again (Y/N)"; Z$
220 IF Z$="Y" GOTO 30 ELSE IF Z$="y" GOTO 30
230 GOTO 570
240 LET TIN=100*X(1)
250 LET C1=.000001
260 LET R1=TIN/C1
261 LET D=X(2)/100
262 LET E=X(2)
270 CLS
280 PRINT "Sample-and-Hold Circuit Model Parameters"
290 PRINT:PRINT TAB(5)"1. A="; SPC(2) X(3)
300 PRINT TAB(5) "2. B="; SPC(2) R1
310 PRINT TAB(5) "3. C="; SPC(2) C1
320 PRINT TAB(5) "4. D="; SPC(2) D
321 PRINT TAB(5) "5. E="; SPC(2) E
322 PRINT TAB(5) "6. F="; SPC(2) X(4)
330 PRINT:PRINT:
340 INPUT "Do you want your output file in a SPICE sub-circuit format
    (Y/N)"; P$
350 IF P$="N" GOTO 570 ELSE IF P$="n" GOTO 570
360 INPUT "What is the output file name for the Sample-and-Hold
    Circuit model"; CHOI$
370 REM -- Output File for Sample-and-Hold Circuit Model --
380 OPEN CHOI$ FOR OUTPUT AS #1
390 PRINT#1, "* Sample-and-Hold Circuit Model *"
400 PRINT#1, ".SUBCKT"; SPC(1) CHOI$; SPC(1) "100 200"
410 PRINT#1, "* 100= Input node for input signal"
412 PRINT#1, "* 200= Output node for sampled signal"
420 PRINT#1, "RIN 100 0"; SPC(1) X(3)
421 PRINT#1, "RS 3 0 1"
430 PRINT#1, "E1 5 0 100 0 1"
431 PRINT#1, "G1 5 6 POLY(2) 5 6 3 0 0 0 0 0 1"
450 PRINT#1, "R1 6 0"; SPC(1) R1
460 PRINT#1, "C1 6 0"; SPC(1) C1
470 PRINT#1, "VS 3 0 PULSE(0 1E3 0 0 0";SPC(1) D;SPC(1) E;")"
480 PRINT#1, "EOUT 7 0 6 0 1"
490 PRINT#1, "R2 200 0 1G"
500 PRINT#1, "ROUT 7 200";SPC(1) X(4)
510 PRINT#1, ".ENDS"; SPC(1) CHOI$
520 CLOSE #1
530 PRINT:
540 PRINT "Your output file (name:";SPC(1) CHOI$; ") is created."
570 END
```

6.3.6 Bistable Network Macromodel: SCHMITT.BAS

```
10 REM ---- Bistable Network Model ---
20 DIM X(50)
30 CLS
40 T$="Schmitt Trigger Model for SPICE 2G.6"
70 L=LEN(T$)
80 PRINT TAB(INT(80-L)/2+.5); T$
110 PRINT:PRINT:
120 INPUT "1. What are the input switching voltages (X1,X2  unit: V)" ;
    X(1),X(2)
130 INPUT "2. What are the output voltage levels (Y1,Y2  unit: V)" ;
    X(3),X(4)
140 INPUT "3. Input resistance (unit: Ω)"; X(5)
141 INPUT "4. Output resistance (unit: Ω)";X(6)
150 INPUT "4. Is this an inverting type (Y/N)"; A$
160 LET D=-1
170 IF A$="Y" THEN LET D=1 ELSE IF A$="y" THEN LET D=1
180 PRINT:PRINT:
190 PRINT "These are your input data.:"
200 PRINT:PRINT TAB(3) "1. X1:"; TAB(35) X(1)
210 PRINT TAB(3) "2. X2:" TAB(35) X(2)
220 PRINT TAB(3) "3. Y1:" TAB(35) X(3)
230 PRINT TAB(3) "4. Y2:" TAB(35) X(4)
240 PRINT TAB(3) "5. Input & Output resistance:" SPC(3) X(5);SPC(5)
    X(6)
250 IF D=-1 THEN  GOTO 280
260 PRINT TAB(3) "6. Noninverting Type Schmitt Trigger."
270 GOTO 290
280 PRINT TAB(3) "6. Inverting Type Schmitt Trigger."
290 PRINT:PRINT:
300 INPUT "Is your input data correct (Y/N)"; D$
310 IF D$="Y" GOTO 350 ELSE IF D$="y" GOTO 350
320 INPUT "Do you want to try again (Y/N)"; Z$
330 IF Z$="Y" GOTO 30 ELSE IF Z$="y" GOTO 30
340 GOTO 790
350 LET B=(X(1)+X(2))/2
360 LET B=B*1E+07
370 LET C=(X(3)+X(4))/2
380 LET F=X(2)-X(1)
390 IF X(1)>X(2) THEN F=X(1)-X(2)
400 LET G=X(3)-X(4)
410 IF X(4)>X(3) THEN G=X(4)-X(3)
420 LET H=F/G
430 LET H=H*1E+07
440 LET I=G/2
450 CLS
460 PRINT "Schmitt Trigger Model Parameters"
470 PRINT:PRINT TAB(5)"1. B=" B
```

```
480 PRINT TAB(5) "2. X=" H
490 PRINT TAB(5) "3. Y=" I
500 PRINT TAB(5) "4. C=" C
510 PRINT TAB(5) "5. D=" D
520 PRINT:PRINT:
530 INPUT "Do you want the output file in a SPICE sub-circuit format
    (Y/N)"; P$
540 IF P$="N" GOTO 790 ELSE IF P$="n" GOTO 790
550 INPUT "What is the output file name for your Schmitt Trigger
    model"; CHOI$
560 REM -- Output File for Schmitt Trigger Model --
570 OPEN CHOI$ FOR OUTPUT AS #1
580 PRINT#1, "* Schmitt Trigger Model *"
590 PRINT#1, ".SUBCKT"; SPC(1) CHOI$; SPC(1) "100 200"
600 PRINT#1, "* 100= Input node,  200= Output node"
610 PRINT#1, "RIN 100 0"; SPC(1) X(5)
620 PRINT#1, "E1 3 0 POLY(2) 100 0 4 0"
630 PRINT#1, "+"; SPC(1) B; SPC(1) "-1E7"; SPC(1) H
640 PRINT#1, "R1 3 4 1T"
650 PRINT#1, "C1 4 0 10F"
660 PRINT#1, "D1 4 5 SWITCH"
670 PRINT#1, "V1 5 0"; SPC(1) I
680 PRINT#1, "D2 6 4 SWITCH"
690 PRINT#1, "V2 6 0"; SPC(1) -I
700 PRINT#1, ".MODEL SWITCH D (IS=10U N=0.001)"
710 PRINT#1, "EOUT 7 0 POLY(1) 4 0"
720 PRINT#1, "+"; SPC(1) C;SPC(1) D
730 PRINT#1, "R2 200 0 1G"
740 PRINT#1, "ROUT 7 200";SPC(1) X(6)
750 PRINT#1, ".ENDS"; SPC(1) CHOI$
760 CLOSE #1
770 PRINT:
780 PRINT "Your output file (name:";SPC(1) CHOI$; ") is created."
790 END
```

6.3.7 Noise Source Model: NOISE.BAS

```
10 REM ---- Noise Source Model ---
20 DIM X(100)
30 CLS
40 T$="Noise Model Extractor for SPICE 2G.6"
70 L=LEN(T$)
80 PRINT TAB(INT(80-L)/2+.5); T$
110 PRINT:PRINT:
120 INPUT "1.What is En (unit: V/root Hz)" ; X(1)
130 INPUT "2.What is In (unit: A/root Hz)" ; X(2)
140 INPUT "3.What is the noise corner frequency of En (unit: Hz)" ;
    X(3)
```

```
141 INPUT "4.What is the noise corner frequency of In (unit: Hz)" ;
    X(4)
150 PRINT:PRINT:
160 PRINT "These are your input data.:"
170 PRINT:PRINT TAB(3) "1. En:"; TAB(45) X(1)
180 PRINT TAB(3) "2. In:" TAB(45) X(2)
190 PRINT TAB(3) "3. Noise corner frequency of En:" TAB(45) X(3)
191 PRINT TAB(3) "4. Noise corner frequency of In:" TAB(45) X(4)
200 PRINT:PRINT:
210 INPUT "Is your input data correct (Y/N)"; D$
220 IF D$="Y" GOTO 260 ELSE IF D$="y" GOTO 260
230 INPUT "Do you want to try again (Y/N)"; Z$
240 IF Z$="Y" GOTO 30 ELSE IF Z$="y" GOTO 30
250 GOTO 760
260 LET Q=1.6E-19
270 LET EREFA=1E-12
280 LET EREFB=1E-24
290 LET IDCE=EREFB/(2*Q)
300 LET KOE=X(1)/EREFA
310 LET KFE=X(3)*EREFB/IDCE
330 LET IREFA=5E-13
340 LET IDCI=IREFA^2/(2*Q)
350 LET KOI=X(2)/IREFA
360 LET KFI=X(4)*IREFA^2/IDCI
370 CLS
380 PRINT:PRINT "Noise Model Parameters of En and In"
390 PRINT:PRINT TAB(10) "En"; TAB(40) "In"
400 PRINT:PRINT TAB(10) "1. IDC=" IDCE; TAB(40) IDCI
410 PRINT TAB(10) "2. Ko= " KOE ; TAB(40) KOI
420 PRINT TAB(10) "3. KF= " KFE ; TAB(40) KFI
430 PRINT:PRINT:
440 INPUT "Do you want the output files as in a SPICE sub-circuit
    format (Y/N)"; P$
460 IF P$="N" GOTO 760 ELSE IF P$="n" GOTO 760
470 INPUT "What is the output file name for the voltage noise source";
    CHOI$
490 INPUT "What is the output file name for the current noise source";
    PYUNG$
500 REM -- Output File for Voltage Noise Source --
510 OPEN CHOI$ FOR OUTPUT AS #1
520 PRINT#1, ".SUBCKT"; SPC(1) CHOI$; SPC(1) "100 200"
530 PRINT#1, "*                  +   -"
540 PRINT#1, "IDC 0 1"; IDCE
550 PRINT#1, "CNE 1 2 1GF"
560 PRINT#1, "VTEST 2 0 DC=0 AC=0"
570 PRINT#1, "DIODE 2 0 DIODE"
580 PRINT#1, ".MODEL DIODE D (AF=1 KF="; KFE;")"
590 PRINT#1, "HOUT 100 200 VTEST"; KOE
```

```
600 PRINT#1, ".ENDS"; SPC(1) CHOI$
610 CLOSE #1
620 REM -- Output file for Current Noise Source --
630 OPEN PYUNG$ FOR OUTPUT AS #2
640 PRINT#2, ".SUBCKT"; SPC(1) PYUNG$; SPC(1) "300 400"
650 PRINT#2, "*                  +   -"
660 PRINT#2, "IDC 0 1"; IDCI
670 PRINT#2, "CNE 1 2 1GF"
680 PRINT#2, "VTEST 2 0 DC=0 AC=0"
690 PRINT#2, "DIODE 2 0 DIODE"
700 PRINT#2, ".MODEL DIODE D (AF=1 KF="; KFI;")"
710 PRINT#2, "FOUT 300 400 VTEST"; KOI
720 PRINT#2, ".ENDS"; SPC(1) PYUNG$
730 CLOSE #2
740 PRINT:
750 PRINT "Your output files are created as SPICE subcircuit."
760 END
```

6.3.8 Macromodels for Piecewise Linear Circuits and Systems: PWLM.BAS

```
10 REM ---- Piecewise Linear Function Model --- For Input Data -----
20 DIM X(100):DIM Y(100):DIM M(100):DIM R(100):DIM V(100):DIM D(100)
30 CLS
40 T$="Piecewise Linear Function Model Extractor for SPICE 2G.6"
50 L=LEN(T$)
60 PRINT TAB(INT(80-L)/2+.5); T$
70 PRINT:PRINT:
80 INPUT " 1. Enter the starting values (x,y)"; X(1),Y(1)
90 INPUT " 2. Number of breakpoints" ; A
100 FOR K=2 TO A+1
110 PRINT " 3. Breakpoint";K-1;"(x,y)";
120 INPUT X(K),Y(K)
130 NEXT
140 INPUT " 4. Enter the final values (x,y)"; X(A+2),Y(A+2)
150 INPUT " 5. Input Resistance (Unit: Ω)" ; B1
151 INPUT " 6. Output resistance (Unit: Ω)" ;B2
160 CLS
170 PRINT "These are your input data:"
180 PRINT:PRINT:
190 PRINT TAB(40) " Input         Output"
200 PRINT:PRINT "1.   Starting values" ; TAB(40) X(1); TAB(54) Y(1)
210 FOR L=2 TO A+1
220 PRINT "2.";L-1;"Breakpoint values" ; TAB(40) X(L); TAB(54) Y(L)
230 NEXT
240 PRINT "3.   Final Values"; TAB(40) X(A+2); TAB(54) Y(A+2)
250 PRINT "4.   Input and output resistance" ; TAB(40) B1; TAB(54) B2
260 PRINT:PRINT:
```

```
270 INPUT "Is your input data correct (Y/N)"; D$
280 IF D$="Y" GOTO 320 ELSE IF D$="y" GOTO 320
290 INPUT "Do you want to try again (Y/N)"; Z$
300 IF Z$="Y" GOTO 30 ELSE IF Z$="y" GOTO 30
310 GOTO 1400
320 INPUT "Do you want to run this program (Y/N)"; H$
330 IF H$="N" GOTO 1400 ELSE IF H$="n" GOTO 1400
340 INPUT "Do you want an output file in a SPICE subcircuit format
    (Y/N)"; P$
350 IF P$="N" GOTO 1400 ELSE IF P$="n" GOTO 1400
360 INPUT "Enter output file name"; CHOI$
370 REM ---------- Calculation for Each Linear Segment Slope --------
380 PRINT:
390 FOR I=1 TO A+1
400 M(I)=(Y(I+1)-Y(I))/(X(I+1)-X(I))
410 NEXT
420 FOR IA=1 TO A
430 IF M(IA+1)<M(IA) GOTO 500
440 REM -------------------- Type 2 Load ----------------------------
450 R(IA)=1/(M(IA+1)-M(IA))
460 V(IA)=X(IA+1)
470 D(IA)=2
480 NEXT
490 GOTO 560
500 REM -------------------- Type 3 Load ----------------------------
510 R(IA)=1/(M(IA)-M(IA+1))
520 V(IA)=X(IA+1)
530 D(IA)=3
540 IF IA=A GOTO 560 ELSE IF IA<>A THEN IA=IA+1
550 GOTO 430
560 REM -------------------- Type 1 Load ---------------------------
570 XA=0:XB=0:XO=0:
580 IF A=1 GOTO 810
590 FOR PA=1 TO A
600 XO=XO+D(PA)
610 NEXT
620 IF XO/A=2 GOTO 930
630 IF XO/A=3 GOTO 1020
640 REM --- Combination of +/- Slope  (More than 2 Breakpoints) ---
650 FOR IB=1 TO A
660 IF D(IB)<>2 GOTO 700
670 XC=(X(A+2)-V(IB))/R(IB)
680 XA=XA+XC
690 XX=Y(A+2)-XA
700 NEXT
710 FOR IC=1 TO A
720 IF D(IC)<>3 GOTO 760
730 XD=(X(1)-V(IC))/R(IC)
```

```
740 XB=XB+XD
750 XY=Y(1)-XB
760 NEXT
770 RC=(X(A+2)-X(1))/(XX-XY)
780 VC=X(A+2)-(XX*RC)
790 GOTO 1100
800 REM ----------- Just 1 Breakpoint  (Type 2 Load)  --------------
810 IF D(1)<>2 GOTO 870
820 XC=(X(A+2)-V(1))/R(1)
830 XX=Y(A+2)-XC
840 RC=(X(A+2)-X(1))/(XX-Y(1))
850 VC=X(A+2)-(XX*RC)
860 GOTO 1100
870 REM ----------- Just 1 Breakpoint (Type 3 Load) ----------------
880 XD=(X(1)-V(1))/R(1)
890 XY=Y(1)-XD
900 RC=(X(A+2)-X(1))/(Y(A+2)-XY)
910 VC=X(1)-(XY*RC)
920 GOTO 1100
930 REM -------------------- All Positive Slopes --------------
940 FOR PB=1 TO A
950 XC=(X(A+2)-V(PB))/R(PB)
960 XA=XA+XC
970 XX=Y(A+2)-XA
980 NEXT
990 RC=(X(A+2)-X(1))/(XX-Y(1))
1000 VC=X(A+2)-(XX*RC)
1010 GOTO 1100
1020 REM ------------------- All Negative Slopes -------------
1030 FOR PC=1 TO A
1040 XD=(X(1)-V(PC))/R(PC)
1050 XB=XB+XD
1060 XY=Y(1)-XB
1070 NEXT
1080 RC=(X(A+2)-X(1))/(Y(A+2)-XY)
1090 VC=X(1)-(XY*RC)
1100 REM ------- Output File as SPICE Sub-circuit Type ------------
1110 OPEN CHOI$ FOR OUTPUT AS #1
1120 PRINT#1, "**-- PIECEWISE LINEAR MODEL --**"
1130 PRINT#1, ".SUBCKT"; SPC(1) CHOI$; SPC(1) "100 500 101"
1140 PRINT#1, "*Node 100=Input, Node 101=Output, Node 500= GND"
1150 PRINT#1, ".MODEL DIODE D(N=1E-6)"
1160 PRINT#1, "RIN 100 500";B1
1170 PRINT#1, "ROUT 103 101";B2
1180 PRINT#1, "RDUM 101 0 1T"
1190 PRINT#1, "E1 102 0 POLY(1) 100 500 0 1"
1200 PRINT#1, "VTEST 102 105 DC=0 AC=0"
1210 PRINT#1, "H1 103 0 POLY(1) VTEST 0 1"
```

```
1220 IF RC>1E+20 THEN RC=1E+20
1230 PRINT#1, "RC 105 106"; SPC(1) RC
1240 PRINT#1, "VC 106 0"; SPC(1) VC
1250 FOR KF=1 TO A
1260 IF D(KF)<>2 GOTO 1320
1270 PRINT#1, "R";-KF;"105";KF;R(KF)
1280 PRINT#1, "D";-KF;KF;200+KF;"DIODE"
1290 PRINT#1, "V";-KF;200+KF;"0";SPC(1) V(KF)
1300 NEXT
1310 GOTO 1370
1320 PRINT#1, "R";-KF;"105";300+KF;R(KF)
1330 PRINT#1, "D";-KF;400+KF;300+KF;"DIODE"
1340 PRINT#1, "V";-KF;400+KF;"0";SPC(1) V(KF)
1350 IF KF=A GOTO 1370 ELSE IF KF<>A THEN KF=KF+1
1360 GOTO 1260
1370 PRINT#1, ".ENDS"; SPC(1) CHOI$
1380 CLOSE #1
1390 PRINT "Output file (named:"; SPC(1) CHOI$; ") is successfully
     created."
1400 END
```

6.3.9 Macromodels for Discrete-Time Domain Transfer Functions: ZFFM.BAS

```
10 REM *- Macromodel Extractor Program for Discrete Filtering
   Functions -*
20 DIM DE(100): DIM DF(100): DIM NU(100): DIM NV(100)
30 CLS
40 T$="Macromodel for Discrete Time domain Transfer Function."
50 P=LEN(T$)
60 PRINT TAB(INT(80-P)/2+.5); T$
70 PRINT:PRINT:INPUT " 1. Enter the highest degree of z in your
   transfer function"; L
80 PRINT
90 LET HO=L: LET MO=L
100 LET LA=2*L+1
110 LET MA=2*L+1
120 PRINT " 2. Enter the coefficients of the denominator polynomial,
    highest power first."
130 FOR LB=1 TO LA
140 PRINT TAB(7) "Coefficient of z**"; SPC(1) HO;"
150 INPUT DE(LB)
160 LET HO=HO-.5
170 NEXT
180 PRINT
190 PRINT " 3. Enter the coefficients of the numerator polynomial,
    highest power first."
200 FOR MB=1 TO MA
```

```
210 PRINT TAB(7) "Coefficient of z**"; SPC(1) MO;
220 INPUT NU(MB)
230 LET MO=MO-.5
240 NEXT
250 PRINT:INPUT " 4. What is the clock frequency (unit: Hz)"; X
251 CLS : PRINT "These are your input data for the denominator
    polynomial:":PRINT:PRINT
252 HO = L
253 FOR LB = 1 TO LA
254 PRINT TAB(7); "Coefficient of z**";:PRINT USING "##.#"; HO; :
    PRINT":";
256 PRINT " "; DE(LB)
257 LET HO = HO - .5
258 NEXT
260 PRINT : PRINT "These are your input data for the numerator
    polynomial:":PRINT : PRINT
261 MO = L
262 FOR MB = 1 TO MA
263 PRINT TAB(7); "Coefficient of z**";:PRINT USING "##.#"; MO; :
    PRINT":";
264 PRINT "";NU(MB)
265 LET MO = MO -.5
266 NEXT
267 PRINT : PRINT "CLOCK FREQUENCY (Unit Hz):";X:INPUT"Is your input
    data correct"; YN$
268 IF YN$ = "N" OR YN$ = "n" THEN 30
269 LET T = 1 / (2 * X)
270 REM *** Calculation of Feedback Coefficients ***
280 FOR LB=1 TO LA
290 LET DF(LB)=DE(LB)/DE(1)
300 NEXT
310 REM *** Calculation of Forward Gain Coefficients ***
320 FOR MB=1 TO MA
330 LET NV(MB)=NU(MB)/DE(1)
340 NEXT
350 REM *** SPICE 2G6 Output File as Subcircuit Format ***
360 PRINT: INPUT " Enter the output file name"; CHOI$
370 OPEN CHOI$ FOR OUTPUT AS #1
380 PRINT#1, "**-- Discrete Time Filtering Function Macromodel --**"
390 PRINT#1, " "
400 PRINT#1, "* Node 100= Input, Node 200= Output"
410 PRINT#1, "RIN 100 0 1G"
420 PRINT#1, "ESUM1 101 0 POLY (";LA;")"
430 PRINT#1, "+   100 0"
440 FOR LB=2 TO LA
450 PRINT#1, "+"; SPC(2) 100+LB; "0"
460 NEXT
470 PRINT#1, "+" SPC(3) "0"
```

```
480 PRINT#1, "+"  SPC(2) DF(1)
490 FOR LB=2 TO LA
500 LET DF(LB)=-1*DF(LB)
510 PRINT#1, "+" SPC(2) DF(LB)
520 NEXT
530 LET XX=0
540 FOR LB=1 TO 2*L
550 LET XX=XX+.5
560 PRINT#1, "XZ";-XX 100+LB 100+LB+1;"DINT"
570 IF XX=L GOTO 590
580 NEXT
590 PRINT#1, "ESUM2 200 0";SPC(1) "POLY("; MA;")"
600 PRINT#1, "+   101 0"
610 FOR MB=1 TO 2*L
620 PRINT#1, "+"; SPC(2) 101+MB;"0"
630 NEXT
640 PRINT#1, "+"; SPC(3) "0"
650 FOR MB=1 TO MA
660 PRINT#1, "+"; SPC(2) NV(MB)
670 NEXT
680 PRINT#1, "RSUM2 200 0 1G"
690 PRINT#1, ".SUBCKT DINT 10 50"
700 PRINT#1, "TDELAY1 10 0 20 0 ZO=1K TD="; T
710 PRINT#1, "RDELAY1 20 0 1K"
720 PRINT#1, "EDELAY1 30 0 20 0 1"
730 PRINT#1, "RDELAY2 30 40 1"
740 PRINT#1, "RDELAY3 40 0 1T"
750 PRINT#1, "RDELAY4 40 50 1"
760 PRINT#1, "EDELAY2 50 0 40 0 -1T"
770 PRINT#1, ".ENDS DINT"
780 PRINT#1, " "
790 CLOSE#1
800 PRINT:PRINT:
810 PRINT "Your output file (name:"; SPC(1) CHOI$;") is successfully
    created."
820 END
```

6.3.10 Macromodel for Operational Amplifier: AMP.BAS

```
10 REM --- Amplifier Macromodel ---
20 DIM X(25)
30 CLS
40 T$ = "Amplifier Macromodel"
50 L = LEN(T$)
60 PRINT TAB(INT(80 - L) / 2 + .5); T$
70 PRINT : PRINT
80 INPUT " 1. What is the differential mode input resistance
```

```
    (unit: Ω)"; X(1)
90 INPUT " 2. What is the common mode input resistance (unit: Ω)"; X(2)
100 INPUT " 3. What is the DC differential mode gain (unit: dB)"; X(3)
110 INPUT " 4. What is the common mode rejection ratio (unit: dB)";
    X(4)
120 INPUT " 5. What is the input offset current (unit: A)"; X(5)
130 INPUT " 6. What is the input bias current (unit: A)"; X(6)
140 INPUT " 7. What is the output resistance (unit: Ω)"; X(7)
150 INPUT " 8. What is the positive slew rate (unit: V/us)"; X(8)
160 INPUT " 9. What is the negative slew rate (unit: V/us)"; X(9)
170 INPUT "10. What is the input capacitance (unit: F)"; X(10)
180 INPUT "11. What is the input offset voltage (unit: V)"; X(11)
190 INPUT "12. What is the frequency of the first pole (unit: Hz)";
    X(12)
200 INPUT "13. What is the frequency of the second pole (unit: Hz)";
    X(13)
230 INPUT "14. What is the frequency of the third pole (unit: Hz)";
    X(14)
240 INPUT "15. What is the frequency of the zero (unit: Hz)"; X(15)
250 IF X(15) > X(14) THEN PRINT "The frequency of the third pole must
    be greater than the frequency of the zero.": GOTO 230
260 INPUT "16. What is the frequency of the fourth pole (unit: Hz)";
    X(16)
270 INPUT "17. What is the maximum source current (unit: A)"; X(17)
280 INPUT "18. What is the maximum sink current (unit: A)"; X(18)
310 CLS
320 PRINT "These are your input data:"
330 PRINT
340 PRINT TAB(3); " 1. Differential input resistance:"; TAB(43); X(1);
    TAB(55); "Ω"
350 PRINT TAB(3); " 2. Common mode input resistance:"; TAB(43); X(2);
    TAB(55); "Ω"
360 PRINT TAB(3); " 3. DC differential mode gain:"; TAB(43); X(3);
    TAB(55); "dB"
370 PRINT TAB(3); " 4. Common mode rejection ratio:"; TAB(43); X(4);
    TAB(55); "dB"
380 PRINT TAB(3); " 5. Input offset current:"; TAB(43); X(5); TAB(55);
    "A"
390 PRINT TAB(3); " 6. Input bias current:"; TAB(43); X(6); TAB(55);
    "A"
400 PRINT TAB(3); " 7. Output resistance:"; TAB(43); X(7); TAB(55);
    "Ω"
410 PRINT TAB(3); " 8. Positive slew rate:"; TAB(43); X(8); TAB(55);
    "V/us"
420 PRINT TAB(3); " 9. Negative slew rate:"; TAB(43); X(9); TAB(55);
    "V/us"
430 PRINT TAB(3); "10. Input capacitance:"; TAB(43); X(10); TAB(55);
    "F"
```

```
440 PRINT TAB(3); "11. Input offset voltage:"; TAB(43); X(11); TAB(55);
    "V"
450 PRINT TAB(3); "12. Frequency of the first pole:"; TAB(43); X(12);
    TAB(55); "Hz"
460 PRINT TAB(3); "13. Frequency of the second pole:"; TAB(43); X(13);
    TAB(55); "Hz"
470 PRINT TAB(3); "14. Frequency of the third pole:"; TAB(43); X(14);
    TAB(55); "Hz"
480 PRINT TAB(3); "15. Frequency of the zero:"; TAB(43); X(15);
    TAB(55); "Hz"
490 PRINT TAB(3); "16. Frequency of the fourth pole:"; TAB(43); X(16);
    TAB(55); "Hz"
500 PRINT TAB(3); "17. Maximum source current:"; TAB(43); X(17);
    TAB(55); "A"
510 PRINT TAB(3); "18. Maximum sink current:"; TAB(43); X(18);
    TAB(55); "A"
540 PRINT : PRINT
550 INPUT "Is your input data correct (Y/N)"; D$
560 IF D$ = "Y" GOTO 600 ELSE IF D$ = "y" GOTO 600
570 INPUT "Do you want to try again (Y/N)"; Z$
580 IF Z$ = "Y" GOTO 30 ELSE IF Z$ = "y" GOTO 30
590 GOTO 10000
600 LET PI = 3.141592654#
610 LET A = X(2)
620 LET B = X(1)
630 LET C = X(7)
640 IF X(14) = X(15) THEN D = 1E+12 ELSE D = 10000 / (X(14) /
    X(15) - 1)
650 LET E = X(6) + X(5) / 2
660 LET F = X(6) - X(5) / 2
670 LET G = 1 / (2000 * PI * X(12))
680 LET H = 1 / (2000 * PI * X(13))
690 LET I = X(10)
700 LET J = 1 / (20000 * PI * X(15))
710 LET K = 1 / (2000 * PI * X(16))
720 LET X3 = EXP(X(3) * LOG(10) / 20)
730 LET X4 = EXP(X(4) * LOG(10) / 20)
740 LET L = X3 / 1000
750 LET M = X3 / (2000 * X4)
760 LET N = 1 / X(7)
770 LET O = -X(8) * G * 1000000!
780 LET P = -X(9) * G * 1000000!
790 LET Q = -1 * X(17)
800 LET R = -1 * X(18)
810 LET S = (D + 10000) / D
820 LET T = X(11)
830 CLS
840 PRINT "Amplifier Macromodel Parameters"
```

```
850 PRINT : PRINT TAB(5); " 1.  A="; A
860 PRINT TAB(5); " 2.  B="; B
870 PRINT TAB(5); " 3.  C="; C
880 PRINT TAB(5); " 4.  D="; D
890 PRINT TAB(5); " 5.  E="; E
900 PRINT TAB(5); " 6.  F="; F
910 PRINT TAB(5); " 7.  G="; G
920 PRINT TAB(5); " 8.  H="; H
930 PRINT TAB(5); " 9.  I="; I
940 PRINT TAB(5); "10.  J="; J
950 PRINT TAB(5); "11.  K="; K
960 PRINT TAB(5); "12.  L="; L
970 PRINT TAB(5); "13.  M="; M
980 PRINT TAB(5); "14.  N="; N
990 PRINT TAB(5); "15.  O="; O
1000 PRINT TAB(5); "16.  P="; P
1010 PRINT TAB(5); "17.  Q="; Q
1020 PRINT TAB(5); "18.  R="; R
1030 PRINT TAB(5); "19.  S="; S
1040 PRINT TAB(5); "20.  T="; T
1050 PRINT
1060 INPUT "Do you want output file in a SPICE sub-circuit format
     (Y/N)"; P$
1070 IF P$ = "N" GOTO 10000 ELSE IF P$ = "n" GOTO 10000
1080 INPUT "What is your output file name for the AMP macromodel";
     NAME$
1090 OPEN NAME$ FOR OUTPUT AS #1
1100 PRINT #1, "* Amplifier Macromodel *"
1110 PRINT #1, "*  2 =  V+ Input Node,          4 = V- Input Node,"
1120 PRINT #1, "*  9 =  V+ Power Supply        11 = V- Power Supply,"
1130 PRINT #1, "* 14 =  Output Node"
1140 PRINT #1, "*"
1145 PRINT #1, "* NODES 2-19 ARE USED."
1147 PRINT #1, "*"
1150 PRINT #1, "R1 3 0"; SPC(1); A
1160 PRINT #1, "R2 3 4"; SPC(1); B
1170 PRINT #1, "R3 4 0"; SPC(1); A
1180 PRINT #1, "R4 6 0 1E3"
1190 PRINT #1, "R5 12 0"; SPC(1); C
1200 PRINT #1, "R6 13 0 1E3"
1210 PRINT #1, "R7 17 18 10E3"
1220 PRINT #1, "R8 18 0"; SPC(1); D
1230 PRINT #1, "R9 19 0 1E3"
1231 PRINT #1, "*USE THE FOLLOWING IF A NPN INPUT IS DESIRED:"
1240 PRINT #1, "I1 3 0"; SPC(1); E
1250 PRINT #1, "I2 4 0"; SPC(1); F
1251 PRINT #1, "*"
1252 PRINT #1, "*USE THE FOLLOWING IF A PNP INPUT IS DESIRED:"
```

```
1255 PRINT #1, "I1 0 3"; SPC(1); E
1257 PRINT #1, "I2 0 4"; SPC(1); F
1258 PRINT #1, "*"
1260 PRINT #1, "C1 7 0"; SPC(1); G
1270 PRINT #1, "C2 13 0"; SPC(1); H
1280 PRINT #1, "C3 3 4"; SPC(1); I
1290 PRINT #1, "C4 17 18"; SPC(1); J
1300 PRINT #1, "C5 19 0"; SPC(1); K
1310 PRINT #1, "G1 0 6 19 0"; SPC(1); L
1320 PRINT #1, "G2 0 6 POLY(2) (3,0) (4,0) 0"; SPC(1); M; SPC(1); M
1330 PRINT #1, "G3 0 12 7 0"; SPC(1); N
1340 PRINT #1, "G4 0 13 3 4 .001"
1350 PRINT #1, "G5 0 19 18 0 .001"
1360 PRINT #1, "VA 12 14 DC 0"
1370 PRINT #1, "VB 6 7 DC 0"
1380 PRINT #1, "F1 8 9 POLY(1) VB"; SPC(1); O; SPC(1); "1"
1390 PRINT #1, "F2 11 10 POLY(1) VB"; SPC(1); P; SPC(1); "-1"
1400 PRINT #1, "F3 15 9 POLY(1) VA"; SPC(1); Q; SPC(1); "1"
1410 PRINT #1, "F4 11 16 POLY(1) VA"; SPC(1); R; SPC(1); "-1"
1420 PRINT #1, "E1 17 0 13 0"; SPC(1); S
1430 PRINT #1, "VC 2 3"; SPC(1); T
1435 PRINT #1, "*USE THE FOLLOWING IF A MORE ACCURATE MODEL OF A
     NPN INPUT IS DESIRED:"
1440 PRINT #1, "*D1 1 2 DX"
1450 PRINT #1, "*D2 5 4 DX"
1451 PRINT #1, "*"
1452 PRINT #1, "*USE THE FOLLOWING IF A MORE ACCURATE MODEL OF A
     PNP INPUT IS DESIRED:"
1455 PRINT #1, "*D1 2 1 DX"
1457 PRINT #1, "*D2 4 5 DX"
1458 PRINT #1, "*"
1460 PRINT #1, "D3 7 8 DX"
1470 PRINT #1, "D4 8 9 DX"
1480 PRINT #1, "D5 10 7 DX"
1490 PRINT #1, "D6 11 10 DX"
1500 PRINT #1, "D7 7 9 DX"
1510 PRINT #1, "D8 11 7 DX"
1520 PRINT #1, "D9 14 15 DX"
1530 PRINT #1, "D10 15 9 DX"
1540 PRINT #1, "D11 16 14 DX"
1550 PRINT #1, "D12 11 16 DX"
1560 PRINT #1, ".MODEL DX D(N=.001)"
1580 CLOSE #1
1590 PRINT
1600 PRINT "Your output file (name:"; SPC(1); NAME$; ") is
     successfully created."
10000 END
```

APPENDIX

Overcoming Convergence Problems in SPICE

Convergence problems in SPICE simulations usually occur in the bias point calculation or transient analysis. These problems are usually caused by two main factors: (1) discontinuities in the device equations that make solving the matrices difficult, and (2) constraints on the analysis technique that prevent the SPICE algorithms from solving the equations. The methods presented here do not change the circuit topology or element values but, when applied properly, significantly increase the number of circuits that SPICE can successfully analyze.

A.1 BIAS POINT CALCULATION

A frequent use of SPICE is to perform small-signal analyses. SPICE starts with a bias point calculation, which it requires for other analyses including the small-signal transfer function, DC sensitivity, noise, and frequency response calculations. Circuits with multiple feedback paths or hysteresis sometimes fail to converge or take considerable time to do so when used with SPICE's default conditions. To overcome convergence problems during the bias point calculation, use the following options:

.OPTION LIST NODE—Before making changes in the input file, check the summary of all circuit elements to verify that each is connected to the correct node; check the table of nodes and devices to verify that each node is connected to the correct elements. The LIST flag option outputs a summary table of all circuit elements. The NODE flag produces a table of nodes and devices attached to each node.

.NODESET—This command allows the user to set the initial voltage or "best guesses" at designed nodes at the beginning of the analysis procedure that solves for node voltages and branch currents. Hand calculations or partial results from other simpler simulations can serve as good starting-point, best-guess values. Specifying accurate initial voltages at only a few nodes often allows SPICE to converge on difficult circuits. It also significantly reduces the time needed to solve for voltages and currents on almost any simulation.

OFF option in active devices—When SPICE starts the numerical analysis that calculates voltages and currents for the simulation run, it assumes that all diodes and transistors are ON. Initially specifying a few key active devices as being OFF so they are not operating in their active region will often allow SPICE to converge on difficult circuits that have hysteresis or multiple feedback paths.

.OPTION ITL1—This option controls the total number of iterations in the bias point calculation. If the analysis has failed to converge within the default maximum number of iterations (= 100), increase the ITL1 value to double or triple its initial value.

A.2 TRANSIENT ANALYSIS

SPICE simulations sometime suffer convergence problems in transient analysis like the bias point calculation. Often these can be overcome by using analysis options, limits, and control parameters with other **.OPTION** commands.

.OPTION ITL4—This option sets the maximum number of iterations (default value = 10) allowed at any point in time during the transient analysis. If the iteration count required for convergence rises above the value assigned to ITL4, the analysis is automatically terminated without any error messages. This event typically happens when a voltage or current in the circuit changes significantly in a short time period. Increasing ITL4 (ITL4 = 50 or 100) often allows transient analysis convergence. However, there will be a significant increase in the simulation time.

.OPTION ITL5—This sets the maximum number of total iterations for the transient analysis. If the maximum number of iterations required for the transient analysis exceeds the value set with ITL5 (default value = 5000), the analysis stops. Setting ITL5 = 0 removes any limitation on the maximum number of iterations.

.OPTION RELTOL—If the convergence problems cannot be solved by using the previous two options, increase the relative error tolerance of the program using the RELTOL option. The default value of RELTOL is 0.001 (0.1%). Although the RELTOL value actually depends on the accuracy that is desired; a recommended value is 0.01 (1%). This option is very useful in solving the convergence problems caused by the small time step of the numerical integration.

A.3 EXAMPLE OF AN ORDERLY PROCEDURE FOR CORRECTING PROBLEMS

Consider a design example of producing a macromodel of a frequency synthesizer that will multiply the input frequency by a factor of 1.5. Suppose we choose to utilize a cascaded arrangement of two drop-in macromodels: a frequency divider (÷ 2) and a frequency multiplier (× 3). The equivalent circuit structure macromodel for this combination is given in Figure A.1.

When this SPICE file is simulated, a convergence problem occurs with the DC bias point calculation. To solve this problem, insert the following option statement into the input control file:

```
.OPTION ITL1=200
```

This corrects the DC bias convergence problem by increasing the number of iterations. Also, there will be no convergence problems in the transient analysis. However, the simulation result produces an incorrect output waveform as shown in Figure A.2.

To correct this error, decrease the RELTOL value from the default value of 0.001 to 1E-5 by modifying the previous options statement to

```
.OPTIONS  ITL1=200  RELTOL=1E-5
```

```
SPICE MACROMODEL FOR FREQUENCY SYNTHESIZER
.SUBCKT FDIV  90     91    92
*             In+   In-   Out
R1  90  91  1T
G1  0  93  POLY(1) 90  91  1  0  -1
G2  93  0  POLY(1) 93  0  0  0  1
R2  93  0  1G
G3  92  0  POLY(1)  93  0  0.5  -0.5
G4  0  92  POLY(1)  92  0  0  0  1
R3  92  0  1G
.ENDS FDIV
.SUBCKT FMUL  90     91     92
*             In+   In-   Out+
R1  90  91  1T
G1  0  94  POLY(1) 90  91  1  0  -1
G2  94  0  POLY(1) 94  0  0  0  1
R2  94  0  1G
E1  92  0  POLY(2)  90  91  94   0
+  0 3  0  0  0  0  -4  0  0
RD  92  0  1G
.ENDS FMUL

VIN  1 0  SIN(0 1 1MEG)
X1  1  0  2  FDIV
X2  2  0  3  FMUL
.TRAN  10N  2U
.PRINT  TRAN V(1) V(2) V(3)
.PROBE
.END
```

Figure A.1 Input control file for the frequency synthesizer

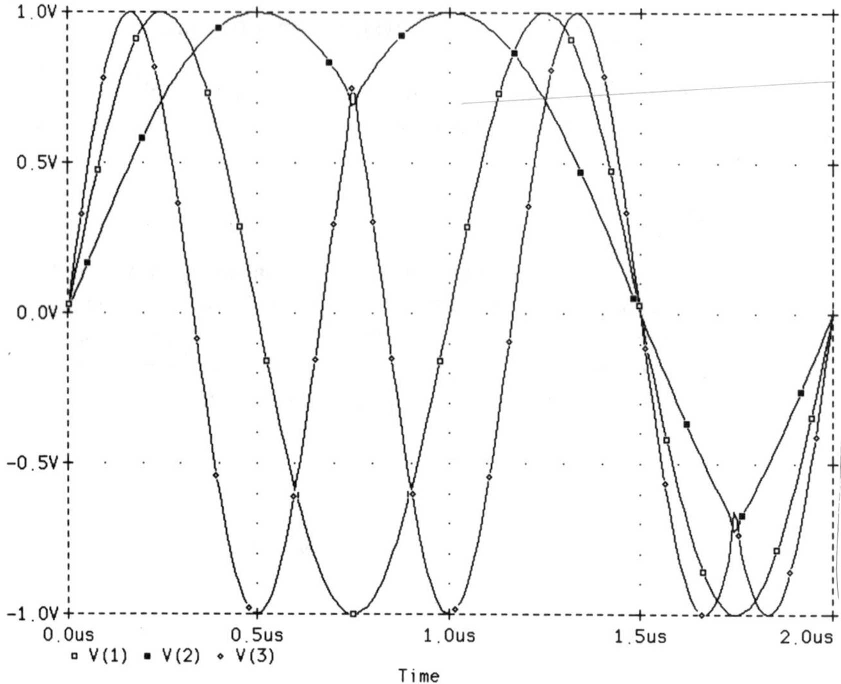

Figure A.2 Simulation of frequency synthesizer showing incorrect output frequency.

When resimulated, a convergence problem occurs during the transient analysis. This can be overcome by increasing the number of iterations at one time point by means of the ITL4 option. If convergence still fails to occur with this increase in ITL4, then increase the RELTOL option value. Again, this may correct the convergence problem but produces an incorrect simulation result. There are two ways to correct this. One way is to decrease the transient step size by making the following change to the transient analysis control:

```
.TRAN  5N  2U
```

For this case a longer simulation time and a larger memory space for the output will be required. In making this change, be sure that the maximum internal time step does not exceed the printing step size. The second way to correct the incorrect output problem is to decrease the internal time step instead of the printing step. Alter the transient analysis control to

```
.TRAN  10N  2U  0  5N
```

The simulation time is about the same as before, but the required memory space is much smaller and the correct result is obtained, as shown in Figure A.3.

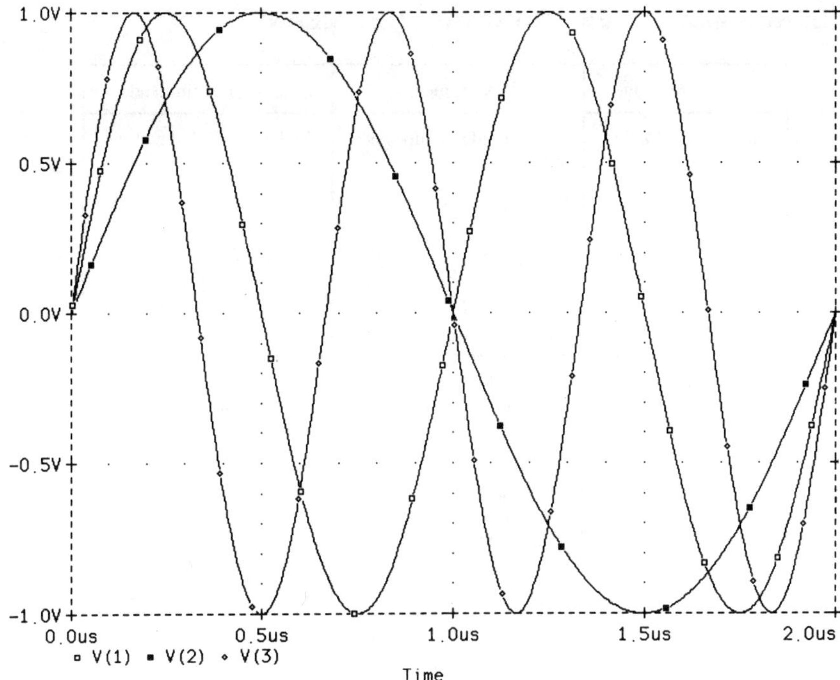

Figure A.3 Simulation of frequency synthesizer showing correct output frequency.

To summarize the procedure, follow these steps:

Step 1.
Run the SPICE input control file without any options or limitations on the time step. Check the simulation results.

Step 2.
If there is a DC convergence problem, use the .OPTION ITL1 = 200 control. Again, check the simulaton results for correctness.

Step 3.
If the results are incorrect , reduce the relative tolerance by adding RELTOL = 1E-5 to the .OPTIONS control. Recheck the simulation results.

Step 4.
If there is no improvement, change the transient analysis command by reducing the step size.

A.4 ERROR MESSAGES AND WHAT THEY MEAN

Error Message	What it means	Solution/Recommendation
Node [x] is floating.	DC closed path does not exist. (Example) 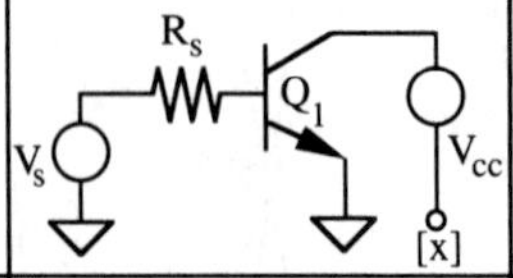 R_s, Q_1, V_s, V_{cc}, [x]	Make DC closed path. (Solution) 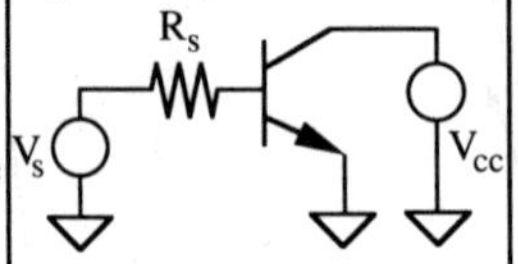 R_s, V_s, V_{cc}
Less than 2 connections at node [x].	Node [x] is not connected with any elements including ground. (Example) 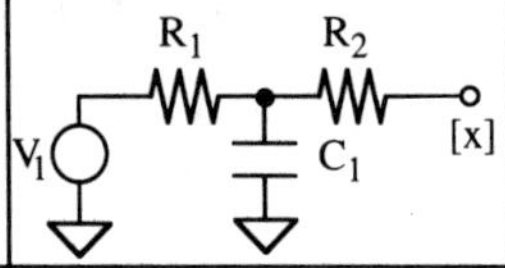 R_1, R_2, V_1, C_1, [x]	Check your circuit description file and connections at node [x]. (Solution) 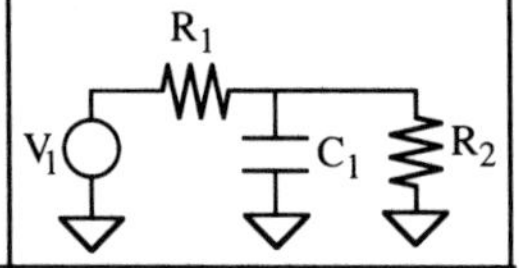 R_1, V_1, C_1, R_2
Duplicate name: [x].	Device name [x] is used more than one time. (Example) R1 2 3 1MEG **R1** 5 9 1K	Change the device name that is used more than one time. (Solution) R1 2 3 1MEG **R2** 5 9 1K
Missing gain/ transconductance/ transresistance.	The gain of dependent sources (E, G, H, and F) is not defined. (Example) ESUM 7 5 2 3	Define the gain of dependent sources. (Solution) ESUM 7 5 2 3 **100**

Error Message	What it means	Solution/Recommendation
Missing model.	The model name is not defined. (Example) QTR 9 8 17	Define model name of the problem element. (Solution) QTR 9 8 17 **NPN**
Invalid card.	. xxxx line is not valid in SPICE program. (Example) . DEC 20 1 100K	Check . xxxx line and redefine. (Solution) **. AC** DEC 20 1 100K
Maximum entry in this column at step [x] (<value>) is less than pivtol.	The (<value>) at step [x] is less than default pivtol value.	.OPTION **PIVTOL=1E -30**
Model <Model name> referenced by <Device name> is undefined. (Example) Model **MOS** referenced by **M1** is undefined.	Missing . MODEL line. (Example) M1 2 3 3 0 MOS	Define . MODEL line. (Solution) M1 2 3 3 0 MOS **. MODEL MOS NMOS**
Only . MODEL valid in subcircuit.	. xxxx line used in subcircuit is not valid. In subcircuit . IC, . NODESET, . TEMP, etc. are not valid.	Remove . xxxx line from subcircuit.
Invalid sweep type.	Does not specify the type of sweep on . AC line.	Specify the type of sweep by using DEC, OCT, or LIN.

RECOMMENDED REFERENCES

J. R. Hines, "Reduce Simulation Times and Costs with SPICE Convergence Aids," *Personal Engineering and Instrumentation News*, May 1987, pp. 47–51.

MicroSim Corporation, *PSpice*, Laguna Hills, California, 1988.

W. Banzhaf, *Computer-Aided Circuit Analysis Using SPICE*, Englewood Cliffs, N.J.: Prentice Hall, 1989.

APPENDIX

B

Getting the Most Out of Probe

B.1 PSPICE PROBE

Probe in PSpice is the graphics postprocessor that calculates and displays results of a simulation. Both screen displays and hard copies are available. In effect, Probe functions as a "software" oscilloscope, calculator, and spectrum analyzer.

B.2 VERSIONS AVAILABLE

MicroSim Corporation released PSpice in January 1984. Two major versions have been available: the commercial or industrial version and the evaluation version, also known as the educational, classroom, or demonstration version. Over the years, MicroSim has developed commercial versions to function on various operational platforms such as the IBM and Macintosh PCs, and the Sun, MicroVAX, and VAX workstations. The latest version as of this writing is 4.05, which incorporates the Windows 3.0 features into the Probe operations. Readers are referred to MicroSim's Users' Manual for specific Probe information for the particular version being used, as there are subtle differences in cursor and function controls among the various versions.

The evaluation version has been distributed without cost to educational institutions and is widely accepted and used. Though limited in the size and complexity of circuits it can simulate, the educational version contains the useful Probe postprocessor. The most recent educational version as of this writing is also 4.05.

B.3 ARITHMETIC FUNCTIONS

Arithmetic operations on output variables are allowed in PSpice Probe. The available operators and functions are as follows:

Function	Actual Meaning	Example
+	Addition of voltage or current	V(3)+V(5)+V(10)
−	Subtraction of voltage or current	I(VSEN1)-I(VSEN2)
*	Multiplication of voltage or current	V(1)*I(VSEN1)
/	Division of voltage or current	V(10)/V(1)
ABS(X)	|X|—Absolute value of X	ABS(V(10))
SQRT(X)	$X^{1/2}$—Square root of X	SQRT(I(VSEN1))
EXP(X)	e^x	EXP(V(5))
LOG(X)	ln(X)—log base e	LOG(V(1))
LOG10(x)	$\log_{10}(x)$—log base 10	LOG10(V(8))
DB(X)	$20^* \log_{10}(\|X\|)$	VDB(6)
PWR(X,Y)	$\|X\|^Y$—X to the power of Y	PWR(V(5),2)
SIN(X)	sin (X)—X in radians	SIN(6.28*V(1))
COS(X)	cos (X)—X in radians	COS(6.28*V(1))
TAN(X)	tan (X)—X in radians	TAN(V(3))
ATAN(X)	$\tan^{-1}$ (X)—X in radians	ATAN(V(5)/V(1))
D(X)	Derivative of X with respect to the x-axis variable	d(V(3))
S(X)	Integral of X over the x-axis variable	s(I(VSEN1))
AVG(X)	Running average of X	AVG(V(5))
RMS(X)	Running RMS average of X	RMS(I(VSENSE))

B.4 CONTROL KEYS

Add trace. The Add Trace command adds one or more traces (up to eight traces) to the selected node voltage or branch current display. The traces can be any output variable(s) in the data file or any combination of output variables modified with the arithmetic functions. Incorporating .PROBE into the input control file will cause all node voltage and branch currents to be saved in the PROBE.DAT file and available for display. If the input control file contains the statement

```
.PROBE V(1) V(4) V(7)
```

only voltages at nodes 1, 4, and 7 will be stored in the data file, and only these three voltages will be available for plotting. However, all various postprocessing functions involving these voltages are available, such as DB, ABS, and so forth.

Remove trace. The Remove Trace key removes one or all traces from the selected display. The traces are listed using the text, which is typed in when adding traces to the plot with the **Add Trace** command.

X-Axis/Y-Axis	The X Axis/Y Axis command enters the x-axis/y-axis menu and applies those commands that are listed below to the x-axis/y-axis.
Exit	Returns to the main menu.
Log	Sets the x-axis/y-axis to be logarithmic scale (base 10).
Linear	Sets the x-axis/y-axis to be linear scale.
Auto-Range	Sets the range of the x-axis/y-axis to be the range of its variable, which is defined in the circuit description file.
Set Range	Sets the range of the x-axis/y-axis variable from the user specification.
X Variable	Sets the variable for the x-axis. The default variable of the x-axis is the sweep variable of the analysis defined in the circuit description file. However, the user can set it to any variable or expression that the user specifies.
Fourier	Switches to Fourier transform mode. If the x-axis variable is Time, the x-axis variable is changed to Frequency. If the x-axis variable is Frequency, an inverse Fourier transform is done.
End Fourier	Ends the Fourier transform mode and goes back to the original domain.
Plot Control	Enters the plot control menu.
Exit	Exits the plot control menu and returns to the main menu.
Add plot	Adds a plot to the display.
Remove plot	Removes the selected plot from the display.
Select plot	Allows the user to change which is selected.
Use symbols	Enables or disables the trace symbols.

Cursors. The Cursor command in Probe allows the tracking of the traces. Two cursors, which may be attached to one or two traces, are activated. To move to the first point of a trace, use the < Home > key. To move to the last point of a trace, use the < End > key.

For tracking one trace,

1. Select the trace desired by combination of the < Ctrl > <→> or < Ctrl > <←> keys.
2. Track the trace by using <→> or <←>.

For tracking two traces,

1. Move one cursor from the first trace to the second trace by using < Ctrl > <→>.
2. Use just the <→> or <←> key for tracking the first trace.
3. Use < Shift > <→> or < Shift > <←> keys for tracking the second trace.

B.5 SPECIAL TRICKS

Conversion from Hz to radians per second (rps) domain. To establish rps values of the x-axis, just multiply 2π times the x-axis in the Hz-domain. The procedure is as follows:

1. Select **X-Axis** from the main Probe menu.
2. Choose **X Variable** under the **X-Axis** command.
3. Type **6.2832*Frequency**.

The result is that the x-axis in Hz changes to rps, although the frequency value on the x-axis will still be labeled as Hz.

Drawing asymptotes. Asymptotic lines can be drawn to check gain, attenuation (dB/decade), and the like.

1. Select **Add Trace** from the main probe menu.
2. If the filter function is low-pass, as shown in Example 1, type

$$-20 * \mathrm{n} * \mathrm{LOG10(Frequency/} < -3 \text{ dB frequency} >)$$

where n is an integer and is the same as the order of low-pass filter function. Example 2 shows how to use the graphic capabilities of PROBE to determine the midband gain (27.98dB) plus the upper (34 KHz) and lower (300 Hz) -3 dB break frequencies of an amplifier.

Example 1

Consider a third-order Butterworth low-pass filter of the generalized behavioral type described in Section 2.5. When denormalized to a cut-off frequency of 1 KHz, the input control file is

```
Third-Order Butterworth Filter
RIN 1 0 1MEG
G1  0 102  POLY(4) 100 0 102 0 103 0 104 0
+0  1 -1.2566E4 -7.8957E7 -2.4805E11
R1  102  0 1T
C1  102  0 1
G2  0 103 102 0 1
R2  103  0 1T
C2  103  0 1
G3  0 104 103 0 1
R3  104 0 1T
C3  104 0 1
EOUT  105  0 104  0 2.4805E11
ROUT  105  106  1K
RD  106  0  1G
VIN 100 0 AC=1
.AC DEC 20 0.1 100K
.PRINT AC VDB(106)
.PROBE
.END
```

Simulate the circuit and plot the output node 106 voltage in dB, VDB(106). Then draw the asymptotic line by following the procedure above, typing

```
-20*3*LOG10(Frequency/1000)
```

The results of this simulation and the asymptote plot are shown in Figure B.1.

Example 2

This is a more complicated example for drawing three asymptotes of a bandpass filter. The SPICE input file is as follows:

```
Bandpass Filter
RIN 1 0 1MEG
G1  0 102  POLY(3) 100 0 102 0 103 0
+0  1  -2.155E5  -4.0268E8
R1  102  0 1T
C1  102  0 1
G2  0 103 102 0 1
R2  103  0 1T
C2  103  0 1
EOUT  104  0 102  0  5.4E6
```

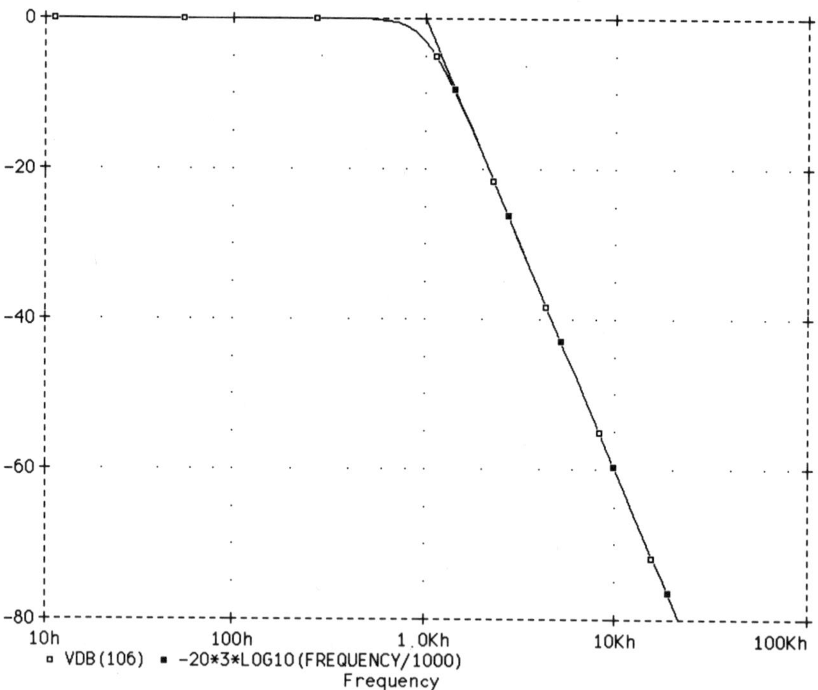

Figure B.1 Simulation of a third-order Butterworth low-pass filter with 1 KHz cut-off frequency and unity gain with asymptotic slope plot.

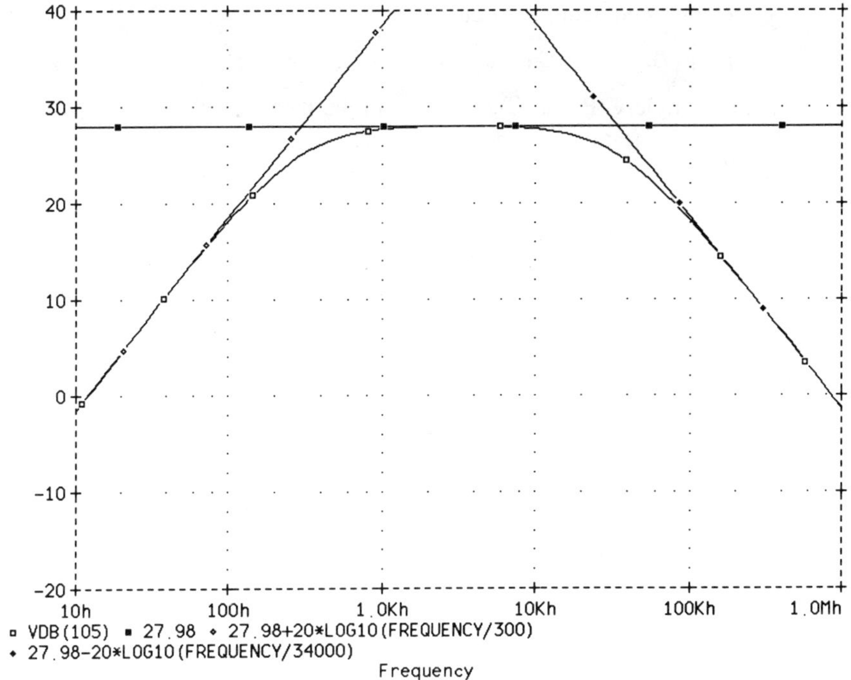

Figure B.2 Simulation of a bandpass filter with three asymptotes.

```
ROUT  104  105  1K
RD  105  0  1G
VIN 100 0 AC=1
.AC DEC  20  1  1MEG
.PRINT AC VDB(105)
.PROBE
.END
```

After simulating the circuit, plot the output node voltage, VDB(105). The pass band gain is 27.98 dB, and the lower and upper -3dB frequencies are located at 300 Hz and 34 KHz, respectively. Figure B.2 shows the equations used to produce the three asymptotic lines.

B.6 COMBINING PROBE PLOTS

PSpice Probe can merge two or more different PSpice simulation results. If three different PSpice simulations are saved under separate names—exampleA.dat, exampleB.dat, and exampleC.dat—then they can be concatenated into one file. For MS-DOS or PC-DOS, use the **copy** command according to:

```
>copy/b  exampleA.dat+exampleB.dat+exampleC.dat  test.dat
```

where exampleA (B and C).dat is the Probe data file of the three different PSpice simulations. The test.dat is a new Probe data file, which merges the three different Probe data files. The saved data format of test.dat is as follows:

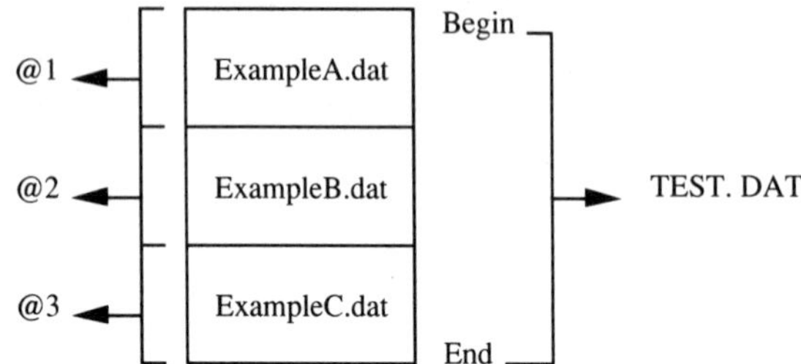

If the test.dat file contains more than one of the same type of analysis (e.g., three transient analyses), the first probe menu will allow the user to choose all transient analyses or to choose just one transient analysis. If all analyses are chosen to view, then the **Add Trace** command will cause all curves for each expression to be displayed. For example, if all transient analyses are chosen to be viewed, then entering the expression

```
V(3)
```

will result in Probe drawing three curves, which are the voltages at node 3 of exampleA.dat, B.dat, and C.dat files.

Just one transient analysis can be called up by entering the expression

```
V(3)@1
```

This will draw the voltage at node 3 with respect to time from the first data saved in the test.dat file, namely exampleA.dat. For a trace of voltage from exampleC.dat, enter the expression

```
V(5)@3
```

In general, voltage and current have "@n" appended where "n" is a number designating which data file is to be selected. Leaving off "@n" gives plots of all the analyses. If a comparison between the voltage at node 3 of exampleB.dat and the voltage at node 5 of exampleA.dat is desired, enter the expression

```
V(3)@2-V(5)@1
```

RECOMMENDED REFERENCES

W. Banzhaf, *Computer-Aided Circuit Analysis Using SPICE*, (Englewood Cliffs, N.J.: Prentice Hall, 1989).

MicroSim Corporation, *PSpice*, Laguna Hills, California, 1988.

Muhammad H. Rashid, *SPICE for Circuits and Electronics Using PSpice* (Englewood Cliffs, N.J.: Prentice Hall, 1990).

Paul W. Tuinenga, *SPICE: A Guide to Circuit Simulation and Analysis Using PSpice* (Englewood Cliffs, N.J.: Prentice Hall, 1988).

APPENDIX

C

Quick SPICE

This section provides a quick reference in the form of examples for using the commands in SPICE. The material presented is a condensation of material given in the SPICE Users' Guide. Figures, equations, and examples have been added for quick reference.

C.1 ELEMENTS

Resistors

```
RXXXXXXX N1 N2 VALUE <TC=TC1<,TC2>>
```

```
R1 1 2 100
RC1 12 17 1K TC=0.001,0.015
```

R_1 R_2

1 —/\/\/— 2 12 —/\/\/— 17

100 1K

```
value(TEMP) = value(TNOM)*[1+TC1*(TEMP-TNOM) +
TC^2 *(TEMP-TNOM)^2]
```

```
RC1(TEMP) = 1K[1+ .001(TEMP-TNOM) + .015(TEMP-TNOM)^2]
```

Capacitors and inductors

```
CXXXXXXX  N+  N-  VALUE  <IC+INCOND>
LXXXXXXX  N+   N- VALUE  <IC+INCOND>
```

```
CBYP 13 0 1UF
COSC 17 23 10U IC=3V
LLINK 42 69 1UH
LSHUNT 23 51 10U IC=15.7MA
```

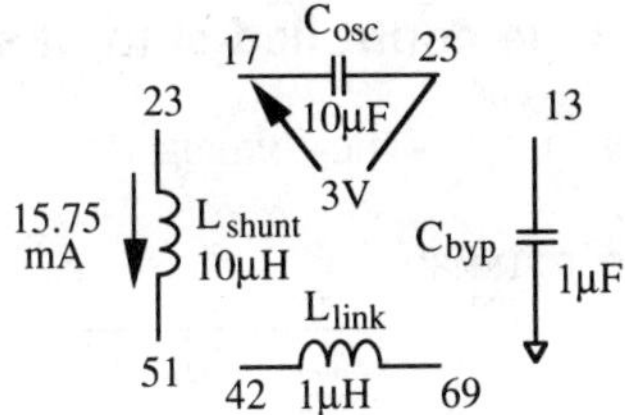

IC is the initial condition. For the capacitor, it is the time zero value of the capacitor voltage in volts. For the inductor, it is the initial value of inductor current in amps flowing from the N+ node to the N− node.

Nonlinear capacitors and inductors

```
CXXXXXXX N+ N- POLY C0 C1 C2 ... <IC=INCOND>
LYYYYYYY N+ N- POLY L0 L1 L2 ... <IC=INCOND>
```

C0 C1 C2 . . .(and L0 L1 L2 . . .) are the coefficients of a polynomial describing the element value.

Coupled (mutual) inductors

```
KXXXXXXX LYYYYYYY LZZZZZZZ VALUE

K43 LAA LBB 0.999
LAA 18 27 1MH
LBB 12 15 1MH

KXFRMR L1 L2 0.87
L1 1 7 1H
L2 9 24 1H
```

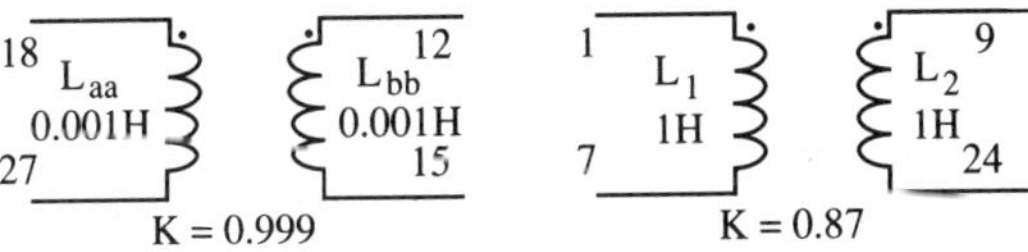

Transmission lines (lossless)

```
TXXXXXXX N1 N2 N3 N4 Z0=VALUE <TD=VALUE> <F=FREQ
<NL=NRMLEN>> <IC=V1,I1,V2,I2>

T1 1 0 2 0 Z0=50 TD=10NS
```

Linear dependent sources

```
i = g*v          v = e*v          i = f*i          v = h*i
```

Linear voltage controlled current sources

```
GXXXXXXX N+ N- NC+ NC- VALUE

G1 2 0 5 0 0.1MMHO
```

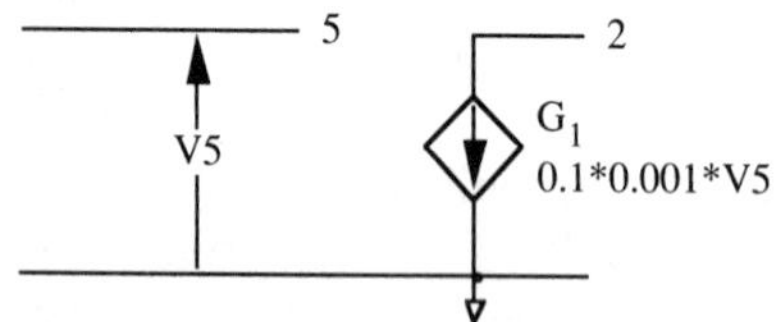

Linear voltage controlled voltage sources

```
EXXXXXXX N+ N- NC+ NC- VALUE

E1 2 3 14 1 2.0
```

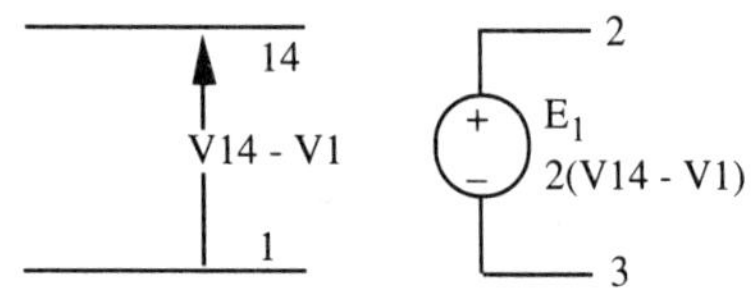

Linear current controlled current sources

```
FXXXXXXX N+ N- VNAM VALUE

F1 13 5 VSENS 5
VSENS 12 8 DC 0 AC 0
```

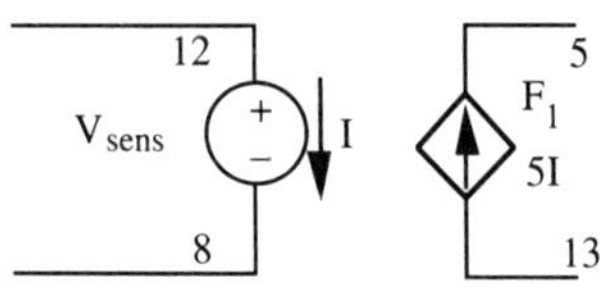

Linear current controlled voltage sources

```
HXXXXXXX N+ N- VNAM VALUE

HX 5 17 VZ 0.5K
VZ 11 16 DC 0 AC 0
```

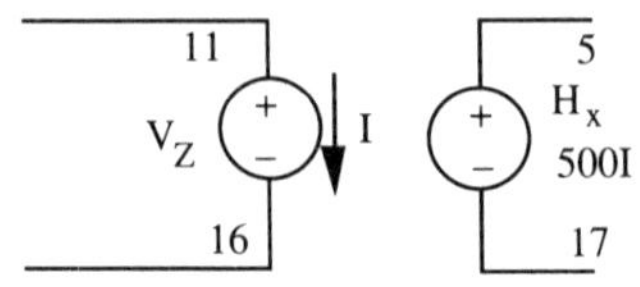

Independent sources

```
VXXXXXXX N+ N- <<DC> DC/TRAN VALUE> <AC <ACMAG
<ACPHASE>>>
```

```
IYYYYYYY N+ N- <<DC> DC/TRAN VALUE> <AC <ACMAG
<ACPHASE>>>
```

```
VCC 10 0 DC 6
VIN 13 2 0.001 AC 1 SIN(0 1 1MEG)
ISRC 23 21 AC 0.333 45.0 SFFM(0 1 10K 5 1K)
```

ISRC $= [0.333\angle 45]1\text{V}\sin[(2\pi \times 10^4\text{time}) + 5\sin(2\pi \times 10^3\text{time})]$

```
VMEAS 12 9
```

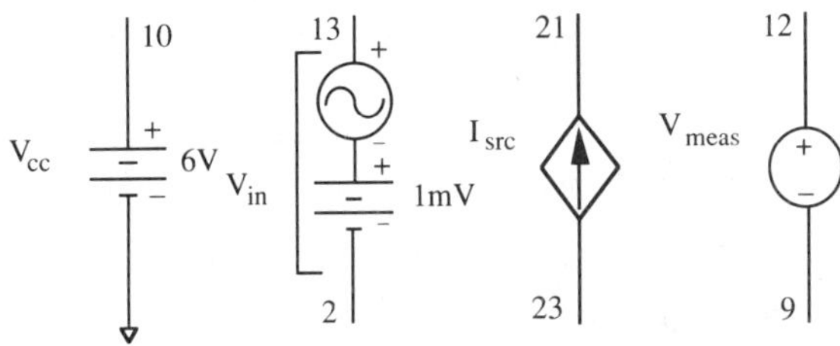

Pulse

```
PULSE(V1 V2 TD TR TF PW PER)
```

```
VIN 3 0 PULSE(-1 1 2NS 2NS 2NS 50NS 100NS)
```

Parameters	Default Values	Units
V1 (initial value)		Volts or amps
V2 (pulsed value)		Volts or amps
TD (delay time)	0.0	Seconds
TR (rise time)	TSTEP	Seconds
TF (fall time)	TSTEP	Seconds
PW (pulse width)	TSTOP	Seconds
PER (period)	TSTOP	Seconds

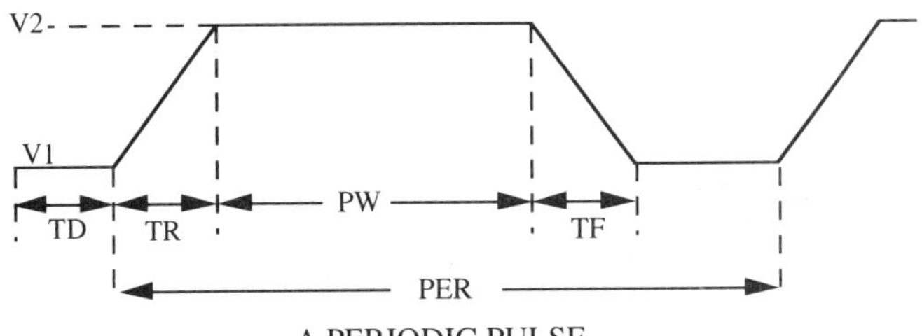

A PERIODIC PULSE

A single pulse is described as follows:

Time	Value
0	V1
TD	V1
TD + TR	V2
TD + TR + PW	V2
TD + TR + PW + TF	V1
TSTOP	V1

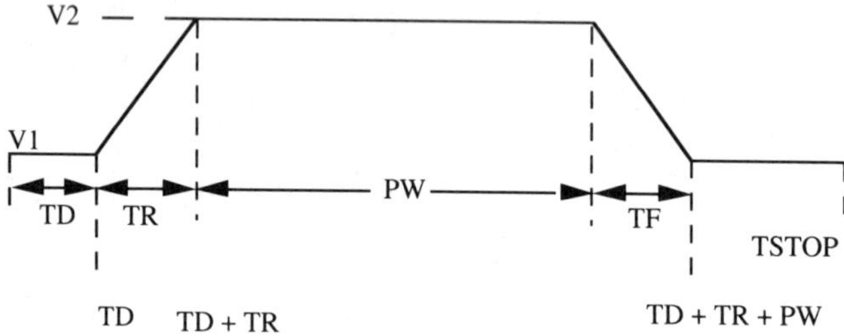

A SINGLE PULSE

Sinusoidal

```
SIN(VO VA FREQ TD THETA)
```

```
VIN 3 0 SIN(0 1 100MEG 1NS 1E10)
```

Parameters	Default Values	Units
VO (offset)		Volts or amps
VA (amplitude)		Volts or amps
FREQ (frequency)	1/TSTOP	Hz
TD (delay)	0.0	Seconds
THETA (damping factor)	0.0	1/Seconds

3

+

$$e^{-(\text{time}-10^{-9})(10^{-10})} \sin\left[2\pi(10^{8})(\text{time}+10^{-9})\right]$$

−

Exponential

```
EXP(V1 V2 TD1 TAU1 TD2 TAU2)
```

```
VIN 3 0 EXP(-4 -1 2NS 30NS 60NS 40NS)
```

Parameters	Default Values	Units
V1 (initial value)		Volts or amps
V2 (pulsed value)		Volts or amps
TD1 (rise delay time)	0.0	Seconds
TAU1 (rise time constant)	TSTEP	Seconds
TD2 (fall delay time)	TD1+TSTEP	Seconds
TAU2 (fall time constant)	TSTEP	Seconds

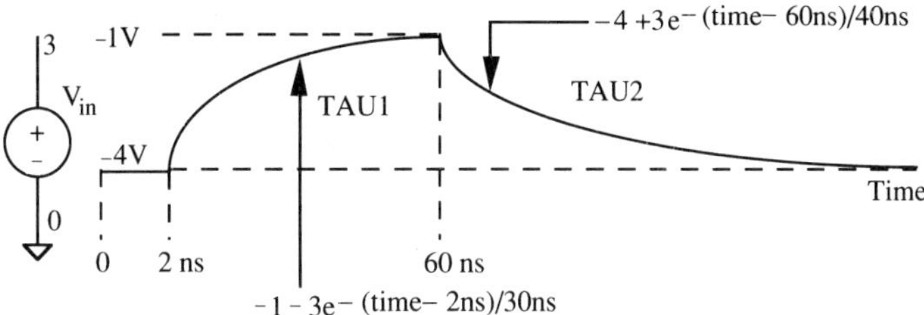

Piecewise linear

```
PWL(T1 V1 <T2 V2 T3 V3 T4 V4 . . . >)
```

```
VCLOCK 7 5 PWL(0 -7 10NS -7 11NS -3 17NS -3 18NS -7 50NS -7)
```

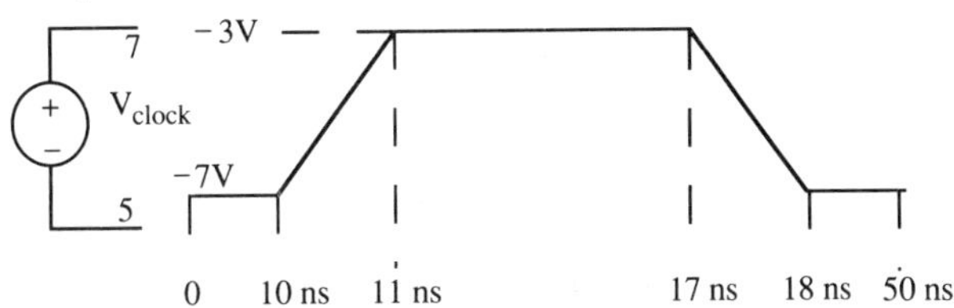

Single-frequency FM

```
SFFM(VO VA FC MDI FS)
```

```
V1 12 0 SFFM(0 1M 20K 5 1K)
```

Parameters	Default Values	Units
VO (offset)		Volts or amps
VA (amplitude)		Volts or amps
FC (carrier frequency)	1/TSTOP	Hz
MDI (modulation index)		
FS (signal frequency)	1/TSTOP	Hz

12

V_1 +

−

$$V_1 = 10^{-3}\sin[((2\pi)(20K)(time)) + 5\sin((2\pi)(1K)(time))]$$

C.2 SEMICONDUCTOR DEVICES

Junction diodes

```
DXXXXXXX N+ N- MNAME <AREA> <OFF> <IC=VD>

DBRIDGE 2 10 DIODE1
DCLMP 3 7 DMOD 3.0 IC=0.2
```

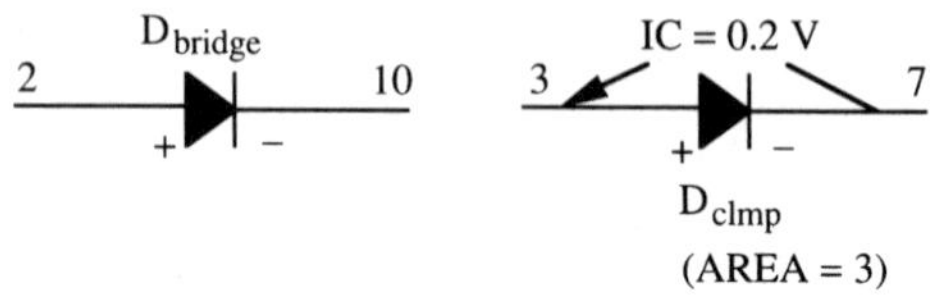


```
.MODEL MNAME D (PNAME1=PVAL1 PNAME2=PVAL2 . . . )
```

Diode model

Name	Parameter	Units	Default	Example
IS	Saturation current	A	1.0E-14	1.0E -14
RS	Ohmic resistance	Ohm	0	10
N	Emission coefficient	—	1	1.0
TT	Transit time	sec	0	0.1Ns
CJO	Zero-bias junction capacitance	F	0	2PF
VJ	Junction potential	V	1	0.6
M	Grading coefficient	—	0.5	0.5
EG	Activation energy	eV	1.11	1.11 Si 0.69 Sbd 0.67 Ge
XTI	Saturation current temperature exponent	—	3.0	3.0 jn 2.0 Sbd
KF	Flicker noise coefficient	—		
AF	Flicker noise exponent			
FC	Coefficient for forward bias Depletion capacitance formula	—	0.5	
BV	Reverse breakdown voltage	V	Infinite	40.0
IBV	Current at breakdown voltage	A	1.0E-3	

Bipolar junction transistors (BJTs)

```
QXXXXXXX NC NB NE <NS> MNAME <AREA> <OFF>
<IC=VBE,VCE>

Q23 10 24 13 QMOD IC=0.6,5.0
Q504 11 26 4 20 MOD1
```

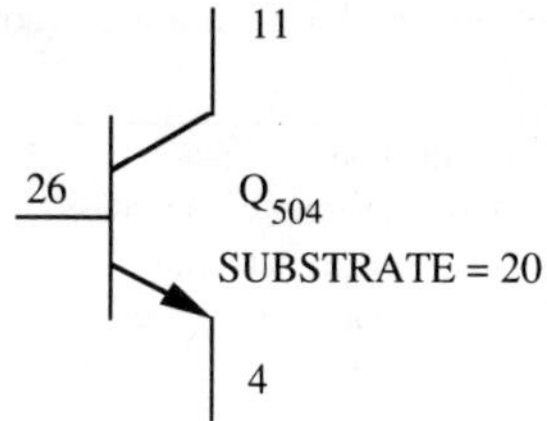


```
.MODEL MNAME TYPE(PNAME1=PVAL1 PNAME2=PVAL2 . . . )

.MODEL MOD1 NPN BF=50 IS=1E-13 VBF=50
```

NPN NPN BJT model

PNP PNP BJT model

BJT models (both NPN and PNP)

Modified Gummel-Poon BJT Parameters.

Name	Parameter	Units	Default	Example
IS	Transport saturation current	A	1.0E-16	1.0E-15
BF	Ideal maximum forward beta	-	100	100
NF	Forward current emission coefficient	—	1.0	
VAF	Forward Early voltage	V	Infinite	200
IKF	Corner for forward beta high-current roll-off	A	Infinite	0.01
ISE	B-E leakage saturation current	A	0	1.0E-13
NE	B-E leakage emission coefficient	—	1.5	2
BR	Ideal maximum reverse beta	—	1	0.1
NR	Reverse current emission coefficient	—	1	1
VAR	Reverse Early voltage	V	Infinite	200
IKR	Corner for reverse beta high-current roll-off	A	Infinite	0.01
ISC	B-C leakage saturation current	A	0	1.0E-13
NC	B-C leakage emission coefficient	—	2	1.5
RB	Zero-bias base resistance	Ohms	0	
IRB	Current where base resistance falls halfway to its minimum value	A	Infinite	0.1
RBM	Minimum base resistance at high currents	Ohms	RB	10
RE	Emitter resistance	Ohms	0	1
RC	Collector resistance	Ohms	0	0
CJE	B-E zero-bias depletion capacitance	F	0	2PF
VJE	B-E built-in potential	V	0.75	0.6
MJE	B-E junction exponent factor	—	0.33	0.33
TF	Ideal forward transit time	sec	0	0.1Ns
XTF	Coefficient for bias dependence of TF	—	0	
VTF	Voltage describing VBC dependence of TF	V	infinite	
ITF	High-current parameter for effect on TF	A	0	
PTF	Excess phase at frequency = $1.0/(TF*2\pi)$Hz	deg	0	
CJC	B-C zero-bias depletion capacitance	F	0	2PF
VJC	B-C built-in potential	V	0.75	0.5
MJC	B-C junction exponent factor	—	0.33	0.5
XCJC	Fraction of B-C depletion capacitance connected to internal base node	—	1	

Modified Gummel-Poon BJT Parameters. (continued)

Name	Parameter	Units	Default	Example
TR	Ideal reverse transit time	sec	0	10Ns
CJS	Zero-bias collector substrate capacitance	F	0	2PF
VJS	Substrate junction built-in potential	V	0.75	
MJS	Substrate junction exponent factor	—	0	0.5
XTB	Forward and reverse beta temperature exponent	—	0	
EG	Energy gap for temperature effect on IS	eV	1.11	
XTI	Temperature exponent for effect on IS	—	3	
KF	Flicker-noise coefficient	—	0	
AF	Flicker-noise exponent	—	1	
FC	Coefficient for forward-bias depletion capacitance formula	—	0.5	

Junction field-effect transistors (JFET's)

```
JXXXXXXX ND NG NS MNAME <AREA> <OFF> <IC=VDS,VGS>

J1 7 2 3 JM1 OFF
```

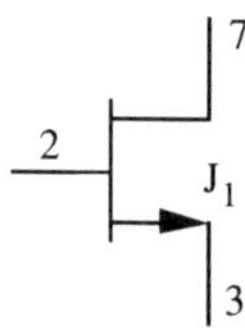


```
.MODEL MNAME TYPE(PNAME1=PVAL1 PNAME2=PVAL2 . . . )
```

NJF N-channel JFET model

PJF P-channel JFET model

```
.MODEL JM1 NJF(VTO = -1.5, BETA = 100)
```

JFET models (both N and P channel)

Name	Parameter	Units	Default	Example
VTO	Threshold voltage	V	-2.0	-2.0
BETA	Transconductance parameter	A/V^2	1.0E-4	1.0E-3
LAMBDA	Channel length modulation parameter	1/V	0	1.0E-4
RD	Drain ohmic resistance	Ohm	0	100
RS	Source ohmic resistance	Ohm	0	100
CGS	Zero-bias G-S junction capacitance	F	0	5PF
CGD	Zero-bias G-D junction capacitance	F	0	1PF
PB	Gate junction potential	V	1	0.6
IS	Gate junction saturation current	A	1.0E-14	1.0E-14
KF	Flicker noise coefficient	—	0	
AF	Flicker noise exponent	—	1	
FC	Coefficient for forward-bias depletion capacitance formula	—	0.5	

MOSFETs

```
MXXXXXXX ND NG NS NB MNAME <L=VAL> <W=VAL>
+ <AD=VAL> <AS=VAL> <PD=VAL> <PS=VAL> <NRD=VAL>
+ <NRS=VAL>    <OFF>   <IC=VDS,VGS,VBS>

M16 24 2 0 20 TYPE1
M31 2 17 6 10 MODM L=5U W=2U
M31 2 16 6 10 MODM 5U 2U
M1 2 9 3 0 MOD1 L=10U W=5U AD=100P AS=100P PD=40U PS=40U
M1 2 9 3 0 MOD1 10U 5U 2P 2P
```

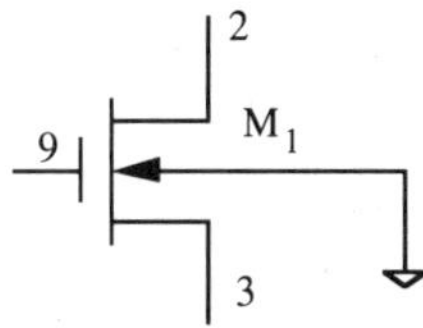


```
.MODEL MNAME TYPE(PNAME1=PVAL1 PNAME2=PVAL2 ... )
```

NMOS N-channel MOSFET model
PMOS P-channel MOSFET model

```
.MODEL MOD1 NMOS(LEVEL=2)
```

MOSFET models (both N and P channel)

LEVEL=1 − > Shichman-Hodges
LEVEL=2 − > MOS2 (as described by Vladimirescu et al.[1])
LEVEL=3 − > MOS3, a semiempirical model[2]

Name	Parameter	Units	Default	Example
LEVEL	Model index	—	1	
VTO	Zero-bias threshold voltage	V	0.0	1.0
KP	Transconductance parameter	A/V^2	2.0E-5	3.1E-5
GAMMA	Bulk threshold parameter	$V^{0.5}$	0.0	0.37
PHI	Surface potential	V	0.6	0.65
LAMBDA	Channel-length modulation (MOS1 and MOS2 only)	1/V	0.0	0.02
RD	Drain ohmic resistance	Ohm	0.0	1.0
RS	Source ohmic resistance	Ohm	0.0	1.0
CBD	Zero-bias B-D junction capacitance	F	0.0	20FF
CBS	Zero-bias B-S junction capacitance	F	0.0	20FF
IS	Bulk junction saturation current	A	1.0E-14	1.0E-15
PB	Bulk junction potential	V	0.8	0.87
CGSO	Gate source overlap capacitance per meter channel width	F/m	0.0	4.0E-11
CGDO	Gate drain overlap capacitance per meter channel width	F/m	0.0	4.0E-11
CGBO	Gate bulk overlap capacitance per meter channel length	F/m	0.0	2.0E-10

(continued)

Name	Parameter	Units	Default	Example
RSH	Drain and source diffusion sheet resistance	Ohm/sq. 0.0	10.0	
CJ	Zero-bias bulk juncton bottom capacitance per square meter of junction area	F/m^2	0.0	2.0E-4
MJ	Bulk junction bottom grading coefficient	—	0.5	0.5
CJSW	Zero-bias bulk junction sidewall capacitance per meter of junction perimeter	F/m	0.0	1.0E-9
MJSW	Bulk junction sidewall grading coefficient	—	0.33	
JS	Bulk junction saturation current per square meter of junction area	A/m^2	1.0E-8	
TOX	Oxide thickness	Meter	1.0E-7	1.0E-7
NSUB	Substrate doping	$1/cm^3$	0.0	4.0E15
NSS	Surface state density	$1/cm^2$	0.0	1.0E10
NFS	Fast surface state density	$1/cm^2$	0.0	1.0E10
TPG	Type of gate material: +1 opposite to substrate -1 same as substrate 0 Al gate	—	1.0	
XJ	Metallurgical junction depth	Meter	0.0	1U
LD	Lateral diffusion	Meter	0.0	0.8U
UO	Surface mobility	cm^2/V-s	600	700
UCRIT	Critical field for mobility degradation (MOS2 only)	V/cm	1.0E4	1.0E4
UEXP	Critical field exponent in mobility degradation (MOS2 only)	—	0.0	0.1
UTRA	Transverse field coefficient (mobility deleted for MOS2)	—	0.0	0.3
VMAX	Maximum drift velocity of carriers	m/s	0.0	5.0E4
NEFF	Total channel charge (fixed and mobile) coefficient (MOS2 only)	—	1.0	5.0
XQC	Thin oxide capacitance model flag and coefficient of channel charge share attributed to drain (0–0.5)	—	1.0	0.4
KF	Flicker noise coefficient	—	0.0	1.0E-26
AF	Flicker noise exponent	—	1.0	1.2
FC	Coefficient for forward-bias depletion capacitance formula	—	0.5	
DELTA	Width effect on threshold voltage (MOS2 and MOS3)	—	0.0	1.0
THETA	Mobility modulation (MOS3 only)	1/V	0.0	0.1
ETA	Static feedback (MOS3 only)	—	0.0	1.0
KAPPA	Saturation field factor (MOS3 only)	—	0.2	0.5

.MODEL

```
.MODEL MNAME TYPE(PNAME1=PVAL1 PNAME2=PVAL2 . . . )
```

NPN	NPN BJT model
PNP	PNP BJT model
D	Diode model
NJF	N-channel JFET model
PJF	P-channel JFET model
NMOS	N-channel MOSFET model
PMOS	P-channel MOSFET model

C.3 SUBCIRCUITS

.SUBCKT

```
.SUBCKT subnam N1 <N2 N3 ...>

.SUBCKT OPAMP 1 2 3 4
```

.ENDS

```
.ENDS <SUBNAM>

.ENDS OPAMP
```

Subcircuit calls

```
XYYYYYYY N1 <N2 N3 ...> SUBNAM

X1 2 4 17 3 1 MULTI
```

C.4 CONTROL STATEMENTS

.TEMP

```
.TEMP T1 <T2 <T3 ...>>

.TEMP -55.0 25.0 125.0
```

.WIDTH

```
.WIDTH IN=COLNUM OUT=COLNUM

.WIDTH IN=72 OUT=133
```

.OPTIONS

General form:

```
.OPTIONS OPT1 OPT2 ... (or OPT=OPTVAL ...)
```

Examples:

```
.OPTIONS ACCT LIST NODE
```

This option allows the user to reset the program control and user options for specific simulation purposes. Any combination of the following options may be included, in any order. In the listing that follows, x represents some positive number.

Option	Effect
ACCT	Causes accounting and run time statistics to be printed.
LIST	Causes the summary listing of the input data to be printed.
NOMOD	Suppresses the printout of the model parameters.
NOPAGE	Suppresses page ejects.
NODE	Causes the printing of the node table.
OPTS	Causes the option values to be printed.
GMIN = x	Resets the value of GMIN, the minimum conductance allowed by the program. The default value is 1.0E-12.
RELTOL = x	Resets the relative error tolerance of the program. The default value is 0.001 (0.1 percent).
ABSTOL = x	Resets the absolute current error tolerance of the program. The default value is 1 picoamp.
VNTOL = x	Resets the absolute voltage error tolerance of the program. The default value is 1 microvolt.
TRTOL = x	Resets the transient error tolerance. The default value is 7.0. This parameter is an estimate of the factor by which SPICE overestimates the actual truncation error.
CHGTOL = x	Resets the charge tolerance of the program. The default value is 1.0E-14.
PIVTOL = x	Resets the absolute minimum value for a matrix entry to be accepted as a pivot. The default value is 1.0E-13.
PIVREL = x	Resets the relative ratio between the largest column entry and an acceptable pivot value. The default value is 1.0E-3. In the numerical pivoting algorithm, the allowed minimum pivot value is determined by EPSREL=AMAX1(PIVREL*MAXVAL,PIVTOL).
NUMDGT = x	Resets the number of significant digits printed for output variable values. X must satisfy the relation $0 < x < 8$. The default value is 4. *Note:* This option is independent of the error tolerance used by SPICE (i.e., if the values options RELTOL, ABSTOL, etc., are not changed, then one may be printing numerical "noise" for NUMDGT > 4.
TNOM = x	Resets the nominal temperature. The default value is 27 C (300 K).
ITL1 = x	Resets the DC iteration limit. The default is 100.
ITL2 = x	Resets the DC transfer curve iteration limit. The default is 50.
ITL3 = x	Resets the lower transient analysis iteration limit. The default value is 4.
ITL4 = x	Resets the transient analysis time point iteration limit. The default is 10.
ITL5 = x	Resets the transient analysis total iteration limit. The default is 5000. Set ITL5=0 to omit this test.
ITL6 = x	Resets the DC iteration limit at each step of the source stepping method. The default is 0, which means not to use this method.
CPTIME = x	The maximum cpu-time in seconds allowed for this job.
LIMTIM = x	Resets the amount of cpu time limit and causes job termination. The default value is 2 (seconds).
LIMPTS = x	Resets the total number of points that can be printed or plotted in a DC, ac, or transient analysis. The default value is 201.
LVLCOD = x	If x is 2 (two), then machine code for the matrix solution will be generated. Otherwise, no machine code is generated. The default value is 2. Applies only to CDC computers.
LVLTIM = x	If x is 1 (one), the iteration time step control is used. If x is 2 (two), the truncation-error time step is used. The default value is 2. If method = Gear and MAXORD > 2, then LVLTIM is set to 2 by SPICE.
METHOD = name	Sets the numerical integration method used by SPICE. Possible names are Gear or trapezoidal. The default is trapezoidal.
MAXORD = x	Sets the maximum order for the integration method if Gear's variable-order method is used. X must be between 2 and 6. The default value is 2.
DEFL = x	Resets the value for MOS channel length. The default is 100.0 micrometer.
DEFW = x	Resets the value for MOS channel width. The default is 100.0 micrometer.
DEFAD = x	Resets the value for MOS drain diffusion area. The default is 0.0.
DEFAS = x	Resets the value for MOS source diffusion area. The default is 0.0.

.OP control functions

.OP—operating point. This command forces SPICE to determine the DC operating point with inductors shorted and capacitors opened. Also, the Q-point information will be listed in the output file for any diodes and transistors in the circuit.

.DC—DC sweep. This command defines the dc sweeping source and its limits.

.DC SRCNAM VSTART VSTOP VINCR [SRC2 START2 STOP2 INCR2]

.DC VIN 0.25 5.0 0.25
Sweeps VIN from 0.25 V to 5 V in 0.25 V steps.

.DC VDS 0 10 .5 VGS 0 5 1
Sweeps VDS from 0 to 10 V in 0.5 V steps with VGS = 0. Then repeats VDS sweep with VGS = 1V. Repeats each VDS sweep until VGS = 5V.

.DC VCE 0 10 .25 IB 0 10U 1U
Sweeps VCE from 0 to 10 V in 0.25 V steps for IB = 0. Then sets IB = 1 μA and repeats VCE sweep. It continues until IB = 10 μA with 1 μA steps for IB.

.NODESET. This command helps determine the DC or initial transient solution by making an initial pass with the indicated nodes set to the specified voltages.

```
.NODESET V(NODNUM)=VAL V(NODNUM)=VAL . . .

.NODESET V(12)=4.5 V(4)=2.23
```

.IC—initial condition. This sets initial transient conditions.

```
.IC V(NODNUM)=VAL V(NODNUM)=VAL ...

.IC V(11)=5 V(4)=-5 V(2)=2.2
```

.TF—transfer function. This command defines the DC, small-signal, input, and output analysis.

```
.TF OUTVAR INSRC

.TF V(5,3) VIN
.TF I(VLOAD) VIN
```

.SENS—sensitivity. This command causes calculation of the DC small-signal sensitivity of each specified output variable with respect to every circuit parameter.

```
.SENS OV1 <OV2 ... >

.SENS V(9) V(4,3) V(17) I(VCC)
```

.AC—AC sweep. This command sweeps the signal frequency.

```
.AC DEC ND FSTART FSTOP
.AC OCT NO FSTART FSTOP
.AC LIN NP FSTART FSTOP
.AC DEC 10 1 10K
.AC DEC 10 1K 100MEG
.AC LIN 100 1 100HZ
```

.DISTO—distortion. This command computes the small-signal distortion present in the circuit as a part of the AC steady-state analysis.

```
.DISTO RLOAD <INTER <SKW2 <REFPWR <SPW2>>>>

.DISTO RL 2 0.95 1.0E-3 0.75
```

.NOISE—noise. This command causes SPICE to compute the equivalent input and output noises as well as the contribution from every noise generator in the circuit.

```
.NOISE OUTV INSRC NUMS

.NOISE V(5) VIN 10
```

.TRAN—transient analysis. This command controls the transient analysis.

```
.TRAN TSTEP TSTOP <TSTART <TMAX>> <UIC>

.TRAN 1NS 100NS
.TRAN 1NS 1000NS 500NS
.TRAN 10NS 1US UIC
```

.FOUR—Fourier analysis. This command, performed in conjunction with the transient analysis, finds the Fourier frequency components at the specified nodes.

```
.FOUR FREQ OV1 <OV2 OV3 ...>

.FOUR 100K  V(5)
```

.PRINT—print. This command defines the contents of the tabular listing of up to eight output variables.

```
.PRINT PRTYPE OV1 <OV2 ... OV8>

.PRINT TRAN V(4) I(VIN)
.PRINT AC VM(4,2) VR(7) VP(8,3)
.PRINT DC V(2) I(VSRC) V(23,17)
.PRINT NOISE INOISE ONOISE
.PRINT DISTO HD3 SIM2(DB)
```

VR	Real part
VI	Imaginary part
VM	Magnitude
VP	Phase
VDB	20 * log 10 (magnitude)

.PLOT—plot. This command controls the plotting of up to eight output variables.

```
.PLOT PLTYPE OV1 <(PLO1,PHI1)> <OV2 <(PLO2,PHI2)> ...OV8>

.PLOT DC V(4) V(5) V(1)
.PLOT TRAN V(17,5) (2,5) I(VIN) V(17) (1,9)
.PLOT AC VM(5) VM(31,24) VDB(5) VP(5)
.PLOT DISTO HD2 HD3(R) SIM2
.PLOT TRAN V(5,3) V(4) (0,5) V(7) (0,10)
```

REFERENCES

1. A. Vladimirescu, Kaihe Zhang, A. R. Newton, D. O. Pederson, and A. Sangiovanni-Vincentelli, *SPICE Version 2G User's Guide*, Department of Electrical Engineering and Computer Sciences, University of California, Berkeley, August 1981.
2. Ibid.

RECOMMENDED REFERENCES

Paolo Antognetti and Giuseppe Massobrio, *Semiconductor Device Modeling with SPICE*, (New York: McGraw-Hill, 1988).

W. Banzhaf, *Computer-Aided Circuit Analysis Using SPICE* (Englewood Cliffs, N.J.: Prentice Hall, 1989).

Index

Absolute value, 64
.AC (AC sweep), 97, 270
AC analysis, 2
Activation energy, 262
Active filter, 7
Add trace, 254
ADVICE, 2
AF, 101
AM detector, 134
Ammeter, 3
Amplifiers, 120
  differential, 120
  log , 91
  microwave, 120
  operational, 120
  voltage, 120
Amplitude modulator, 88
ASPEC, 2
Asymptotes, 251
Auto-range, 250
Average
  arithmetic, 104

Balanced-modulator, 138, 140
Bandpass filter, 253
Beat note, 173, 177, 192
BETA, 263, 264
Bias point, 241
Bipolar junction transistors, 262
Butterworth low-pass filter, 38, 251, 252

Capacitors, 256
Capture range, 135, 167, 169, 182, 187, 193, 200
Carrier frequency, 88, 90
Characteristic impedance, 79
Chebyshev low-pass filter, 30
Circuit structure, 182
  equivalent, 182
  mathematical, 182
Clock regenerators, 134
Common mode gain, 122
Common mode input resistance, 120
Comparator, 117, 210
Convergence, 241
Conversion gain, 168, 192, 196, 197
Coupled inductors, 257
Current at breakdown voltage, 262
Current Controlled Current Source (CCCS), 5, 17, 258
Current Controlled Voltage Source (CCVS), 4, 17, 258
Cursors, 250

Damping factor, 168, 177, 193, 206, 208
.DC (DC sweep), 7, 269
DC analysis, 2
DC operating point, 2
DC small signal sensitivity analysis, 2
DC transfer function analysis, 2
Delay line, 79
Dependent sources, 257
Depletion capacitance, 263, 266
Differential input impedance, 98
Differential mode gain, 122
Differentiation, 20
Differentiator, 62
Diode, 23, 262
Diode subcircuit, 101
Discriminator, 117
.DISTO (distortion analysis), 270
Distortion, 2
Division
  voltage, 19, 47
Dominant pole, 121
Doppler radar, 57
Drift velocity, 266
Dynamic range, 123, 137, 144, 168, 201

Early voltage, 263
End fourier, 250
.ENDS (end of subcircuit), 6, 267
Energy gap, 264
Emission coefficient, 23, 24, 154, 262, 263
Equivalent circuit structure, 135
Excess phase, 263
Exit, 250
Exponential, 260

Fall time, 154
Feedback, positive, 80
Flicker noise, 97
  coefficient, 99, 262, 264, 266
  exponent, 99, 262, 264, 266
FM detector, 134
.FOUR (Fourier analysis) , 2, 156, 270
Fourier, 174, 250
Fourier frequency spectrum, 91
Free-running frequency, 135, 137, 147, 182, 192, 193, 196, 197
Frequency
  −3dB, 253
Frequency adder/subtracter, 73
Frequency counter: frequency-to-voltage converter, 111
Frequency divider, 71, 242
Frequency multiplier, 68, 243
Frequency response, 121, 241
Frequency shift key (FSK), 209
Frequency synthesizer, 134, 242
FSK demodulator, 134
Full wave rectifier, 64
Function generator, 159

Gain margin, 194
Gain stage, 121, 123
GAMMA, 265
Grading coefficient, 262
Gummel-Poon FET model, 263

HSPICE, 2
Hybrid phase-locked loop, 134
Hystersis, 241

.IC (initial conditions), 269
IG-SPICE, 2
Independent sources, 259
Inductors, 256
Input bias current, 120, 121
Input capacitance, 120, 121
Input offset current, 120, 121
Input offset voltage, 120
Integration , 21
Integrator, 59, 142, 144, 146, 148, 151, 152, 153
Integrator, discrete , 33
I-SPICE, 2
IS-Spice, 2
Iterations, 242

Jitter, 80
Junction field-effect transistor (JFET), 264
Junction potential, 262

KF, 101
KP, 265

LAMBDA, 264, 265
Laplace transform, 30, 162
Log amplifier, 91
Linear, 250
Linear small signal analysis, 2
Lock range, 167, 182, 193, 200
Locked mode, 135
Lock-up, 170, 171
Lockup time, 135, 177

Log, 250
Loop conversion gain, 135
Loop damping factor, 135
Loss of lock, 135
Low-pass filter, 134, 135, 159, 163, 169

Macromodel development, 16
Mason's gain formula, 31
Mathematical structure, 135
Mean value, 105
Mobility, 266
.MODEL- 6, 266
Modem, 134
Modulation signal, 88, 90
MOSFET, 265
Multiplier, voltage, 6, 45, 142, 151

Nagel, Lawrence, 1
Natural frequency, 206, 208
.NODESET, 269
Noise, 80, 241
  corner frequency, 99
  excess, 100
  flicker, 97, 100, 102
  Gaussian, 103
  index, 100
  1/f, 100
  random noise generator, 103
  shot, 97, 102
  thermal, 97, 101
.NOISE (noise analysis), 2, 97, 103, 270
Noise sources
  current, 97
  random, 103
  voltage, 97
Nyquist rate, 33, 38

Offset voltage, 120
Operational amplifier, 120
.OP (operating point), 269
.OPTIONS, 267
Oscillation, 170
Oscillator, 117
Output offset voltage, 142
Output stage, 123
Overshoot, 135, 170, 171, 195
Oxide capacitance, 266
Oxide thickness, 266

Peak detector, 66, 76
Phase comparator, 134
Phase detector, 135, 136, 138, 140, 142, 159, 168
Phase error, 201
Phase locked loop, 57, 134
Phase margin, 194
Phase modulation, 57
Phase shifter, 57
Piecewise linear, 23, 261
Plot, 250
  add, 250
  remove, 250
  select, 250
.PLOT (plot control), 250, 271
Pole, 121
Polynomial dependent source, 19
Power supplies, 123
.PRINT (print control), 270
Probe, 248
PSpice, 2
Pull-out range, 177
Pulse, 259
Pulse width modulator, 86

Ramp, 86
Ramp of frequency, 167, 182
Random noise generator, 103
Relative error tolerance, 242
Resistors, 256
Reverse breakdown voltage, 23, 262
Reverse current, 23
Reverse saturation current, 154
Rise time, 154
Root extractor, 94

Sampled data filter, s-domain, 33
Sample-hold, 83, 111
Saturation current, 23, 262, 263, 264, 265
Schmitt Trigger, 80, 86, 117, 140, 142, 144, 146, 148, 152–154, 156
S-domain, 30
.SENS, 269
Sensitivity, 241
Set range, 250
Sheet resistance, 266
Signal acquisition, 135
Signal-to-noise ratio (SNR), 175
Single-frequency FM, 261
Sinusoidal, 260
Slew rate stage, 123
SLICE, 2
Sources
  independent, 3
  dependent, 4
  polynomial, 5
  voltage, 3
Square root extractor, 52
Squarer, 49
Stability, 194
Standard derivation, 104, 106
Step of frequency, 173
Step of phase, 170
Stereo demodulator, 134
Storistor, 33
Subcircuit, 7, 267
.SUBCKT, 6, 267
Substrate capacitance, 264
Summer, 5, 43, 86, 142, 151
Sweeping component values, 7
Switched capacitor, 134
Symbols, 250

.TEMP (temperature analysis), 2, 267
Temperature, 10
.TF (transfer function analysis), 269
Thermal voltage, 23
Threshold voltage, 139, 142, 144, 151, 264, 265
Time delay, 79, 210
Time-dependent elements, 10
Timing capacitors, 135
Timing resistors, 135
Tone decoder, 134
Tracking range, 135
.TRAN (transient analysis), 103, 270
Transfer characteristic, 81, 117
Transfer function, 241
Transformer, 51
Transient analysis, 2, 103, 242
Transient response, 169, 170, 187, 207
Transit time, 262, 263
Transmission line , 33, 79, 257
Triangle-to-sine-wave converter, 54

Unlocked state, 135

Voltage comparator, 117
Voltage Controlled Current Source (VCCS), 4, 17, 258
Voltage controlled oscillator (VCO), 107, 114, 134, 135, 137, 142, 145, 159, 168
Voltage Controlled Voltage Source (VCVS), 4, 17, 258
Voltage level detector, 117
Voltage limiter, 142, 148

Waveform generator, 134
.WIDTH (width control), 267

X-axis, 250
X variable, 250

Y-axis, 250

Zener diode, 10, 13
Zero, 121

Disks to accompany MACROMODELING WITH SPICE
by J. Alvin Connelly and Pyung Choi

Prentice Hall

YOU SHOULD CAREFULLY READ THE FOLLOWING TERMS AND CONDITIONS BEFORE OPENING THIS DISKETTE PACKAGE. OPENING THIS DISKETTE PACKAGE INDICATES YOUR ACCEPTANCE OF THESE TERMS AND CONDITIONS. IF YOU DO NOT AGREE WITH THEM, YOU SHOULD PROMPTLY RETURN THE PACKAGE UNOPENED, AND YOUR MONEY WILL BE REFUNDED.

IT IS A VIOLATION OF COPYRIGHT LAW TO MAKE A COPY OF THE ACCOMPANYING SOFTWARE EXCEPT FOR BACKUP PURPOSES TO GUARD AGAINST ACCIDENTAL LOSS OR DAMAGE.

Prentice Hall, Inc. provides this program and licenses its use. You assume responsibility for the selection of the program to achieve your intended results, and for the installation, use, and results obtained from the program. This license extends only to use of the program in the United States or countries in which the program is marketed by duly authorized distributors.

LICENSE

You may:

a. use the program;
b. copy the program into any machine-readable form without limit;
c. modify the program and/or merge it into another program in support of your use of the program.

LIMITED WARRANTY

THE PROGRAM IS PROVIDED "AS IS" WITHOUT WARRANTY OF ANY KIND, EITHER EXPRESSED OR IMPLIED, INCLUDING, BUT NOT LIMITED TO, THE IMPLIED WARRANTIES OF MERCHANTABILITY AND FITNESS FOR A PARTICULAR PURPOSE. THE ENTIRE RISK TO THE QUALITY AND PERFORMANCE OF THE PROGRAM IS WITH YOU. SHOULD THE PROGRAM PROVE DEFECTIVE, YOU (AND NOT PRENTICE HALL, INC. OR ANY AUTHORIZED DISTRIBUTOR) ASSUME THE ENTIRE COST OF ALL NECESSARY SERVICING, REPAIR, OR CORRECTION.

SOME STATES DO NOT ALLOW THE EXCLUSION OF IMPLIED WARRANTIES, SO THE ABOVE EXCLUSION MAY NOT APPLY TO YOU. THIS WARRANTY GIVES YOU SPECIFIC LEGAL RIGHTS AND YOU MAY ALSO HAVE OTHER RIGHTS THAT VARY FROM STATE TO STATE.

Prentice Hall, Inc. does not warrant that the functions contained in the program will meet your requirements or that the operation of the program will be uninterrupted or error free.

However, Prentice Hall, Inc., warrants the diskette(s) on which the program is furnished to be free from defects in materials and workmanship under normal use for a period of ninety (90) days from the date of delivery to you as evidenced by a copy of your receipt.

LIMITATIONS OF REMEDIES

Prentice Hall's entire liability and your exclusive remedy shall be:

1. the replacement of any diskette not meeting Prentice Hall's "Limited Warranty" and that is returned to Prentice Hall with a copy of your purchase order, or
2. if Prentice Hall is unable to deliver a replacement diskette or cassette that is free of defects in materials or workmanship, you may terminate this Agreement by returning the program, and your money will be refunded.

IN NO EVENT WILL PRENTICE HALL BE LIABLE TO YOU FOR ANY DAMAGES, INCLUDING ANY LOST PROFITS, LOST SAVINGS OR OTHER INCIDENTAL OR CONSEQUENTIAL DAMAGES ARISING OUT OF THE USE OR INABILITY TO USE SUCH PROGRAM EVEN IF PRENTICE HALL OR AN AUTHORIZED DISTRIBUTOR HAS BEEN ADVISED OF THE POSSIBILITY OF SUCH DAMAGES, OR FOR ANY CLAIM BY AN OTHER PARTY.

SOME STATES DO NOT ALLOW THE LIMITATION OR EXCLUSION OF LIABILITY FOR INCIDENTAL OR CONSEQUENTIAL DAMAGES, SO THE ABOVE LIMITATION OR EXCLUSION MAY NOT APPLY TO YOU.

GENERAL

You may not sublicense, assign, or transfer the license or the program except as expressly provided in this Agreement. Any attempt otherwise to sublicense, assign, or transfer any of the rights, duties, or obligations hereunder is void.

This Agreement will be governed by the laws of the State of New York.

NEITHER PRENTICE HALL NOR THE AUTHORS OFFER ANY TECHNICAL SUPPORT FOR THIS PROGRAM.

Should you have any question concerning this Agreement, you may contact Prentice Hall, Inc., by writing to:

Prentice Hall
College Division
Englewood Cliffs, N.J. 07632

YOU ACKNOWLEDGE THAT YOU HAVE READ THIS AGREEMENT, UNDERSTAND IT, AND AGREE TO BE BOUND BY ITS TERMS AND CONDITIONS. YOU FURTHER AGREE THAT IT IS THE COMPLETE AND EXCLUSIVE STATEMENT OF THE AGREEMENT BETWEEN US THAT SUPERCEDES ANY PROPOSAL OR PRIOR AGREEMENT, ORAL OR WRITTEN, AND ANY OTHER COMMUNICATIONS BETWEEN US RELATING TO THE SUBJECT MATTER OF THIS AGREEMENT.

ISBN 0-13-544941-3